Environmental Impacts of Pasture-based Farming

Environmental Impacts of Pasture-based Farming

Edited by

Richard W. McDowell

AgResearch
Invermay Agricultural Centre
Mosgiel
New Zealand

www.cabi.org

CABI is a trading name of CAB International

CABI Head Office
Nosworthy Way
Wallingford
Oxfordshire OX10 8DE
UK

Tel: +44 (0)1491 832111
Fax: +44 (0)1491 833508
E-mail: cabi@cabi.org
Website: www.cabi.org

CABI North American Office
875 Massachusetts Avenue
7th Floor
Cambridge, MA 02139
USA

Tel: +1 617 395 4056
Fax: +1 617 354 6875
E-mail: cabi-nao@cabi.org

A catalogue record for this book is available from the British Library, London, UK.

Library of Congress Cataloging-in-Publication Data

Environmental impacts of pasture-based farming / edited by: R.W. McDowell.
p. cm.
Includes bibliographical references and index.
ISBN 978-1-84593-411-8 (alk. paper)
1. Pastures--Environmental aspects. 2. Grazing--Environmental aspects. I. McDowell, Richard W. II. Title.

SB199.E58 2009
636.08'45--dc22

2008018126

ISBN-13: 978 1 84593 411 8

Typeset by SPi, Pondicherry, India.
Printed and bound in the UK by MPG Biddles Ltd, Kings Lynn.

Contents

Contributors

Kirsten Barlow, *Victorian Department of Primary Industries, Rutherglen, RMB 1145, Chiltern Valley Road, Rutherglen, Victoria 3685, Australia. E-mail: Kirsten.barlow@dpi.vic.govt.au*

Denise Bewsell, *AgResearch Limited, Lincoln Research Centre, Corner Springs Road and Gerald Street, Private Bag 4749, Christchurch 8140, New Zealand. E-mail: denise.bewsell@agresearch.co.nz*

Steve P. Cuttle, *Institute of Grassland and Environmental Research, Plas Gogerddan, Aberystwyth, Ceredigion SY23 3EB, UK. E-mail: steve.cuttle@bbsrc.ac.uk*

Chris Dake, *AgResearch Limited, Grasslands Research Centre, Tennent Drive, Private Bag 11008, Palmerston North 4442, New Zealand. E-mail: chris.dake@agresearch.co.nz*

Cecile A.M. de Klein, *AgResearch Invermay, Private Bag 50034, Mosgiel, New Zealand. E-mail: cecile.deklein@agresearch.co.nz*

Peter J.A. Kleinman, *US Department of Agriculture, Agricultural Research Service, Pasture Systems and Watershed Management Research Unit, University Park, PA 16802, USA. E-mail: peter.kleinman@ars.usda.gov*

Ross M. Monaghan, *AgResearch, Invermay Agricultural Centre, Private Bag 50034, Mosgiel, New Zealand. E-mail: ross.monaghan@agresearch.co.nz*

David Nash, *Victorian Department of Primary Industries, Ellinbank, RMB 2460 Hazeldean Road, Ellinbank, Victoria 3821, Australia and eWater CRC, University of Canberra, GPO Canberra 2601, ACT, Australia. E-mail: david.nash@dpi.vic.govt.au*

Cesar Pinares-Patino, *AgResearch Grassland, Private Bag 11008, Palmerston North, New Zealand. E-mail: Cesar.pinares@agresearch.co.nz*

Patrick E. Reece, *University of Nebraska-Lincoln, Department of Agronomy and Horticulture, Lincoln, NE 68583-0915, USA. E-mail: preece@unlnotes.unl.edu*

Fiona A. Robertson, *Future Farming Systems Research, Department of Primary Industries, Hamilton Centre, Mount Napier Road, Private Bag 105, Hamilton, Victoria 3300, Australia. E-mail: fiona.robertson@dpi.voc.gov.au*

Walter H. Schacht, *University of Nebraska-Lincoln, Department of Agronomy and Horticulture, Lincoln, NE 68583-0915, USA. E-mail: wshacht@unlnotes.unl.edu*

John T. Scott, *Landcare Research, Manaaki Whenua, Private Bag 3127, Waikato Mail Centre, Hamilton 3240, New Zealand. E-mail: scottJ@candcareresearch.co.nz*

Andrew N. Sharpley, *Department of Crop, Soil and Environmental Sciences, University of Arkansas, Fayetteville, Arkansas, USA. E-mail: sharpley@uark.edu*

Kathy Soder, *US Department of Agriculture, Agricultural Research Service, Pasture Systems and Watershed Management Research Unit, University Park, PA 16802, USA. E-mail: kathy.soder@ars.usda.gov*

Garry C. Waghorn, *DairyNZ, Private Bag 3221, Hamilton, New Zealand. E-mail: Garry.waghorn@DairyNZ.co.nz*

Charles West, *Department of Crop, Soil and Environmental Sciences, University of Arkansas, Fayetteville, Arkansas, USA. E-mail: west@uark.edu*

Robert J. Wilcock, *National Institute of Water and Atmospheric Research Ltd, PO Box 11–115, Hamilton, New Zealand. E-mail: r.wilcock@niwa.co.nz*

Foreword

At the heart of this book reside the tensions that people face when managing grazed ecosystems throughout the world. Sustaining the natural capital of our grazed lands is crucial, as these areas represent a very large part of our global terrestrial ecosystems. This challenge is grounded in the socio-economic expectations of land users and nations.

The biophysical elements that need to be managed will vary, depending on the ecosystem involved. For rangelands, the maintenance of perennial vegetations cover is critical from a soil erosion perspective. For more intensive grasslands, the emission of nutrients and gases is of great concern. But in all cases, the connections between land, water and air are becoming increasingly evident.

While work involving singular disciplines of study provides knowledge of greater depth, we also need a better understanding of the interactions and emergent properties of our grazed ecosystems. The long-standing principles of ecology, armed with the analytical power of simulation modelling, have a major role to play in understanding and designing sustainable systems for the future.

People must be an integral part of any future system design. They are not observers and their expectations will shape the way grazed ecosystems will be managed. While some people seek high-quality food and a pristine environment, we must recognize that many other communities simply seek a little more food and economic wealth to survive.

It will be interesting to see whether the desired changes we seek in people will occur voluntarily, or whether they will require incentives and regulations. Like most things in life, I expect that a mix will be required to ensure the necessary knowledge and motivation is in place.

Good science must not take a political position with regard to resolving these tensions and managing our grazed lands in a better way. Rather, it must inform the various communities of interests. Herein lies the value of this book. In the end, wise solutions will be a balance of trade-offs that are based on informed decisions and actions.

Gavin W. Sheath
Chair, International Grasslands Congress

Preface

The principal objective of this text is to raise awareness among scientists and policy makers of the impact that grazed grasslands have on our environment. It is common for scientists and policy makers to have an area of speciality. However, while focusing on one area can undoubtedly increase our knowledge, it must be put in a wider context. Too often are our decisions and conclusions based on a narrow view of the world dictated by our field of expertise. Unfortunately, the processes within the environment, defined here as the effect of grasslands on land, air and water, are linked and can have many repercussions on one another. For instance, a quantity of nutrient may be lost from land to water or air. This will be quantified and management practices suggested for mitigating the loss. However, these practices are rarely evaluated in terms of economic or social impact. We must remember that while preventing damage to the environment, humans have for millennia formed part of the landscape – after all it is our values that define what an acceptable environment is.

While collating this book I have tried to do two things, and have split the book into two parts as a consequence. First, was to establish a base of knowledge that the reader (and indeed I found myself) could use to interpret how land, air and water interact within grazed grasslands. Coupled to this is the socio-economic impact, too often neglected in environmental analyses. The second task recognized that use of grassland is becoming increasingly specialized as farms become larger, more intensive and profitable. Furthermore, specific systems have their own problems. For instance, intensive pastoral grazing by dairy cattle can result in soil damage by treading or poaching that would not occur in an extensive sheep operation. As a consequence, one land-use not only requires specific management practices to mitigate effects on the environment, but it also needs to account for regional influences like climate, topography, economic and social factors.

It is predicted that by 2050, production from grazed grassland will double to just over 1 billion t of milk and 465 million t of meat (Steinfield *et al.*, 2006). Much of this is driven by demand from areas like South-east Asia. While local production in these areas has huge potential for growth, largely because little has been produced in the past, until production can be increased (if possible), traditional producers of grassland-based protein and milk products will supply the bulk of the demand. This brings with it the problem of how to increase production in areas with a long history of grassland production given that they contain generally affluent societies, which are increasingly aware of their environment.

This book brings together authors from countries with large livestock industries. Each chapter in the second part of the book is written by a specialist in the field, but with cross references to areas of basic knowledge in the first part where appropriate. The choice of authors was an indication of the importance of the issue in their country or region. For instance, the air quality chapter was written by three New Zealand authors largely because unlike other countries, where industry or transport may have the most effect on air pollution and climate change, the production of nitrous oxide and methane by grazing ruminants is estimated to be the main influence on air quality in New Zealand (Chapter 1). In contrast, a chapter focusing on hybrid dairies where cows graze pastures for a few months during the growing season and are housed otherwise was written by experts in the north-east of the USA where the practice is becoming increasingly common (Fales *et al.*, 1993).

There is increasing realization that grazed grassland and intensively managed pastures are impacting on our environment. Citations that mention pastures or grassland have almost doubled in many electronic databases since the late 1990s (e.g. ARICOLA, SCOPUS, CABI Abstracts). However, as mentioned earlier, an approach that considers the many facets of environment and people is missing. We still continue to struggle with the issue of improving environmental metrics without compromising productivity or social constructs. To this end I hope that the following text helps to meet this goal.

R.W. McDowell, April 2008
Editor

References

Fales, S.L., McMurry, S.A. and McSweeny, W.T. (1993) The role of pasture in north-eastern dairy farming: historical perspectives, trends and research imperatives for the future. *Agricultural Research* 677, 111–132.

Steinfeld, H., Gerber, P., Wassenaar, T., Castel, V., Rosales, M. and de Haan, C. (2006) *Livestock's Long Shadow: Environmental Issues and Options*. FAO, Rome. 390 pp.

Acknowledgements

The editor wishes to thank the authors for their contributions and the following people for helping to maintain quality during the reviewing process: Keith Cameron, Owen Carton, Harry Clark, Leo Condron, Rob Davies-Colley, Hong Di, John Drewry, David Horne, Steve Jarvis, Bob Knight, Stewart Ledgard, Alex MacKay, Roger Parfitt, John Quinn, Caroline Saunders, Rob Sherlock, Paul Withers and Vic Wright.

1 Greenhouse Gas Emissions

C.A.M. de Klein,[1] C. Pinares-Patino[2] and G.C. Waghorn[3]

[1]AgResearch Invermay, Private Bag 50034, Mosgiel, New Zealand; [2]AgResearch Grassland, Private Bag 11008, Palmerston North, New Zealand; [3]DairyNZ, Private Bag 3221, Hamilton, New Zealand

Introduction

Greenhouse gases (GHGs) are gaseous components of the atmosphere that absorb solar energy reflected from the earth's surface as infrared radiation. This energy is transferred to the major non-GHGs (nitrogen (N) and oxygen) resulting in an overall temperature increase in the lower atmosphere. GHGs are critically important for regulating the earth's surface temperature, as without any atmospheric GHGs the average temperature would be −6°C instead of 15°C (Steinfeld *et al.*, 2006). Since the pre-industrial period, the global emissions of several major GHGs have increased exponentially and by 70% between 1970 and 2004 as a result of industrial and agricultural activities (IPCC, 2007). Consequently, the average temperature of the earth's surface has increased by 0.6°C since the late 1800s, with further increases of 1–5°C projected by 2100 (Steinfeld *et al.*, 2006). These increases in temperature are projected to affect the earth's climate and weather patterns, and extreme events such as droughts, floods and storms are expected to occur more frequently.

The main anthropogenic or human-induced GHGs are carbon dioxide (CO_2), methane (CH_4) and nitrous oxide (N_2O). These GHGs each have a different 'global warming potential' (GWP) based on the gases' ability to absorb solar energy and on their atmospheric lifetime. The GWP for CO_2, CH_4 and N_2O is currently calculated to be 1, 25 and 298 (Solomon *et al.*, 2007), respectively, indicating that 1 kg of CH_4 is 25 times and 1 kg of N_2O 298 times as potent as 1 kg of CO_2. Weighted by their GWP, CO_2, CH_4 and N_2O currently contribute *c.*75%, 15% and 10% of the global GHG emissions (Fig. 1.1; IPCC, 2007). The main source of CO_2 emissions is fossil fuel use, while agricultural practices contribute 40% and 90% of the global CH_4 and N_2O emissions, respectively. CH_4 and N_2O are the two main GHGs from agriculture, contributing over 90% of the emissions (Fig. 1.1). Of the total agricultural emission, pastoral grazing systems have been estimated to contribute 20% of the CH_4 emissions and between 16% and 33% of the N_2O emissions (Clark *et al.*, 2005).

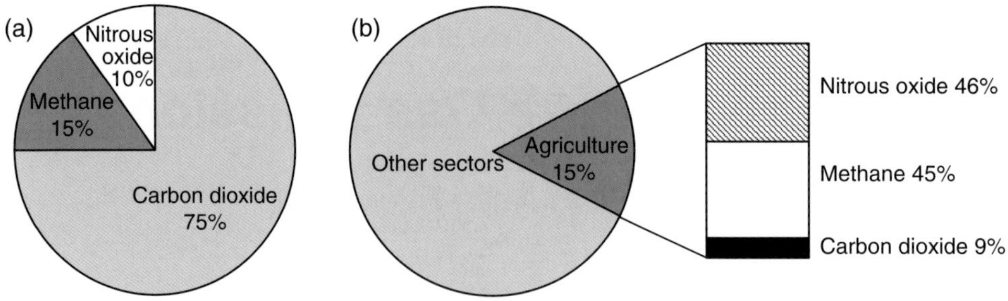

Fig. 1.1. (a) Global GHG emission profile in CO_2 equivalents. (b) Agricultural contribution to global GHG emissions and agricultural GHG emission profile in CO_2 equivalents.

In 1997, the Kyoto Protocol was adopted by industrialized countries to reduce GHG emissions (UNFCCC, 1997). The Kyoto Protocol is an international agreement that came into force on 16 February 2005 and sets legally binding targets for reducing GHG emissions for member countries of the United Nations Framework Convention on Climate Change (UNFCCC) that ratified the protocol. The targets vary per country and range from a reduction of 8% to an increase of 10% relative to 1990 emissions. The overall aim of the targets is to reduce global GHG emissions by at least 5% relative to 1990 levels, by 2008–2012 (UNFCCC, 1997). The Kyoto Protocol also requires countries to submit a GHG inventory each year, and the Intergovernmental Panel on Climate Change (IPCC) provides guidelines for compiling these inventories (IPCC, 1997, 2006). For agriculture, these guidelines include generic inventory methodologies for estimating CH_4 and N_2O emissions from the size or the strength of a given source of these emissions (e.g. number of animals or amount of N fertilizer applied), multiplied by a CH_4 or N_2O emission factor associated with that source. The guidelines provide default values for these emission factors that are based on the best available knowledge.

Pastures have diverse topography, productivity and management, but are usually dominated by grasses and grazed by ruminants (cattle, sheep, buffalo, goats) as well as camelids and other species (horses, kangaroos, etc.). Annual pasture growth varies from little more than 1000 kg dry matter (DM)/ha in arid or cold environments to 20,000 kg DM/ha in warm, fertile temperate situations with higher yields for tropical grasses. Pastures can also contain legume species that provide N to the system through biological N fixation. High levels of animal production are dependent on substantial input of N, either through N fixation, or, if legumes are a minor sward component, through inputs of N fertilizers (McGrath *et al.*, 1998). GHG emissions from grazed pastoral systems increase with increased DM production and hence animal production, and are mainly affected by forage species, pasture yield, animal production, climate, soil type and fertilizer application.

This chapter describes the sources and sinks of each of the main GHGs in pastoral grazing systems, and focuses mainly on productive temperate environments. It also describes the factors affecting these emissions, the latest knowledge on N_2O emission factors and the methods for estimating or modelling emissions. Finally, examples of the GHG profile for different grazing systems and the threats of increasing GHG emissions for pastoral farming are discussed.

Sources and Sinks in Pastoral Grazing

Methane sources and sinks

CH_4 emissions from managed grasslands can derive from: (i) enteric fermentation; (ii) emissions from manure and dung deposited on pastoral land; (iii) grassland soils; and (iv) forage plants. The single most important source of CH_4 from grazing systems is ruminant CH_4 arising from the fermentation of feed in the reticulo-rumen (rumen). Enteric CH_4 emissions account for about 44Tg/year (Clark *et al.*, 2005). Excretion from this source is mostly via the lungs, rather than flatus (Murray *et al.*, 1976; Torrent and Johnson, 1994). Flatus CH_4 accounts for less than 2% of the total enteric emissions in sheep (Murray *et al.*, 1976). CH_4 can also be produced following manure and dung deposition in the field but these sources are relatively insignificant compared to enteric fermentation. For example, in New Zealand pastoral agriculture, 99% of CH_4 emissions arise from enteric sources and only 1% from faecal material (Ministry for the Environment, 2007). CH_4 fluxes from soil and plants in grazed pastures are also minor, compared to enteric fermentation. Although net CH_4 emission from soil might occur in anoxic micro-environments that are present in otherwise aerobic, well-drained soils (Nicol *et al.*, 2003), grasslands are generally considered sinks for CH_4 (van den Pol-van Dasselaar *et al.*, 1999). Reports that CH_4 is emitted by plants in aerobic environments (Keppler *et al.*, 2006) remain contentious and even if plants were found to emit CH_4, estimates suggest these will be less than 3% of that arising from enteric sources (Kelliher *et al.*, 2006).

Enteric methane production

Ruminant digestion is primarily a microbial process that enables the fibre fraction of forages to be digested, yielding by-products that are absorbed and used by the ruminant (acetate, propionate and butyrate). Microbial digestion in the rumen is followed by hydrolysis of un-degraded plant (and microbial) material in the small intestine and absorption of protein and lipid components, while the remaining residues are subjected to further bacterial digestion in the large intestine (Van Nevel and Demeyer, 1996). Both rumen and large intestinal digestion yield hydrogen (H_2) as well as CO_2, volatile fatty acids (VFAs – acetate, propionate, butyrate), ammonia and heat (McAllister *et al.*, 1996). A final step in the process is reduction of CO_2 to CH_4 with H_2 as the energy source. The CH_4 formation acts as the most important ruminal electron sink for the H_2 produced by ruminal micro-organisms (McAllister and Newbold, 2008). CH_4 and heat production represent a loss of dietary energy, whereas ammonia represents a loss of dietary N that can ultimately yield N_2O. Energy loss to CH_4 is often 6–7% of feed gross energy intake (GEI) from temperate pasture, or about 10% of absorbed energy (Waghorn and Woodward, 2006). CH_4 represents 20–30% of intra-ruminal gas and is eructated via the oesophagus to the lungs and is lost in respired gas exchange.

Methanogens belong to the Euryarchaeota kingdom within the domain Archaea (Nicol *et al.*, 2003) and constitute a fundamental component of the rumen microbiota, becoming established very early in the life of the ruminant (Morvan *et al.*, 1994). The most common species of methanogens isolated from the rumen are strains of *Methanobrevibacter*, *Methanomicrobium*, *Methanobacterium* and *Methanosarcina* (Jarvis *et al.*, 2000). A feature of methanogens is that they are frequently associated with protozoa and although protozoa do not seem to be essential for efficient functioning of the rumen, their presence facilitates fibre degradation, CH_4 production (Finlay *et al.*, 1994; Van Nevel and Demeyer, 1996; Hegarty, 2004) and the efficient removal and disposal of hydrogen. Although protozoa are associated with about a quarter of CH_4 production (Newbold *et al.*, 1995; Morgavi *et al.*, 2008), the underlying source of CH_4 is from the utilization of hydrogen arising from digestion of plant matter by methanogens.

Enteric CH_4 production is affected by diet, the amount of feed intake and rumen microbial diversity. Typical CH_4 yields from digestion of temperate pastures are about 21 g/kg dry matter intake (DMI; Ministry for the Environment, 2007). This is equivalent to annual emissions of about 12 kg from an adult sheep and 70–90 kg from lactating cows fed temperate pastures. With stocking rates (SRs) of 20 sheep/ha or 3 cows/ha, enteric CH_4 emissions are about 240 kg/ha/year.

Factors affecting enteric methane emissions

CH_4 emissions (per unit of land area) from grazed pastoral systems are determined mainly by livestock numbers and animal productivity, which are dependent on pasture growth and total DMI. Higher-yielding (improved) pastures utilized by grazing animals will yield most CH_4, but also highest production, which is an essential aspect of food supply for the human population. The balance between CH_4 emissions per hectare and animal production per hectare is determined to a large extent by human inputs. These include establishment of forage species able to sustain higher levels of production (legumes, herbs, annual grasses), provision of fertilizer, water for irrigation, appropriate management to utilize a high proportion of the forage grown and provision of supplemental feeds in times of scarcity. Improved feeding and animal production will increase absolute emissions per hectare, but can lower CH_4 emissions per unit of product. Factors able to affect changes in methanogenesis include: forage type and composition, level of intake, animal species and genotype, grazing management and animal productivity.

FORAGE TYPE High-quality forages such as legumes or herbs generally yield lower CH_4 emissions per unit of DMI compared to poor-quality grasses. Waghorn and Woodward (2006) summarized feeding experiments with sheep in New Zealand involving fresh forages (Table 1.1) and found that white clover or *Lotus pedunculatus* fed to sheep yielded the lowest CH_4 emissions per unit of DMI: ~12 g/kg DMI compared to up to 26 g/kg DMI for ryegrass and other forages. Legume-feeding results in a lower rumen pH and higher ammonia, total VFAs concentrations and proportion of propionate compared to cows grazing perennial ryegrass pasture (Williams *et al.*, 2005), all of which lower H_2 excess and methanogenesis. In addition to these feed-quality effects of legumes, legume diets containing condensed tannins (CT) can further reduce CH_4 emissions by 15% (Waghorn and Woodward, 2006; Waghorn, 2007), particularly if fed as a sole feed. The cause of the CT effect on reducing CH_4 emissions is not known.

However, pastures are dominated by grasses rather than legumes, resulting in relatively high CH_4 per unit of feed digested and moderate levels of animal

Table 1.1. Methane emissions from cattle and sheep fed good-quality forages. All data are based on group means, typically of 6–30 individuals with measurements determined over a 4-day period. (Data from Waghorn and Woodward, 2006, with updates [a,b,c])

	No. of groups	CH_4 (g/kg DMI)	DM intake (kg/day)	NDF (fibre)	Crude protein
		Mean; range	Mean; range	Mean; range	Mean; range
Cattle					
Ryegrass	24[a]	20.4; 15.1–26.1	14.1; 3.6–19.1	44.6; 35–57	23.1; 15–30
Ryegrass + other forages	11[b]	20.0; 16.6–26.4	17.4; 14.1–20.5	40; 35–44	21; 15–29
Sheep					
Ryegrass	25[c]	21.7; 15.1–26.6	1.05; 0.51–1.89	47.3; 33–55	20.0; 11–29
Legumes and herbs	9[d]	17.0; 11.5–20.6	1.47; 0.93–2.0	26; 20–34	22; 15–27

[a]Woodward *et al.*, 2004, 2006; Robertson and Waghorn, 2002; Woodward, 2003; Waghorn *et al.*, 2008; Van Vugt *et al.*, 2005; Molano *et al.*, 2006; Clark *et al.*, 2005; Kolver and Aspin, 2006.
[b]Waghorn *et al.*, 2003a,b,c; Lee *et al.*, 2004; Waugh *et al.*, 2005; Grainger *et al.*, 2008.
[c]Knight *et al.*, 2008; Cosgrove *et al.*, 2008; Molano and Clark, 2008; Ulyatt *et al.*, 2002; Lassey *et al.*, 1997.

production. Including small quantities of legumes (e.g. 15–30%) with grasses has little effect on CH_4 emissions (g/kg DMI; Lee *et al.*, 2004; van Dorland *et al.*, 2007). As grass matures, cell wall content increases, which is more methanogenic than non-structural carbohydrates (Moe and Tyrrell, 1979), and the rate of passage through the digestive tract decreases (Mambrini and Peyraud, 1994), which can further increase CH_4 emissions. Tropical forages (C4 grasses) usually are less digestible than temperate forages (Minson, 1990), and energy lost as CH_4 is about 8.6, and 9.6% of GEI for cattle (Hunter, 2007) and 7.5% of GEI for sheep (Margen *et al.*, 1988), whereas values rarely exceed 7% of GEI with temperate grasses (Clark *et al.*, 2005).

Aside from the general relationships showing lowest CH_4 production from digestion of legumes, reductions due to CT and higher emissions from tropical than temperate grasses, there are no clear associations between CH_4 yields (g/kg DMI) and the dietary components such as fibre, non-structural carbohydrates, crude protein, lipid or ash (Waghorn and Woodward, 2006; Clark *et al.*, 2007). A number of studies with fresh temperate grasses also suggest that CH_4 yield (% of GEI) is relatively insensitive to either forage maturity (quality) or intake (e.g. Boadi and Wittenberg, 2002; Mbanzamihigo *et al.*, 2002; Molano and Clark, 2008).

Even though temperate grass quality has little effect on CH_4 yield, quality does affect how much feed is needed to achieve a given level of production. Increasing forage quality could be used to decrease emissions per animal simply because less feed is processed in the rumen to achieve a given level of production (Table 1.2). However, if feed quality is increased without reducing the quantity available, intakes of individual animals and/or the number of animals kept per unit area will increase so that CH_4 emissions increase either per animal or per unit area (Clark *et al.*, 2005).

Table 1.2. Methane emissions associated with production from growing 40 kg lambs fed contrasting diets *ad libitum* or restricted pasture intakes, and Friesian/Holstein lactating cows fed pasture at days 60, 150 and 240 of lactation.

	Diet/days of lactation	Diet ME[a] (MJ/kg DM)	Production	Methane (g/kg DMI)	Methane (g/production)
Lambs			Daily gain (g)		g/kg gain
	Pasture	10	100	24	330
	Pasture	11	150	22	210
	Pasture	12	200	21	160
	Lucerne	11.5	250	20	135
	Lotus	12	250	12	80
	White clover	12	300	10	100
	Rest pasture[b,c]	11	100	22.5	280
		11	50	22.5	560
Cows[d]			kg milk/day		g/kg milk
	Day 60	12.7	26.5	18.0	11.7
	Day 150	11.1	19.5	22.2	19.4
	Day 240	11.3	14.7	23.8	24.3

[a]ME, Metabolizable energy (mega joule) content of the dry matter.
[b]Lamb data calculated from measured daily gain (Burke *et al.*, 2004) and methane production from different diets (Waghorn and Woodward, 2006); maintenance requirements 11 MJ ME.
[c]Rest, restricted.
[d]From Robertson and Waghorn (2002).

LEVEL OF FEED INTAKE As a ruminant's feed intake increases above its maintenance requirements, CH_4 yield (g/kg DMI or % of GEI) decreases by 5–15% for each multiple of the amount of intake above maintenance requirements (Blaxter and Clapperton, 1965). The reduction in CH_4 per unit of DMI is greatest with high-quality diets, so high intakes of high-quality pasture will reduce emissions per unit of DMI as well as per unit of production (Tables 1.2 and 1.3; Ulyatt and Lassey, 2000). The decreased emissions have been associated with a more rapid passage of digesta through the rumen (less time for digestion and CH_4 production) resulting in a lower ruminal digestion (Benchaar *et al.*, 2001). However, the effects of increased intakes appear less important with forage diets than with mixed or concentrate diets (Blaxter and Clapperton, 1965; Mathison *et al.*, 1998). Molano and Clark (2008) did not find an effect of the level of DMI or feed quality (reproductive versus vegetative ryegrass) on CH_4 emission rates when lambs were fed at 0.75, 1.0, 1.25 and 1.5 times their maintenance requirements.

Feed intake can also increase through the provision of supplementary feed, such as conserved forage (silage or hay) or cereal grains, brought to animals on a pasture diet. Not only will this increase emissions above that from pasture alone due to increased DMI, but CH_4 emissions per unit of DMI can also increase. For example, substitution of 36% ryegrass pasture with maize silage increased CH_4 emissions from 16.3 to 19.0 g/kg DMI (Waugh *et al.*, 2005). On the other hand, provision of 20–30% grain in a forage diet had little effect on rumen CH_4 yield or the CH_4 per kg DMI (Boadi *et al.*, 2002; Lovett *et al.*, 2005).

Table 1.3. Calculated proportions of methane emissions attributable to maintenance or production in 40 kg growing lambs and 500 kg lactating cows fed ryegrass dominant pasture, with a metabolizable energy content of 11 MJ/kg. Data are per day unless indicated.

		Methane	Percentage of CH_4 associated with		
DMI (kg)	Production	G	Maintenance	Production	CH_4/production
Lambs	Daily gain (g)				g/kg gain
1.0	0	20.9	100	0	0
1.4	100	30.3	68	32	303
1.9	200	39.7	52	48	199
2.4	300	49.5	42	58	165
Cows	Milk (kg)				g/kg milk
5.2	0	112	100	0	0
12.0	12	259	43	57	21.6
15.3	18	330	34	66	18.3
18.7	24	404	28	72	16.8

Assumes 20.9 g CH_4 emitted for each kg DMI by growing lambs and 21.6 g CH_4 emitted for each kg DMI by dairy cows (New Zealand Climate Change Office, 2003).

ANIMAL SPECIES AND GENOTYPES There are very few direct comparative studies of CH_4 emission between ruminant species. Terada *et al.* (1985) reported the CH_4 yield (% of GEI) of cattle was higher than that of sheep or goats, but Simpson *et al.* (1978) found no differences between sheep and red deer in CH_4 yields. Studies conducted by Pinares-Patiño *et al.* (2003b) revealed much higher CH_4 yields for alpaca than for sheep fed under controlled conditions. Recently, Swainson *et al.* (2007) indicated that both daily CH_4 emissions and emissions per unit of intake varied between ruminant species on the same diet. Daily CH_4 production (g CH_4/day) significantly reduced as follows: cattle > deer > sheep, while the order of CH_4 emission per unit of DMI was: cattle > sheep > deer.

CH_4 emissions are similar for different breeds of cows fed temperate forages (Boadi and Wittenberg, 2002; Münger and Kreuzer, 2006), but Robertson and Waghorn (2002) reported higher CH_4 emissions from New Zealand Friesian compared to North American Holstein cows in early and mid-lactation when fed either pasture or grain-based rations. The difference disappeared in late lactation.

Within-species variation in CH_4 emitted per unit of feed intake is a common feature of grazing animals (Ulyatt *et al.*, 1997; Pinares-Patiño, 2000), with a range from 11 to 31 g/kg DMI found in 300 lactating dairy cows (Clark *et al.*, 2005). However, evidence for persistent differences over time is equivocal (Pinares-Patiño, 2000; Pinares-Patiño *et al.*, 2003b, 2005; Waghorn *et al.*, 2001). Recently, Vlaming *et al.* (2008) suggested that the commonly used SF_6 tracer technique for measuring ruminant CH_4 emissions may exaggerate apparent within-species variation.

Finally, the age of grazing animals may affect enteric CH_4 emissions due to differences in digestive function and forage selection. Knight *et al.* (2008) reported CH_4 emission from lambs to be 8% lower than from ewes on the same diet (21.9

versus 23.8 g/kg DMI). In contrast, Graham (1980) found no differences in CH_4 emission rates (per unit of DMI) in sheep between weaning (12 weeks of age) and maturity when mixed diets were fed. Under grazing, the effect of the age on CH_4 emissions may arise from selection of more succulent (low-fibre) diets by young animals compared to adults. However, under intensive management when animals have a high drive to eat, most edible forage will be eaten and diet selection is not likely to influence CH_4 emissions.

GRAZING MANAGEMENT The efficiency of herbage utilization, individual animal performance and production per hectare is largely determined by SR (the number of animals per unit of land area) and grazing practices (rotational grazing versus set-stocking). Rotational grazing is more common with improved pastures and requires a high SR for short (1–3 days) periods to achieve a high pasture utilization, production per hectare and maintenance of sward quality. In contrast, continuous (set) stocking involves a lower and more sustainable SR for several weeks. Improved management often coincides with incorporation of improved cultivars to increase production but SR and grazing management also affect sward composition, diet quality, soil characteristics (Hodgson, 1990) and CH_4 emissions (Pinares-Patiño *et al.*, 2007). For example, modelling cattle grazing in tropical systems, Howden *et al.* (1994) showed that emissions per unit of live weight (LW) gain decreased when SR increased to an optimum. Thereafter, further increases in SR increased CH_4 emissions per unit of LW gain, as individual growth rates declined and the proportion of energy used for maintenance increased. When tropical pasture was over-seeded with annual ryegrass and rotationally grazed, LW gain improved and CH_4 per kg gain was reduced by 22% compared to unimproved pasture under continuous grazing (De Ramus *et al.*, 2003). A similar conclusion was reached by Pavao-Zukerman *et al.* (1999) who incorporated white clover and fertilizer and applied rotational grazing to fescue pasture, which reduced CH_4 emissions per unit of LW gain when compared to low-input conventional management. However, in both these examples the absolute CH_4 emissions increased because pasture and animal production increased. Although gross enteric CH_4 emissions from ruminants fed improved and managed pastures are higher than unimproved pastures, the increase in farm gross margin is greater and, based on the current price of carbon (C), the profit attributable to improvements currently far exceeds the cost of gaseous emissions associated with higher SRs (Alcock and Hegarty, 2006).

ANIMAL PRODUCTIVITY Enhancing ruminant productivity generally requires simultaneous improvements in nutrition, genetics, reproductive efficiency, health and animal welfare. When animal productivity is increased the absolute amount of CH_4 per animal will increase, but the CH_4 yield per unit of animal product will decrease. The benefits of increasing animal productivity on CH_4 emission mitigation will result from the 'dilution' effect of the fixed 'maintenance' CH_4 production over the CH_4 associated with productive functions (Table 1.3; Clark *et al.*, 2005). Improved pasture quality increases production (Table 1.2), so lambs gaining 100, 150 or 200 g/day had 330, 210 and 160 g CH_4 associated with each kg LW gain. Good-quality diets result in lower emissions per kg DMI as well as higher levels of production, so white clover or lotus will generate only 100 or 80 g CH_4/kg LW gain, respectively (Table 1.2). This argument is similar for grazing dairy cows, but data must be interpreted with caution because in early lactation cows mobilize body reserves to provide energy for milk, so CH_4/milk production is underestimated. Conversely, in late lactation some of the feed energy is used to replenish LW, and the CH_4 expressed in terms of milk production is an overestimate. Ideally, CH_4 emissions should be evaluated on an annual rather than a short-term basis to avoid distorting the data. For example, the costs of lamb growth should include the cost of feeding the ewe as well as the lamb. If this approach is adopted, the CH_4 emissions per lamb (or kg lamb) will diminish if ewes have a long productive life (and lamb every year), and if they have multiple lambs per year. In effect, the costs of growing the ewe and her maintenance costs are diluted by high fecundity and a long productive life. Although improvements in production efficiency will reduce the amount of CH_4 emitted per unit of product, they will not necessarily reduce absolute CH_4 emissions if more animals are farmed (Clark *et al.*, 2005). A reduction in gross emissions will only occur if the animal numbers are static or increase more slowly than the decline in CH_4 emitted per unit of product.

Emissions from dung and manure

In pastoral grazing systems, between 20% and 50% of DM eaten is returned to pasture as dung or as manure or slurry in sheds and storage ponds. The percentage of DMI returned depends on feed quality, with lower values associated with higher-quality diets, and higher values with lower feed quality. The CH_4 from animal excreta is from two potential sources: entrapped CH_4 arising from enteric digestion and that arising from microbial fermentation in the excreta itself. The warm and moist conditions of the excreta, as well as the appropriate microflora and the organic substrates contained within it, provide conditions favourable for yielding hydrogen and ultimately CH_4 from fermentation. However, the potential for CH_4 emissions from animal excreta varies depending on physical form (shape, size, solid or sloppy), the amount of digestible matter, climate (temperature, humidity) and the amount of time it remains undisturbed (Saggar *et al.*, 2004b). Holter (1997) showed that CH_4 production from dung patches (*c.*0.1 g/kg dung DM) was, on average, six times lower than that from solid manure and 17 times lower than that from slurry. As dung patches are small in size, often in discrete portions, they can dry out and decompose relatively rapidly, reducing the amount of CH_4 produced (Flessa *et al.*, 2002). For example, emissions from beef dung patches declined rapidly as they dried, and total CH_4 from pasture-based beef dung was about 0.5 g/kg dung DM (Flessa *et al.*, 2002). The effect of forage diet was reported by Jarvis *et al.* (1995), who showed a fivefold range increase in CH_4 yields from dung (0.14–1.10 g CH_4/kg dung DM) from cows fed forage diets with a low C/N ratio. The importance of physical form is demonstrated with sheep dung, which is often deposited as discrete pellets up to about 20 mm in length. Carran *et al.* (2003) showed the duration of emissions did not exceed 10 days from sheep dung and although values ranged from 0.3 to 1.3 g CH_4/kg dung DM, this was mainly entrained CH_4 of enteric origin entrapped in pellets. Small quantities of dung are less likely to be anaerobic, so emissions from sheep dung and from cattle that defaecate while walking (leaving small, thin deposits) are likely to be less than 1.0 g CH_4/kg dung DM (Jarvis *et al.*, 1995; Flessa *et al.*, 2002; Carran *et al.*, 2003). Therefore, if 1 kg dung arose from 3 kg feed DM (i.e. 67% DM digestibility), the CH_4 arising from enteric digestion (about 21 g/kg feed DM) would be 63 g and that from dung <1.0 g – or about 1% of total animal emissions.

Soil and plant sinks and emissions

Soils may utilize or emit CH_4, but grasslands are generally considered sinks for atmospheric CH_4 (Dasselmar *et al.*, 1999) with net emissions occurring only from saturated environments such as swamps or rice fields. When CH_4 is formed in anaerobic soils, most is oxidized by methanotrophs at the aerobic interface with the saturated zone (Le Mer and Roger, 2001). Methanotropic activity in saturated soils is limited mainly by availability of oxygen (enabling formation of CO_2) and methanotrophs oxidize CH_4 from both soil and atmospheric sources (Horz *et al.*, 2002).

Although pastures may have anaerobic zones in wet conditions, with associated methanogenic activity, most evaluations of soil methanogens have been undertaken in rice fields (e.g. Minami, 1995) rather than grazed pastures. In their review, Le Mer and Roger (2001) summarized emissions (g CH_4/ha/day) from several environments and calculated means (and ranges) as: rice fields 1×10^3 ($0–29 \times 10^3$); swamps 720 ($0–17 \times 10^3$); temporarily submerged pasture emitted only 3 (0–216) g CH_4/ha/day. Factors affecting emissions from soils include water content, oxygen availability, organic matter content, pH, as well as temperature, fertilizer and microbial inhibitors (Le Mer and Roger, 2001), but the net emissions from pastures are negligible compared to those from ruminant digestion.

CH_4 oxidation (uptake; sink) is widespread across soil types and environments. Rates are affected by the amount and source of CH_4 (e.g. anaerobic soil conditions or decaying vegetation), oxygen availability and appropriate microflora, but methanotrophs have been isolated in wide-ranging environments. Le Mer and Roger (2001) summarized published data to show mean (with ranges) soil CH_4 oxidation activity (g CH_4/ha/day) to be: cultivated 5.5 (0–866), grassland 6.5 (2–485), non-cultivated upland 8.8 (0–288), forest 9.9 (0–1660) and wetland 172 ($0–7 \times 10^5$). Similarly, in a recent review of New Zealand methanotrophic studies, Saggar *et al.* (2007b) reported a net uptake of CH_4 (g/ha/day) to be about 30 for beech forest, 14 for pine forest, 4 for cropped soils and less than 3 for most New Zealand pastures. Their value for pastures (<1 kg CH_4/ha/year) is similar to reports from the Netherlands where the highest uptake from pasture was 1.1 kg CH_4/ha/year. Saggar *et al.* (2007b) also indicated that CH_4 uptake was 2–3 times higher in dry warm conditions compared to a cool and wet winter.

Recent claims that plants release substantial quantities of CH_4 (Keppler *et al.*, 2006) and subsequent extrapolation to a global situation have been challenged by a number of researchers. Parsons *et al.* (2006) demonstrated mathematical inconsistencies in the original calculations and in the extrapolation of laboratory findings. They concluded emissions from pasture plants could be 3 kg/ha/

annum, which is less than 3% of emissions from ruminants grazing the pasture. Although further measurements are needed to quantify possible CH_4 emissions from pasture plants, the many measurements of methanogenic and methanotropic activity from pastures have used chambers that include plants as well as soil. In nearly all situations these data have shown that pastures represent a net sink for CH_4.

Nitrous oxide

Annual N_2O emission from pastures range from negligible in arid or infertile regions to as much as 15 kg/ha in fertile, wet, pugged anaerobic situations (Velthof and Oenema, 1995; de Klein *et al.*, 2001; Bolan *et al.*, 2004). The availability of N in pastoral soils is an important factor affecting N_2O, and urine from animals can account for the majority of emissions in many environments.

Processes of nitrous oxide production

N_2O is produced during the microbial processes of nitrification and denitrification. Nitrification is the biological oxidation of ammonium (NH_4^+) to nitrite (NO_2^-) and nitrate (NO_3^-), with N_2O being a by-product. Denitrification is the stepwise biological reduction of nitrate to gaseous nitrogen (N_2), with N_2O being an obligatory intermediate (Fig. 1.2). Denitrification is generally accepted as the main source of N_2O from grazed pastoral soils (Stevens *et al.*, 1998). Although N_2O production from nitrification is less significant, nitrification provides the NO_3^- substrate and is often a critical prerequisite for denitrification.

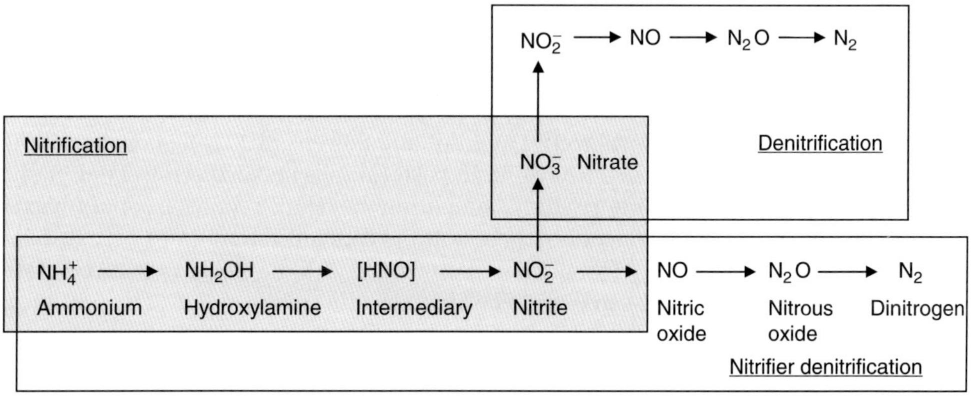

Fig. 1.2. Schematic representation of N_2O production from nitrification and denitrification. (From Wrage *et al.*, 2001.)

The nitrification process is carried out by a range of microorganisms, namely: (i) ammonia-oxidizing autotrophic bacteria (AOB); (ii) nitrite-oxidizing bacteria (NOB); (iii) ammonia-oxidizing archaea-bacteria; and (iv) heterotrophic nitrifying bacteria and fungi (Prosser, 2007). In pastoral soils, the most common forms of nitrifiers are the *Nitrosomonas* genus (AOB) and the *Nitrobacter* genus (NOB). Both groups are autotrophic and express various enzymes that catalyse the *aerobic* oxidation of ammonia or nitrite. The energy released from the oxidation processes is used for cell growth. *Nitrosomonas* oxidizes ammonia (or ammonium cations) into hydroxylamine (NH_2OH) and subsequently into nitrite (Ferguson *et al.*, 2007), whereas *Nitrobacter* oxidizes nitrite into nitrate. Ammonia oxidizing archaea have only recently been discovered. Little is known yet about their relative importance but they could be more abundant than AOBs and be a major contributor to nitrification (Prosser, 2007). Heterotrophic nitrification is less well understood but heterotrophic nitrifiers are believed to be widely distributed in soils (Ferguson *et al.*, 2007). They are physiologically diverse and are active in a range of environmental conditions (Prosser, 2007).

Under some soil conditions, including low oxygen concentrations, *Nitrosomonas* species (AOB) can also express enzymes that catalyse the reduction of NO_2^- and NO. As a result, these nitrifiers are able to reduce NO_2^- and NO to produce N_2O (Ferguson *et al.*, 2007), a process commonly referred to as nitrifier denitrification (Fig. 1.2). The reason for possessing this ability is not fully understood, but may be a mechanism to avoid toxic accumulation of nitrite. Nitrifier denitrification is likely to be greatest at the interface between aerobic and anaerobic conditions because oxygen is required to produce the nitrite (NO_2^-), and anoxic (anaerobic) conditions facilitate NO_2^- and NO reduction (Prosser, 2007). *Nitrosomonas* (AOB) do not contain N_2O reductase and thus are unable to reduce N_2O into N_2 (Ferguson *et al.*, 2007).

Denitrification is a four-step process (Fig. 1.2), carried out by a wide range of facultative aerobic heterotrophic bacteria, archaea and fungi. Each step of the process is catalysed by reductase enzymes that are produced under (near) anaerobic conditions (van Spanning *et al.*, 2007). The most well-defined denitrifying bacteria are *Paracoccus denitrificans* and *Pseudomonas* species, which can express all the enzymes required for the full denitrification pathway (nitrate-, nitrite-, nitric oxide- and N_2O-reductase). Microbial populations of denitrifying archaea are predominantly associated with extreme conditions such as saltwater lakes and hot springs (van Spanning *et al.*, 2007) and are therefore not likely to play a role in denitrification in pastoral soils. Laughlin and Stevens (2002) demonstrated the occurrence of fungal denitrification in pastoral soils in Northern Ireland. Based on results from a laboratory study, comparing the relative contributions of bacteria and fungi to denitrification, these authors concluded that fungi were responsible for most of the N_2O production. Fungi often lack the ability to reduce N_2O to N_2 (Shoun *et al.*, 1992), and could be an important source of N_2O. The ecological importance of fungi N_2O production warrants further investigation.

Both nitrification and denitrification can occur in soils as well as in aquatic systems (Seitzinger *et al.*, 2000), although soils are generally accepted to be the main sources of N_2O.

Factors affecting nitrous oxide emissions

The processes that control N_2O emissions from pastoral soils are: (i) the rate of nitrification and denitrification; (ii) the ratio of the end products of denitrification; and (iii) the diffusion of N_2O through the soil profile (Firestone and Davidson, 1989). These processes are all affected by a range of 'proximal' soil factors that are in turn affected by various more 'distal' factors (Fig. 1.3). As a result, the regulation of N_2O emissions is very complex and numerous field and laboratory experiments have been conducted in an attempt to untangle the complex interactions between the various regulators of N_2O emissions (e.g. Langeveld *et al.*, 1994; Dobbie *et al.*, 1999; Luo *et al.*, 1999; Van Groenigen *et al.*, 2005). However, since the same soil and climatic factors affect the three processes that determine the rate of N_2O release from the soil surface, it is often impossible to distinguish how each influences the N_2O flux. For example, soil aeration has a major influence on N_2O production from both nitrification and denitrification (e.g. Frolking *et al.*, 1998; Luo *et al.*, 1999; Bolan *et al.*, 2004), but it also affects N_2O diffusion rate from the soil. Researchers have shown that reduced soil aeration increases the N_2 to N_2O ratio of denitrification (Bolan *et al.*, 2004). This is probably because of restricted N_2O diffusion though the soil profile increases the chance of N_2O reduction and its subsequent emission as N_2 from the soil surface.

In grazed pastoral soils, N_2O emissions from soils are largely regulated by N inputs from excreta and fertilizer and soil aeration. The latter is a function of rainfall or irrigation, soil compaction and grazing management (de Klein and Eckard, 2007). N_2O emissions from aquatic systems are also largely regulated by N supply and oxygen availability. Liikanen and Martikainen (2003) found that oxygen availability in the water overlying the sediment was a key driver of N_2O emissions through its interaction with both nitrification and denitrification. When N enters the aquatic system as NH_4^+, oxygen is required to ensure nitrification to NO_3^-. However, increased nitrification consumes oxygen, decreasing the oxygen content of the sediment and thus increasing N_2O production from denitrification (Liikanen and Martikainen, 2003).

Nitrous oxide sources

The main source of N_2O in grazed systems is excreta deposited by grazing animals or applied to land as manure collected in the milking parlour or winter-housing system (Oenema *et al.*, 2005). Grazing ruminants utilize relatively little of the N in feed (Whitehead, 1995) and 75–90% of their dietary N (which originates from inputs of N fertilizer and biological N fixation) is recycled back into the system via urine and dung. N_2O emissions also occur immediately following applications of N fertilizer before this N is utilized by the plant/animal system. In addition to these *direct* N_2O emissions following excreta deposition and N fertilizer application, *indirect* N_2O emissions from these sources can occur through N that is lost from the system via NO_3^- leaching or NH_3 volatilization and subsequently emitted from surface waters or following redeposition of NH_3 to land (Fig. 1.4).

Until recently, the IPCC inventory methodology also included biological N fixation as a direct source of N_2O (IPCC, 1997). However, Rochette and Janzen (2005) concluded that the increase in soil N_2O emissions in legume crops was due

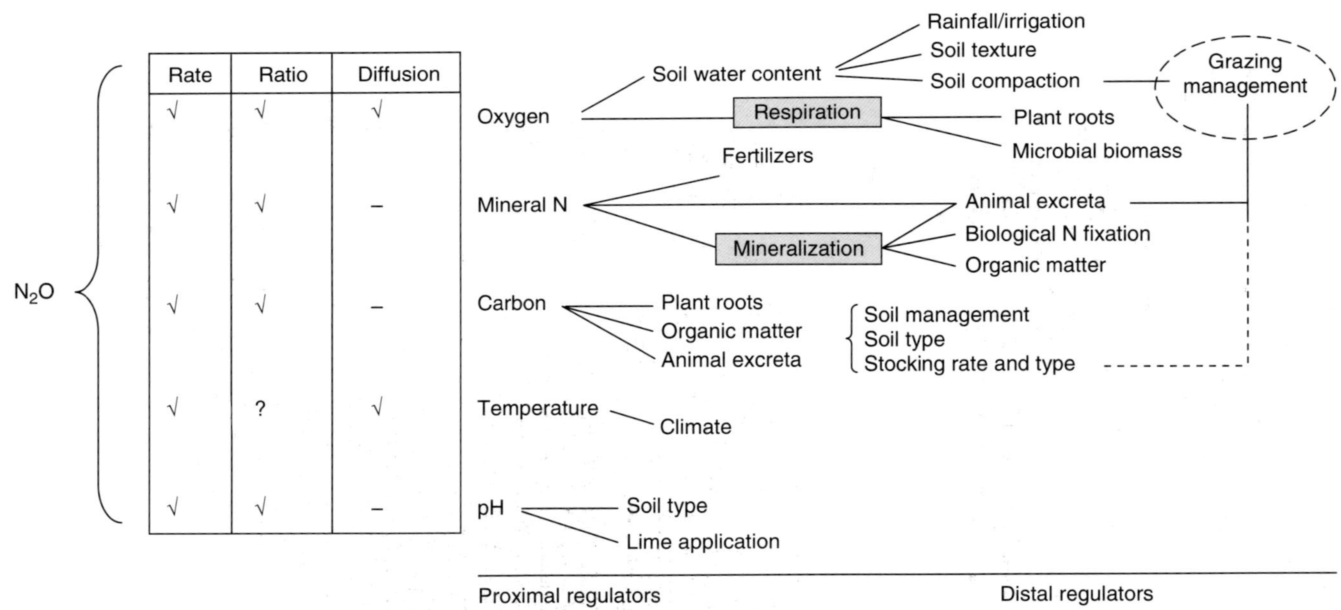

Fig. 1.3. Schematic diagram of factors affecting: (i) the rate of nitrification and denitrification; (ii) the N_2/N_2O ratio of denitrification; and (iii) N_2O diffusion from soil (√, affected by this factor; –, not affected by this factor; ?, unclear if affected by this factor). Shaded boxes represent biological processes. (Adapted from Tiedje, 1988 and de Klein *et al.*, 2001.)

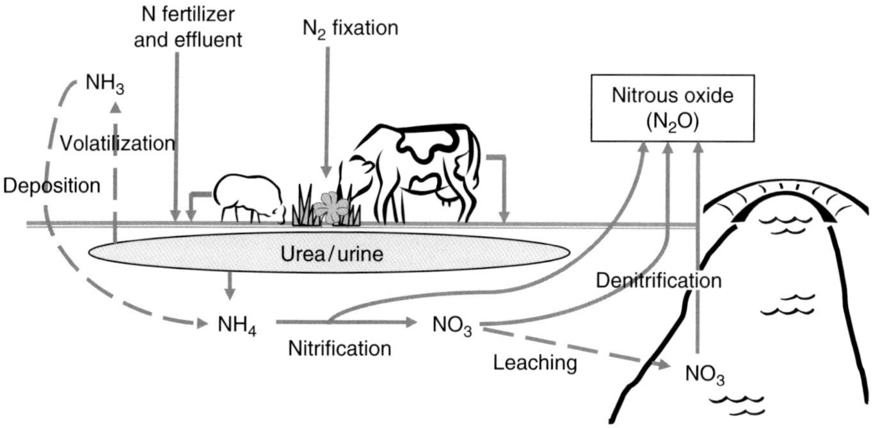

Fig. 1.4. Schematic diagram of N cycle and N_2O emissions in grazed systems. Dashed lines are pathways resulting in indirect losses of N_2O from grazing systems.

to the N release from root exudates and decomposition of residue after harvest, rather than from biological N fixation per se. Similarly, in clover-based pastoral grazing systems, N_2O emissions from biologically fixed N occur following the recycling of dietary N through the animal, and N_2O emissions from the fixation process itself are no longer included in estimates of N_2O emissions (IPCC, 2006).

In systems that are characterized by year-round grazing of grass/clover pastures and limited use of N fertilizer (e.g. on average 100–150 kg N/ha/year, typical of New Zealand and Australia), direct N_2O emissions from urine and dung patches generally contribute about 50–60% of the total N_2O emissions (de Klein *et al.*, 2001). Direct N_2O emissions from N fertilizer account for about 10–15%, while emissions from effluent applied to land generally contribute <5%. Indirect N_2O emissions from NO_3^- leaching and NH_3 volatilization make up about 20–30% of the emissions. It should be noted that indirect emissions originate from excreta as well as fertilizer N. Therefore, direct and indirect N_2O emissions from urine and dung from grazing animals can account for up to 80% of the total emissions and up to 20% from N fertilizer. In European all-grass dairy systems that are characterized by high N fertilizer inputs (over 200 kg N/ha/year) and animal housing for 5–6 months per year, direct deposition of excreta contributes less than 50% of the total emissions. The N fertilizer inputs and excreta N applied as manure or slurry contribute up to 35% and 15% of the emissions, respectively (e.g. Schils *et al.*, 2005).

N_2O EMISSION FACTORS: DIRECT EMISSIONS Direct N_2O emissions from excreta deposited to grazed grasslands largely derive from urine N, rather than dung N. The reason for this is twofold. First, in well-fed animals the amount of N excreted as urine represents 60–80% of the total excreted N (Whitehead, 1995) and the proportion of N excreted in urine increases as the N content of the diet increases. Second, the N_2O emission factor, i.e. N_2O emitted as percentage of N voided, is generally greater for urine than for dung, with median values of 1.5% and 0.2% for cattle urine and dung, respectively (Table 1.4). N_2O emissions from animal

Table 1.4. Summary of estimated N_2O emission factors for the main sources of N in grazed grasslands.

	Emission factors (N_2O~N as a percentage of N input)			
N source	Range	Median	Geomean	References
Excreta deposited during grazing				de Klein *et al.* (C.A.M. de Klein, 2001, unpublished data); Van Groenigen *et al.* (2005)
Grazing cattle (*n* = 12)	1.0–10	2.2	2.5	
Grazing sheep (*n* = 5)	0.2–1.7	1.0	0.8	
Cattle urine (*n* = 43)	0.0–14	1.5	0.9	
Cattle dung (*n* = 15)	0.0–4.0	0.2	0.3	
Sheep urine (*n* = 11)	0.0–2.6	0.4	0.3	
Sheep dung (*n* = 2)	0.0			
Synthetic urine (*n* = 147)	0.1–16	1.4	1.4	
Effluent/manure applied to land				de Klein *et al.* (2001); Saggar *et al.* (2004b)
Cattle slurry (*n* = 20)	0.01–17.4	1.8	1.0	
Injected cattle slurry (*n* = 6)	0.01–0.22	0.2	0.1	
Solid manure (*n* = 8)	0.06–20.0	4.0	1.3	
N Fertilizer				Bouwman *et al.* (2002b)
Ammonium Nitrate (*n* = 117)		0.8		
Ammonium Sulfate (*n* = 59)		1.0		
Calcium Ammonium Nitrate (*n* = 61)		0.7		
Nitrate-based fertilizers (*n* = 53)		0.9		
Urea (*n* = 98)		1.1		
Ammonia volatilization	0.1–22	2.8	2.6	Denier Van Der Gon and Bleeker (2005)
N leaching	0.05–2.5	0.75		IPCC (2006)

excreta collected as effluent or manure in the milking parlour or the winter-housing system and reapplied to land tend to be lower than from urine patches, as the N is applied more evenly to the soil, rather than in concentrated urine patches, although N_2O emissions from solid manure applied to pastures were similar to those of urine patches (Table 1.4).

In grazed systems, the direct losses of N_2O from dung and urine patches can be exacerbated if animal treading reduces soil aeration (Van Groenigen *et al.*, 2005), e.g. in wet soil conditions. Grazing management can therefore have a marked influence on N_2O emissions (Anger *et al.*, 2003; de Klein *et al.*, 2006). In addition, the temporary increase in soil pH in urine patches can increase the release of water-soluble C and this can stimulate N_2O emissions (Monaghan and Barraclough, 1993). As a result, N_2O emissions from grazed grasslands are generally greater than those from cut grassland, with values ranging from 1% to 7% (median 2.1%) and from 0.3% to 4% (median 1.8%) of total N inputs (Fowler *et al.*, 1997).

The N_2O emission factor for N fertilizer has been estimated in many studies summarized by Bouwman *et al.* (2002a; Table 1.4). The emission factor is affected by the rate, type and timing of N fertilizer applications, with N_2O emissions generally increasing exponentially with the amount of N applied (Velthof and Oenema, 1995; Whitehead, 1995; Eckard *et al.*, 2006). The type of fertilizer had a relatively small impact with estimated emission factors ranging from 0.8% to 1.1% (Table 1.4) and the mean global emission factor for N fertilizer used at typical application rates calculated at 0.9% (Bouwman *et al.*, 2002b).

N_2O EMISSION FACTORS: INDIRECT N_2O EMISSIONS Measurements of indirect N_2O emissions from NH_3 that are volatilized and redeposited on to pastoral soil are limited. Studies by Mosier *et al.* (1998a,b) suggested N_2O emission rates of 0.2–2% of the wet and dry deposition N from the atmosphere to a short grass steppe. Emission rates of N_2O from N deposited in semi-natural ecosystems ranged from 0.2% to 15% (Skiba *et al.*, 1998, 2004), with the highest rates found closest to the source of the deposited N. N_2O emission rates from N deposited to forest soils have been studied more widely, and the results from 22 peer-reviewed studies suggested that the N_2O emission factor for deposited N ranged from 0.1% to 22% with a median value of 2.6% (Denier Van Der Gon and Bleeker, 2005). These authors also indicated that the N_2O emission factor for atmospheric N deposition is land-use-specific, and thus depends on where volatilized N is ultimately deposited.

N lost from the system through leaching and runoff enters groundwater and surface water, riparian zones, rivers and eventually the ocean (Mosier *et al.*, 1998a). Although information is available on the amount of nitrate leaching/runoff from grazed soils, relatively little is known about the fraction of this leached N that is converted to N_2O. Clough *et al.* (1999) have demonstrated a mechanism whereby N_2O in the soil profile may be moved down the profile by rainfall or irrigation, but the fate of N_2O in the soil profile is not known and the potential exists for further denitrification in the subsoil (Clough *et al.*, 2000). The uncertainty of the fate of leached N is reflected in the wide uncertainty range around the 1996 IPCC default value of 2.5% (0.2–12% of N leaching/runoff). However, recent studies have indicated that the fraction of leached N that is converted to N_2O is much lower than this default value (e.g. Reay *et al.*, 2003; Sawamoto *et al.*, 2005; Clough *et al.*, 2006). As a result, in the 2006 revision of the IPCC guidelines, this value is reduced to 0.75% with an uncertainty range of 0.05–2.5% (IPCC, 2006).

Nitrous oxide sinks

As explained above, denitrifying bacteria can reduce N_2O to N_2 gas and can act as a source as well as a sink of N_2O in soils. In addition, nitrifier denitrification can also reduce N_2O to N_2 and thus act as an N_2O sink (Wrage *et al.*, 2001; Chapuis-Lardy *et al.*, 2007). Negative fluxes of N_2O (i.e. N_2O consumption by soil) have been reported in various studies and summarized by Chapuis-Lardy *et al.* (2007), who suggested that negative fluxes are too frequent and substantial to be dismissed as experimental error. However, in most studies, gross N_2O production is generally larger than gross N_2O consumption, resulting in net N_2O production. Whether a soil acts as a sink for N_2O depends on the ratio between N_2O production and N_2O reduction to N_2 and is thus regulated by the relative activities of nitric oxide reductase (NIR) and N_2O reductase (NOR; Fig. 1.2). If NOR activity increases relative to NIR activity, the N_2 to N_2O ratio increases and gross N_2O consumption occurs. NOR activity is affected by several factors and net N_2O consumption in pastoral soils has sometimes been measured under high soil water content and low soil N availability but it has not yet been studied systematically (Chapuis-Lardy *et al.*, 2007). However, in grazed pastoral soils where N availability is high due to either excreta N and/or N fertilizer applications, net N_2O consumption is unlikely to exceed N_2O production.

Carbon dioxide

Sources and sinks of CO_2

A detailed description of C flows within grassland soils is provided in Chapter 2 (this volume). Here, we give a brief summary of the key sources and sinks of CO_2 under pastoral grazing systems (Fig. 1.5).

Net CO_2 emissions from pastoral soils occur when CO_2 sequestration into the systems through gross primary production (photosynthesis) is smaller than the CO_2 emitted from the system through plant, soil and animal respiration (Fig. 1.5), and cultivation associated with pasture renewal. In grazed pastures, C sequestered into the above-ground biomass is turned over rapidly, rather than being accumulated as standing biomass such as in trees, and some is incorporated into the soil (Soussana *et al.*, 2004). In grazed systems, organic material is returned to the soil as faeces, and C mineralization due to cultivation is less, or absent, compared to

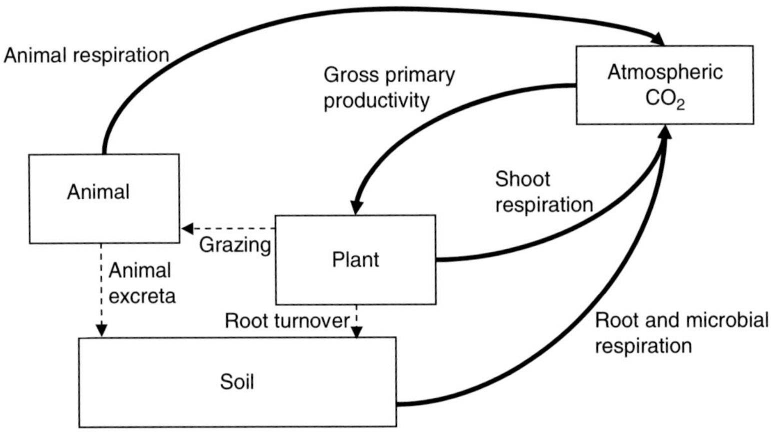

Fig. 1.5. Schematic diagram of carbon cycling in pastoral grazing. (Adapted from Soussana *et al.*, 2004.)

cropped soils. Consequently, pastoral soils contain larger amounts of C per unit area than other ecosystem soils (Soussana *et al.*, 2004). Therefore, the key factor controlling the amount of C returned to pastoral soils is the amount of net primary production (NPP) that is removed by the grazing animal.

In addition to CO_2 derived from plant, soil and animal respiration, CO_2 emissions from fossil fuel use are also part of pastoral grazing systems. In particular, CO_2 emissions associated with the production of fertilizers, machinery or electricity that are used in the farming system, so-called embedded CO_2, can be a major source of fossil fuel-derived CO_2 (Wheeler *et al.*, 2003). However, compared to CH_4 and N_2O, these emissions contribute a relatively small part to total CO_2-equivalent emissions from a pastoral grazing system (Wheeler *et al.*, 2003; Schils *et al.*, 2005).

Factors affecting emissions/sequestration in grazed pastoral systems

Management practices that increase the specific growth rate of the pasture, e.g. through improved fertility, will increase the flow of C to soil. For example, development of unimproved pasture will increase the soil C stock (Soussana *et al.*, 2004). However, further intensification of improved pasture through intensive N fertilizer use can reduce soil C stocks due to increased SRs, and thus increased NPP removal, associated with increased productivity. In addition, N fertilization can also increase mineralization and decomposition of soil organic matter (Soussana *et al.*, 2004). As noted already, increased N fertilizer usage will also increase CO_2 emissions associated with the manufacture of the fertilizer. It has been estimated that 0.4 t CO_2 is emitted per t of N fertilizer produced (Steinfeld *et al.*, 2006).

Estimating/Modelling GHG Emissions from Pastoral Farming

CH_4 models

Most CH_4 research from animals grazing pastures has focused on inventory and opportunities for mitigating enteric CH_4 production. This focus has been brought about by the dominance of enteric CH_4 relative to manure and other fluxes and because of the very significant loss of potentially useful feed energy as CH_4. Mechanistic models have attempted to explain CH_4 production on the basis of rumen biochemistry (Baldwin *et al.*, 1987; Dijkstra *et al.*, 1992), and although Mills *et al.* (2001) reported a high correlation ($r^2 = 0.72$) between predicted and observed CH_4 production, uncertain input values (Benchaar *et al.*, 1998), perhaps associated with unexplained *in vivo* variations (e.g. Table 1.1), are likely to prevent accurate estimation of methanogenesis from ruminants.

Although mechanistic models may provide the most accurate prediction of methanogenesis from contrasting diets, the use of simple empirical equations (e.g. Blaxter and Clapperton, 1965; Yan *et al.*, 2000; Kurihara *et al.*, 2002; Garnsworthy, 2004) will enable strategies to be evaluated for pastoral management to lessen emissions. Recent models have predicted emissions from various grazing systems (e.g. OVERSEER, Wheeler *et al.*, 2003; Moorepark dairy systems model, Lovett *et al.*, 2006; GrassGro, Cohen *et al.*, 2004), but all are hampered to some extent by variation in reported emission rates (Table 1.1) and lack of data concerning effects of supplementation on emissions, which leads to uncertain inputs upon which predictions are based.

Nitrous oxide models

Due to the highly spatial and temporal nature of N_2O emissions, a large research investment of near-continuous measurements on spatially integrated areas is required to accurately measure N_2O emissions from soils. However, field measurements are expensive, so simulation models have become an important means for improving our understanding of the complex interactions between drivers of N_2O emissions and of evaluating practices that can reduce emissions. In recent years, a substantial effort has been put into the development of N_2O emission models (Table 1.5). These models range in scale and complexity from annual time step, annual accounting models such as the N_2O inventory methodology of the IPCC (IPCC, 1997) and the nutrient budgeting model OVERSEER (Wheeler *et al.*, 2003), to farm-scale models such as FASSET (Chatskikh *et al.*, 2005; Hutchings *et al.*, 2007) and Moorepark Dairy Systems Model (MDSM; Lovett *et al.*, 2006), to detailed process-based daily time step models such as the Denitrification/Decomposition (DNDC; Li *et al.*, 1992), DAYCENT (Del Grosso *et al.*, 2005, 2006), DairyMod (Eckard *et al.*, 2006) and the Pasture Simulation Model (PaSim; Calanca *et al.*, 2007).

The annual time step models are generally accounting or inventory models that estimate annual N_2O emissions from farms, regions and/or countries by multiplying N sources in grazed pastures with a single N_2O emission factor for each source. For example, the IPCC methodology, which is used by many countries to estimate their national GHG inventory, estimates N_2O emissions from the annual anthropogenic N inputs into a system, such as N fertilizer use, N excreta deposition on grazed pastures, N in waste, biological N fixation and N deposition to land following ammonia volatilization, multiplied by a (default) emission factor for each source (IPCC, 1997). The default emission factors are generic values and generally do not account for the effect of climate, soil and/or management practices on N_2O emissions. In contrast, the process-based daily time step models tend to simulate changes in the proximal factors affecting the emissions (e.g. N availability, soil moisture content, soil temperature) from information on distal factors (e.g. climate, soil type, management practices). The models apply algorithms to estimate the effect of the proximal factors on N_2O emissions and to improve our understanding of drivers of N_2O emissions at a process level. Farm-scale models tend to use a combination of accounting and dynamic modelling approaches based on either daily, monthly or annual time steps. The annual time step models are more generally applicable but are surrounded by great uncertainties, while the daily time step models may yield more reliable results but are applicable only to the system or the climatic condition for which they were developed.

The different N_2O models have different purposes, from annual accounting to evaluating potential mitigation strategies and/or improving understanding of the drivers of N_2O emissions at a process level. The level of complexity and the detail of the required input data generally increase with the extent to which a model

Table 1.5. Overview of selected N_2O emission models, their scale and application.

Model name	Scale	Time step	Comment	References
IPCC	National	Annual	Generalized GHG accounting/ inventory methodology; does not currently account for farm or regional-scale variation in soil or climatic conditions or management practices	IPCC (1997)
OVERSEER	Farm/ regional	Annual	GHG accounting largely following IPCC inventory approach; some allowance for soil and management impacts; includes options for evaluating mitigation strategies	Ledgard *et al.* (1999); Wheeler *et al.* (2003)
Moorepark Dairy Systems Model	Farm	Annual	GHG accounting largely following IPCC inventory approach; evaluating mitigation strategies	Lovett *et al.* (2006)
FASSET	Farm	Daily	GHG accounting largely following IPCC inventory approach; evaluating mitigation strategies	Chatskikh *et al.* (2005); Hutchings *et al.* (2007)
DNDC	Paddock/ farm/ regional	Daily	Dynamic modelling; biogeochemical model for predicting C sequestration and GHG emissions; farm or regional-scale GHG accounting and evaluating mitigation strategies	Li *et al.* (1992); Saggar *et al.* (2004a; 2007a)
DAYCENT	Paddock/ farm	Daily	Dynamic modelling; biogeochemical model for predicting C sequestration and GHG emissions	Del Grosso *et al.* (2005; 2006)
DairyMod	Paddock/ farm	Daily	Dynamic modelling	Eckard *et al.* (2006)
CASA	Ecosystem	Daily	Dynamic modelling; biogeochemical model for predicting C sequestration and GHG emissions; large-scale modelling. Detailed management practices not accounted for	Potter *et al.* (1996)

accounts for soil, climate and management variability. Depending on the purpose of the model, a balance is often required between available input data and reliability of the estimates. Although N_2O emission models have been developed and improved in recent years, the spatial and temporal variability and complex interactions between the drivers of N_2O emissions and dynamic nature of the emissions challenge our ability to predict emissions (Calanca *et al.*, 2007). Ironically, the spatial and temporal variability of N_2O emissions that underpins the need for these models also hampers their development.

GHG Profiles from Grazing Systems

As already discussed, CH_4 and N_2O are the dominant GHGs from pastoral grazing, although their absolute and relative contribution to total GHG emissions can vary for different systems. Examples of GHG profiles for grazing systems in New Zealand and the Netherlands are given in Table 1.6. These profiles represent the on-farm emissions only, although the C emissions associated with lime and fertilizer manufacturing and use are also included. These data show that total GHG emissions from clover-based pastures are lower than fertilized grass pastures, in part because of the CO_2 emissions associated with the manufacturing of N fertilizer that is required for grass pastures (Steinfeld *et al.*, 2006). Similarly, animals that are raised under year-round grazing generally have lower emissions per hectare than animals that are housed during the autumn/winter months.

However, emissions per unit of product and total GHG emissions from the New Zealand year-round dairy grazing systems are higher, as the milk production per animal is much higher in the Dutch systems (Table 1.6). Emissions intensity, i.e. GHG emissions per unit of product, has recently been suggested as a new approach to reducing GHG emissions. The intensity of GHG emissions can be reduced through

Table 1.6. Examples of GHG emission profiles (in kg CO_2 equivalent/ha) from different pastoral grazing systems in New Zealand (NZ) and the Netherlands (NL). Values in brackets represent the percentage contribution to total GHG emissions.

	NZ dairy[a,b]	NZ sheep/ beef[a,c]	NL dairy[d] Grass N	NL dairy[e] Grass/clover
Methane				
Enteric fermentation	n/a	n/a	5,628	4,809
Effluent/manure management	n/a	n/a	1,932	1,512
Grazing	n/a	n/a	57	50
Total	5,295 (59)	3,579 (71)	7,617 (64)	6,371 (72)
Nitrous oxide				
N Fertilizer use	656	173	974	97
Urine/dung deposited during grazing	1,498	948	1,413	1,267
Effluent applied to land	232	n/a	341	292
Indirect emissions	497	218	390	365
Total	2883 (32)	1339 (26)	3118 (26)	2021 (23)
Carbon				
Energy/fuel use	272	63	263	219
Lime use	174	0		
Embedded C in N fertilizer	300	72	825	207
Total	746 (8)	135 (3)	1,088 (9)	426 (5)
Total GHG	8,924	5,053	11,823	8,818

[a]GHG profiles estimated using OVERSEER (Wheeler *et al.*, 2003).
[b]NZ dairy farm: year-round grazing grass/clover pasture; 100 kg fertilizer N/ha/year; 2.8 cows/ha; 350 kg milk solids/cow, i.e. *c.*4,060 l milk/cow.
[c]NZ sheep/beef farm: year-round grazing on grass/clover pasture; 24 kg fertilizer N/ha/year; 16 stock units/ha.
[d]From Schils *et al.* (2005). NL dairy farm grass N: part housing, part grazing grass only pasture; 275 kg fertilizer N/ha/year; 2.2 cows/ha; 8,095 kg FPCM/cow; i.e. *c.*7,600 l milk/cow.
[e]From Schils *et al.* (2005). NL dairy farm grass/clover: part housing, part grazing grass/clover pasture; 69 kg fertilizer N/ha/year; 1.9 cows/ha; 8,294 kg FPCM/cow; i.e. *c.*7,800 l milk/cow; n/a, not available.

reducing emissions per unit of product, or by increasing productivity per unit of emissions. This is an important consideration for identifying management strategies that can have the largest reduction in environmental emissions for a given level of production. However, to achieve a net reduction in total GHGs, any reduction in GHG per unit of product needs to be greater than the increase in production (products per hectare). The GHG profiles presented in Table 1.6 are on-farm emissions and do not include estimates of emissions associated with processing, transport and the use of products, or any losses of soil C associated with cultivation and pasture replacement. A full assessment of GHG associated with contrasting systems requires a life cycle assessment to allow evaluation of the complete impact of these systems on GHG emissions (van der Nagel *et al.*, 2003; Casey and Holden, 2005; Basset-Mens *et al.*, 2008).

Threats to Pastoral Farming

As a result of ongoing intensification of pastoral grazing systems, GHG emissions from these systems have increased globally (Clark *et al.*, 2005). These represent both a political and a biophysical threat to the ongoing sustainability of pastoral grazing. In the current political climate, countries that have ratified the Kyoto Protocol are legally bound to take responsibility for any GHG emissions in excess of agreed targets in the first commitment period (2008–2012), either by reducing emissions to the agreed targets or by offsetting any excess by purchasing credits on the international market. For most industrialized countries, the contribution of pastoral farming to the national GHG emissions is limited (often less than 20%), and mitigation strategies have largely focused on reducing CO_2 emissions from fossil fuel burning. However, in countries such as New Zealand, Ireland and Uruguay where pastoral agriculture is the dominant sector and the ratio of livestock to human population is relatively high, CH_4 and N_2O emissions contribute up to 50% of the total GHG emissions (de Klein and Ledgard, 2005; Lovett *et al.*, 2006). For these countries, the development of mitigation strategies for reducing GHG emissions from pastoral agriculture is likely to be a priority. Furthermore, less-developed countries have a substantial portion of GHG emissions from grazing ruminants, and improved pastoral production (especially in tropical environments) will increase emissions from these sources. Much of the increase will be associated with application of nitrogenous fertilizer and increased ruminant numbers.

Increased GHG emissions and the subsequent effects on climate change can also have biophysical impacts on pastoral farming including: (i) increase in the unpredictability and frequency of extreme weather events such as floods, drought or storms; (ii) loss of biodiversity in more fragile environments; (iii) sea-level rises submerging coastal agricultural land; (iv) shift in agro-ecological climate zones; (v) increased pasture production due to increased temperatures; and (vi) increased threat of invasions of pest and diseases (Bazzaz *et al.*, 1996). In particular, future increases in tropospheric CO_2 concentrations and the likely influence of the so-called CO_2 fertilization effect could have major implications for pastoral farming. Recent work has confirmed that elevated CO_2 levels can increase both plant production and the resilience of grassland ecosystems to lower levels of precipitation

(Soussana and Lüscher, 2007). However, elevated CO_2 levels can also induce a biogeochemical feedback mechanism resulting in a reduction in N and/or phosphorus (P) availability. This mechanism, referred to as progressive nutrient limitation (PNL), has been examined in recent studies (Finzi *et al.*, 2006; Gill *et al.*, 2006; Hungate *et al.*, 2006) that suggest that the CO_2 fertilization effect on plant growth could be restricted due to the reduced availability of N and/or P. In legume-based pasture, where external N inputs occur via biological N fixation, P is likely to be the main limiting factor on plant growth. Although the PNL effect can be alleviated by external inputs of these nutrients, a better understanding of the full impacts of elevated CO_2 and/or increased temperatures on pastoral grazing systems is required to allow the development of management practices that are adapted to global changes (Soussana and Lüscher, 2007).

In summary, both the development of GHG mitigation strategies and the ability to understand and adapt to the changing environment will be critical for ensuring the ongoing sustainability of pastoral grazing systems. Potential GHG mitigation strategies and management practice adapted to climate change will be discussed for specific pastoral grazing systems in Chapter 2.

References

Alcock, D. and Hegarty, R.S. (2006) Effects of pasture improvement on productivity, gross margin and methane emissions of a grazing sheep enterprise. *International Congress Series* 1293, 103–106.

Anger, M., Hoffmann, C. and Kuhbauch, W. (2003) Nitrous oxide emissions from artificial urine patches applied to different N-fertilized swards and estimated annual N2O emissions for differently fertilized pastures in an upland location in Germany. *Soil Use and Management* 19, 104–111.

Baldwin, R.L., Thornley, J.H.M. and Beever, D.E. (1987) Metabolism of the lactating cow 11. Digestive elements of a mechanistic model. *Journal of Dairy Research* 54, 107–131.

Basset-Mens, C., Ledgard, S.F. and Boyes, M. (2008) Eco-efficiency of intensification scenarios for milk production in New Zealand. *Journal of Ecological Economics*. doi:10.1016/j.ecolecon.2007.11.017.

Bazzaz, F.A., Bassow, S.L., Berntson, G.M. and Thomas, S.C. (1996) Elevated CO_2 and terrestrial vegetation: Implications for and beyond the global carbon budget. In: Walker, B. and Steffen, W. (eds) *Global Change and Terrestrial Ecosystems*. Cambridge University Press, Cambridge, pp. 43–76.

Benchaar, C., Rivest, J., Pomar, C. and Chiquette, J. (1998) Comment on the paper 'prediction of methane production from dairy cows using existing mechanistic models and regression equations'. Reply to a letter by Donovan, K.C. and Baldwin, R.L. *Journal of Animal Science* 76, 2752–2753.

Benchaar, C., Pomar, C. and Chiquette, J. (2001) Evaluation of dietary strategies to reduce methane production in ruminants: a modelling approach. *Canadian Journal of Animal Science* 81, 563–574.

Blaxter, K.L. and Clapperton, J.L. (1965) Prediction of the amount of methane produced by ruminants. *The British Journal of Nutrition* 19, 511–522.

Boadi, D.A. and Wittenberg, K.M. (2002) Methane production from dairy and beef heifers fed forages differing in nutrient density using the sulphur hexafluoride (SF_6) tracer gas technique. *Canadian Journal of Animal Science* 82, 201–206.

Boadi, D.A., Wittenberg, K.M. and McCaughey, W.P. (2002) Effects of grain supplementation on methane production of grazing steers using the sulphur hexafluoride (SF_6) tracer gas technique. *Canadian Journal of Animal Science* 82, 151–157.

Bolan, N.S., Surinder, S., Luo, J., Rita, B. and Jagrati, S. (2004) Gaseous emissions of nitrogen from grazed pastures: processes, measurements and modelling, environmental implications, and mitigation. *Advances in Agronomy* 84, 37.

Bouwman, A.F., Boumans, L.J.M. and Batjes, N.H. (2002a) Emissions of N_2O and NO from fertilized fields: summary of available measurement data. *Global Biogeochemical Cycles* 16, 6-1–6-13.

Bouwman, A.F., Boumans, L.J.M. and Batjes, N.H. (2002b) Modeling global annual N_2O and NO emissions from fertilized fields. *Global Biogeochemical Cycles* 16, 28-1–28-9.

Burke, J.L., Waghorn, G.C. and Brookes, I.M. (2004) An evaluation of sulla (*Hedysarum coronarium*) with pasture, white clover and lucerne for lambs. *Proceedings of the New Zealand Society of Animal Production* 62, 152–156.

Calanca, P., Vuichard, N., Campbell, C., Viovy, N., Cozic, A., Fuhrer, J. and Soussana, J.F. (2007) Simulating the fluxes of CO_2 and N_2O in European grasslands with the Pasture Simulation Model (PaSim). *Agriculture, Ecosystems and Environment* 121, 164–174.

Carran, R.A., Dewar, D. and Theobald, P.W. (2003) *Methane and Nitrous Oxide Emissions from Sheep Dung.* Report prepared for the Ministry of Agriculture and Forestry, p. 14.

Casey, J.W. and Holden, N.M. (2005) Analysis of greenhouse gas emissions from an average Irish milk production system. *Agricultural Systems* 86, 97–114.

Chapuis-Lardy, L., Wrage, N., Metay, A., Chotte, J.-L. and Bernoux, M. (2007) Soils, a sink for N_2O? A review. *Global Change Biology* 13, 1–17.

Chatskikh, D., Olesen, J.E., Berntsen, J., Regina, K. and Yamulki, S. (2005) Simulation of effects of climate, soils and management on N_2O emission from grasslands. *Biogeochemistry* 76, 395–419.

Clark, H., Pinares, P.C. and de Klein, C.A.M. (2005) Methane and nitrous oxide emissions from grazed grasslands. In: McGilloway, D.A. (ed.) *Grassland: A Global Resource. Plenary and Invited Papers from the XX International Grassland Congress Dublin Ireland 26 June–1 July 2005.* Wageningen Academic Publishers, Wageningen, the Netherlands, pp. 279–293.

Clark, H., Kelliher, F.M., Waghorn, G.C., Johnstone, P. and Rys, G. (2007) An updated New Zealand enteric methane inventory calculated using algorithms developed from an analysis of New Zealand experiments conducted between 1997 and 2005. *Greenhouse Gas and Animal Agriculture – Conference Book*, November 2007. Christchurch, New Zealand, p. 163.

Clough, T.J., Jarvis, S.C., Dixon, E.R., Stevens, R.J., Laughlin, R.J. and Hatch, D.J. (1999) Carbon induced subsoil denitrification of 15N-labelled nitrate in 1 m deep soil columns. *Soil Biology and Biochemistry* 31, 31–41.

Clough, T.J., Sherlock, R.R. and Cameron, K.C. (2000) Entrapment and displacement of nitrous oxide in a drained pasture soil. *Nutrient Cycling in Agroecosystems* 57, 191–193.

Clough, T.J., Bertram, J.E., Sherlock, R.R., Leonard, R.L. and Nowicki, B.L. (2006) Comparison of measured and EF5-r-derived N_2O fluxes from a spring-fed river. *Global Change Biology* 12, 352–363.

Cohen, R.H.D., Stevens, J.P., Moore, A.D., Donnelly, J.R. and Freer, M. (2004) Predicted methane emissions and metabolisable energy intakes of steers grazing a grass/alfalfa pasture and finished in a feedlot or at pasture using the Grass Gro decision support tool. *Canadian Journal of Animal Science* 84, 125–132.

Cosgrove, G.P., Waghorn, G.C., Anderson, C.B., Peters, J.S., Smith, A., Molano, G. and Deighton, M. (2008) The effect of oils fed to sheep on methane production and digestion of ryegrass pasture. *Australian Journal of Experimental Agriculture* 28, 189–192.

Dasselmar, A.P., van Beusichem, M.L. and Oenema, O. (1999) Effects of nitrogen input and grazing on methane fluxes of extensively and intensively managed grasslands in the Netherlands. *Biology and Fertility of Soils* 29, 24–30.

de Klein, C.A.M. and Eckard, R.J. (2007) Targeted technologies for nitrous oxide abatement from animal agriculture. *Australian Journal of Experimental Agriculture* 48, 14–20.

de Klein, C.A.M. and Ledgard, S.F. (2005) Nitrous oxide emissions from New Zealand agriculture – key sources and mitigation strategies. *Nutrient Cycling in Agroecosystems* 72, 77–85.

de Klein, C.A.M., Sherlock, R.R., Cameron, K.C. and van der Weerden, T.J. (2001) Nitrous oxide emissions from agricultural soils in New Zealand – a review of current knowledge and directions for future research. *Journal of The Royal Society of New Zealand* 31, 543–574.

de Klein, C.A.M., Smith, L.C. and Monaghan, R.M. (2006) Restricted autumn grazing to reduce nitrous oxide emissions from dairy pastures in Southland, New Zealand. *Agriculture, Ecosystems and Environment* 112, 192–199.

De Ramus, H.A., Clement, T.C., Giampola, D.D. and Dickison, P.C. (2003) Methane emissions of beef cattle on forages: efficiency of grazing management systems. *Journal of Environmental Quality* 32, 269–277.

Del Grosso, S.J., Mosier, A.R., Parton, W.J. and Ojima, D.S. (2005) DAYCENT model analysis of past and contemporary soil N_2O and net greenhouse gas flux for major crops in the USA. *Soil and Tillage Research* 83, 9–24.

Del Grosso, S.J., Parton, W.J., Mosier, A.R., Walsh, M.K., Ojima, D.S. and Thornton, P.E. (2006) DAYCENT national-scale simulations of nitrous oxide emissions from cropped soils in the United States. *Journal of Environmental Quality* 35, 1451–1460.

Denier Van Der Gon, H. and Bleeker, A. (2005) Indirect N_2O emission due to atmospheric N deposition for the Netherlands. *Atmospheric Environment* 39, 5827–5838.

Dijkstra, J., Neal, H.D.StC., Beever, D.E. and France, J. (1992) Simulation of nutrient digestion, absorption and outflow in the rumen: model description. *The Journal of Nutrition* 122, 2239–2256.

Dobbie, K.E., McTaggart, I.P. and Smith, K.A. (1999) Nitrous oxide emissions from intensive agricultural systems: variations between crops and seasons, key driving variables, and mean emission factors. *Journal of Geophysical Research* 104, 26891–26899.

Eckard, R., Johnson, I. and Chapman, D. (2006) Modelling nitrous oxide abatement strategies in intensive pasture systems. *International Congress Series* 1293, 76–85.

Ferguson, S.J., Richardson, D.J. and van Spanning, R.J.M. (2007) Biochemistry and molecular biology of nitrification. In: Bothe, H., Ferguson, S.J. and Newton, W.E. (eds) *Biology of the Nitrogen Cycle*. Elsevier, Amsterdam, the Netherlands, pp. 209–222.

Finlay, B.J., Esteban, G., Clarke, K.J., Williams, A.G., Embley, T.M. and Hirt, R.P. (1994) Some rumen ciliates have endo-symbiotic methanogens. *FEMS Microbiology letters* 117, 157–162.

Finzi, A.C., Moore, D.J.P., DeLucia, E.H., Lichter, J., Hofmockel, K.S., Jackson, R.B., Kim, H.S., Matamala, R., McCarthy, H.R., Oren, R., Pippen, J.S. and Schlesinger, W.H. (2006) Progressive nitrogen limitation of ecosystem processes under elevated CO_2 in a warm-temperate forest. *Ecology* 87, 15–25.

Firestone, M.K. and Davidson, E.A. (1989) Microbiological basis of NO and N_2O production and consumption in soil. In: Andreae, M.O. and Schimel, D.S. (eds) *Exchange of Trace Gases Between Terrestrial Ecosystems and the Atmosphere*, Wiley, Chichester, UK, pp. 7–21.

Flessa, H., Ruser, R., Dorsch, P., Kamp, T., Jimenez, M.A., Munch, J.C. and Beese, F. (2002) Integrated evaluation of greenhouse gas emissions (CO_2, CH_4, N_2O) from two farming systems in southern Germany. *Agricultural Ecosystems and Environment* 91, 175–189.

Fowler, D., Skiba, U. and Hargreaves, K.J. (1997) Emissions of nitrous oxide from grasslands. In: Jarvis, S.C. and Pain, B.F. (eds) *Gaseous Nitrogen Emissions from Grasslands*. CAB International, Wallingford, UK, pp. 147–164.

Frolking, S.E., Mosier, A.R., Ojima, D.S., Li, C., Parton, W.J., Potter, C.S., Priesack, E., Stenger, R., Haberbosch, C., Dorsch, P., Flessa, H. and Smith, K.A. (1998) Comparison of N_2O emissions from soils at three temperate agricultural sites: simulations of year-round measurements by four models. *Nutrient Cycling in Agroecosystems* 52, 77–105.

Garnsworthy, P.C. (2004) The environmental impact of fertility in dairy cows: a modelling approach to predict methane and ammonia emission. *Animal Feed Science and Technology* 112, 211–223.

Gill, R.A., Anderson, L.J., Polley, H.W., Johnson, H.B. and Jackson, R.B. (2006) Potential nitrogen constraints on soil carbon sequestration under low and elevated atmospheric CO_2. *Ecology* 87, 41–52.

Graham N. McC. (1980) Variation in energy and nitrogen utilization by sheep between weaning and maturity. *Australian Journal of Agricultural Research* 31, 335–345.

Grainger, C., Clarke, T., Auldist, M.J., Beauchemin, K.A., McGinn, S.M., Waghorn, G.C. and Eckhard, R.J. (2008) Mitigation of greenhouse gas emissions from dairy cows fed pasture and grain through supplementation with Acacia mearnsii tannins. *Journal of Dairy Science* (submitted).

Hegarty, R.S. (2004) Genotype differences and their impact on digestive tract function of ruminants: a review. *Australian Journal of Experimental Agriculture* 44, 459–467.

Hodgson, J. (1990) *Grazing Management, Science into Practice.* Longman Scientific and Technical, Essex, UK.

Holter, P. (1997) Methane emissions from Danish cattle dung pats in the field. *Soil Biology and Biochemistry* 29, 31–37.

Howden, S.M., White, D.H., McKeon, G.M., Scanlan, J.C. and Carter, J.O. (1994) Methods for exploring management options to reduce greenhouse gas emissions from tropical grazing systems. *Climate Change* 27, 49–70.

Horz, H.-P., Rughubansh, A.S., Heyer, J., Kammann, C., Conrad, R. and Dunfield, P.F. (2002) Activity and community structure of methane oxidising bacteria in a wet meadow soil. *FEMS Microbiology Ecology* 41, 247–257.

Hungate, B.A., Johnson, D.W., Dijkstra, P., Hymus, G., Stiling, P., Megonigal, J.P., Pagel, A.L., Moan, J.L., Day, F., Li, J., Hinkle, C.R. and Drake, B.G. (2006) Nitrogen cycling during seven years of atmospheric CO_2 enrichment in a scrub oak woodland. *Ecology* 87, 26–40.

Hunter, R.A. (2007) Methane production by cattle in the tropics. *British Journal of Nutrition* 98, 657.

Hutchings, N.J., Olesen, J.E., Petersen, B.M. and Berntsen, J. (2007) Modelling spatial heterogeneity in grazed grassland and its effects on nitrogen cycling and greenhouse gas emissions. *Agriculture, Ecosystems and Environment* 121, 153.

IPCC (1997) *Revised 1996 IPCC Guidelines for National Greenhouse Gas Inventories: Volume 3 Reference Manual.* Intergovernmental Panel on Climate Change. IPCC/OECD/IEA, Paris, France.

IPCC (2006) *2006 IPCC Guidelines for National Greenhouse Gas Inventories.* Prepared by the National Greenhouse Gas Inventories Programme, Eggleston H.S., Buendia L., Miwa K., Ngara T. and Tanabe K. (eds) IGES, Japan.

IPCC (2007) *Climate Change 2007: Mitigation.* Contribution of Working Group III to the Fourth Assessment Report of the Intergovernmental Panel on Climate Change. Metz, B., Davidson, O.R., Bosch, P.R., Dave, R. and Meyer, L.A. (eds), Cambridge University Press, Cambridge, p. 852.

Jarvis, G.N., Strömpl, C., Burgess, D.M., Skillman, L.C., Moore, E.R.B. and Joblin, K.N. (2000) Isolation and identification of ruminal methanogens from grazing cattle. *Current Microbiology* 40, 327–332.

Jarvis, S.C., Lovell, R.D. and Panayides, R. (1995) Patterns of methane emission from excreta of grazing animals. *Soil Biology and Biochemistry.* 27, 1581–1588.

Kelliher, F.M., Clark, H., Li, Z., Newton, P.C.D., Parsons, A.J. and Rys, G. (2006) A comment on scaling methane emissions from vegetation and grazing ruminants in New Zealand. *Functional Plant Biology* 33, 613–615.

Keppler, F., Hamilton, J.T.G., Brass, M., Röckman, T. (2006) Methane emissions from terrestrial plants under aerobic conditions. *Nature* 439, 187–191.

Knight, T.W., Molano, G., Clark, H. and Cavanagh, A. (2008) Methane emissions from weaned lambs measured at 13, 17, 25 and 35 weeks of age compared with mature ewes consuming a fresh forage diet. *Australian Journal of Experimental Agriculture* 48, 240–243.

Kolver, E.S. and Aspin, P.W. (2006) Supplemental fumarate did not influence milksolids or methane production from dairy cows fed high quality pasture. *Proceedings of the New Zealand Society of Animal Production* 66, 409–415.

Kurihara, M., Nishida, T., Purnomoadi, A., Shibata, M. and Terada, I. (2002) The prediction of methane conversion rate from dietary factors. In: Takahasi, J. and Young, B.A. (eds) *Greenhouse Gases and Animal Agriculture.* Elsevier Science, The Netherlands.

Langeveld, C.A., Leffelaar, P.A., Goudriaan, J., Van Cleemput, O., Hofman, G. and Vermoesen, A. (1994) Factors affecting nitrous oxide emission from a sandy grassland soil under controlled field conditions. *Progress in Nitrogen Cycling Studies: Proceedings of the 8th Nitrogen Workshop Held at the University of Ghent* 1996, pp. 575–579.

Lassey, K.R., Ulyatt, M.J., Martin, R.J., Walker, C.F. and Shelton, I.D. (1997) Methane emissions measured directly from grazing livestock in New Zealand. *Atmospheric Environment* 31, 2905–2914.

Laughlin, R.J. and Stevens, R.J. (2002) Evidence for fungal dominance of denitrification and codenitrification in a grassland soil. *Soil Science Society of America Journal* 66, 1540–1548.

Ledgard, S.F., Williams, P.H., Broom, F.D., Thorrold, B.S., Wheeler, D.M. and Willis, V.J. (1999) OVERSEER – A nutrient budgeting model for pastoral farming, wheat, potatoes, apples and kiwifruit. In: Currie, L.D., Hedley, M.J., Horne, D.J. and Loganathan, P. (eds) *Best Soil Management Practices for Production*. Vol. Occasional Report No. 12, Fertilizer and Lime Research Centre, Massey University, Palmerston North, New Zealand, pp. 143–152.

Lee, J., Woodward, S.L., Waghorn, G.C. and Clark, D.A. (2004) Methane emissions by dairy cows fed increasing proportions of white clover (Trifolium repens) in pasture. *Proceedings of the New Zealand Grassland Association* 66, 151–155.

Le Mer, J. and Roger, P. (2001) Production, oxidation, emission and consumption of methane by soils: a review. *European Journal of Soil Biology* 37, 25–50.

Li, C., Frolking, S. and Frolking, T.A. (1992) A model of nitrous oxide evolution from soil driven by rainfall events: 1. Model structure and sensitivity. *Journal of Geophysical Research* 97, 9759–9776.

Liikanen, A. and Martikainen, P.J. (2003) Effect of ammonium and oxygen on methane and nitrous oxide fluxes across sediment-water interface in a eutrophic lake. *Chemosphere* 52, 1287–1293.

Lovett, D.K., Shalloo, L., Dillon, P. and O'Mara, F.P. (2006) A systems approach to quantify greenhouse gas fluxes from pastoral dairy production as affected by management regime. *Agricultural Systems* 88, 156–179.

Lovett, D.K., Stack, L.J., Lovell, S., Callan, J., Flynn, B., Hawkins, M. and O'Mara, F.P. (2005) Manipulating enteric methane emissions and animal performance of late-lactation dairy cows through concentrate supplementation at pasture. *Journal of Dairy Science* 88, 2836–2842.

Luo, J., Tillman, R.W. and Ball, P.R. (1999) Factors regulating denitrification in a soil under pasture. *Soil Biology and Biochemistry* 31, 913–927.

Mambrini, M. and Peyraud, J.L. (1994) Mean retention time in digestive tract and digestion of fresh perennial ryegrass by lactating dairy cows: influence of grass maturity and comparison with a maize silage diet. *Reproduction, Nutrition, Development* 34, 9–23.

Margen, D.E., Graham, N.McC., Minson, D.J. and Searle, T.W. (1988) Energy and protein values of four forages, including a comparison between tropical and temperate species. *Australian Journal of Experimental Agriculture* 28, 729–736.

Mathison, G.W., Okine, E.K., McAllister, T.A., Dong, Y., Galbraith, J. and Dmytruk, O.I.N. (1998) Reducing methane emissions from ruminant animals. *Journal of Applied Animal Research* 14, 1–28.

Mbanzamihigo, L., Fievez, V., da Costa Gomez, C., Piattoni, F., Carlier, L. and Demeyer, D. (2002) Methane emissions from the rumen of sheep fed a mixed grass-clover pasture at two fertilisation rates in early and late season. *Canadian Journal of Animal Science* 82, 69–77.

McAllister, T.A., Okine, E.K., Mathison, G.W. and Cheng, K.-J. (1996) Dietary, environmental and microbiological aspects of methane production in ruminants. *Canadian Journal of Animal Science* 76, 231–243.

McAllister, T.A. and Newbold, C.J. (2008) Redirecting rumen fermentation to reduce methanogenesis. *Australian Journal of Experimental Agriculture* 48, 7–13 (in press).

McGrath, J.A., Penno, J.W., Macdonald, K.A., Carter, W.A. (1998) Using nitrogen fertiliser to increase dairy farm profitability. *Proceedings of the New Zealand Society of Animal Production* 58, 117–120.

Mills, J.A.N., Dijkstra, J., Bannink, A., Cammell, S.B., Kebreab, E. and France, J. (2001) A mechanistic model of whole tract digestion and methanogenesis in the lactating dairy cow: model development, evaluation and application. *Journal of Animal Science* 79, 1589–1597.

Minami, K. (1995) The effect of nitrogen fertiliser use and other practices on methane emissions from flooded rice. *Fertilizer Research* 40, 71–84.

Ministry for the Environment (2007) *New Zealand's Greenhouse Gas Inventory 1990–2005*. The National Inventory Report and Common Reporting Format. Ministry for the Environment, New Zealand. p. 158.

Minson, D.J. (1990) *Forages in Ruminant Nutrition*. Academic Press, London.

Moe, P.W. and Tyrrell, H.F. (1979) Methane production in dairy cows. *Journal of Dairy Science* 62, 1583–1586.

Molano, G. and Clark, H. (2008) The effect of level of intake and forage quality on methane production by sheep. *Australian Journal of Experimental Agriculture* 48, 219–222 (in press).

Molano, G., Clark, H., Knight, T.W. and Cavanagh, A. (2006) Methane emissions from growing beef cattle grazing hill country. *Proceedings of the New Zealand. Society of Animal Production* 66, 172–175.

Monaghan, R.M. and Barraclough, D. (1993) Nitrous oxide and dinitrogen emissions from urine-affected soil under controlled conditions. *Plant and Soil* 151, 127–138.

Morgavi, D.P., Jouany J.-P. and Martin, C. (2008) Changes in methane emission and rumen fermentation parameters induced by refaunation in sheep. *Australian Journal of Experimental Agriculture* 48, 69–72 (in press).

Morvan, B., Dore, J., Rieulesme, F., Foucat, L., Fonty, G. and Gouet, P. (1994) Establishment of hydrogen-utilizing bacteria in the rumen of newborn lamb. *FEMS Microbiology Letters* 117, 249–256.

Mosier, A., Kroeze, C., Nevison, C., Oenema, O., Seitzinger, S., Cleemput, O.V., Abrahamsen, G., Bouwman, L., Bockman, O., Drange, H., Frolking, S., Howarth, R., Smith, K. and Bleken, M.A. (1998a) Closing the global N2O budget: nitrous oxide emissions through the agricultural nitrogen cycle: OECD/IPCC/IEA phase II development of IPCC guidelines for national greenhouse gas inventory methodology. *Nutrient Cycling in Agroecosystems* 52, 225–248.

Mosier, A.R., Parton, W.J. and Phongpan, S. (1998b) Long-term large N and immediate small N addition effects on trace gas fluxes in the Colorado shortgrass steppe. *Biology and Fertility of Soils* 28, 44–50.

Münger, A. and Kreuzer, M. (2008) Methane emission as determined in contrasting dairy cattle breeds over the reproduction cycle. *International Congress Series* 1293, 119–122.

Murray, R.M., Bryant, A.M. and Leng, R.A. (1976) Rates of production of methane in the rumen and large intestine of sheep. *British Journal of Nutrition* 36, 1–14.

Newbold, C.J., Lassalas, B. and Jouany, J.P. (1995) The importance of methanogens associated with ciliate protozoa in ruminal methane production *in vitro*. *Letters in Applied Microbiology* 21, 230–234.

New Zealand Climate Change Office (2003) *New Zealand Inventory Report 2001*. New Zealand Climate Change Office, Wellington, New Zealand.

Nicol, G.W., Anne Glover, L. and Prosser, J.I. (2003) Molecular analysis of methanogenic archaeal communities in managed and natural upland pasture soils. *Global Change Biology* 9, 1451–1457.

Oenema, O., Wrage, N., Velthof, G.L., van Groenigen, J.W., Dolfing, J. and Kuikman, P.J. (2005) Trends in global nitrous oxide emissions from animal production systems. *Nutrient Cycling in Agroecosystems* 72, 54–65.

Parsons, A.J., Newton, P.C.D., Clark, H. and Kelliher, F.M. (2006) Scaling methane emissions from vegetation. *Trends in Ecology and Evolution* 21, 423–424.

Pavao-Zukerman, M.A., Waller, J.C., Ingle, T. and Fribourg, H.A. (1999) Methane emissions of beef cattle grazing tall fescue pastures at three levels of endophyte infestation. *Journal of Environmental Quality* 28, 1963–1969.

Pinares-Patiño, C.S. (2000) Methane emission from forage-fed sheep, a study of variation between animals. PhD thesis, Massey University, Palmerston North, New Zealand. p. 226.

Pinares-Patiño, C.S., Baumont, R. and Martin, C. (2003a) Methane emissions by Charolais cows grazing a monospecific pasture of timothy at four stages of maturity. *Canandian Journal of Animal Science* 83, 769–777.

Pinares-Patiño, C.S., Ulyatt, M.J., Lassey, K.R., Barry, T.N. and Holmes, C.W. (2003b) Persistance of the between-sheep variation in methane emission. *The Journal of Agricultural Science* 140, 205–214.

Pinares-Patiño, C.S., Vlaming, B., Cavanagh, A., Molano, G. and Clark, H. (2005) Persistence of dairy cows in animal-to-animal variation in methane emission. In: Soliva, C.R., Takahashi, J. and Kreuzer, M. (eds) *Proceedings of the 2nd International Conference in Greenhouse Gases and Animal Agriculture*, ETH Zurich, Zurich, Switzerland, pp. 401–404.

Pinares-Patiño, C.S., D'Hour, P., Jouany, J.-P. and Martin, C. (2007) Effects of stocking rate on methane and carbon dioxide emissions from grazing cattle. *Agriculture, Ecosystems and Environment* 121, 30–46.

Potter, C.S., Matson, P.A., Vitousek, P.M. and Davidson, E.A. (1996) Process modeling of controls on nitrogen trace gas emissions from soils worldwide. *Journal of Geophysical Research* 101, 1361–1377.

Prosser, J.I. (2007) The ecology of nitrifying bacteria. In: Bothe, H., Ferguson, S.J. and Newton, W.E. (eds) *Biology of the Nitrogen Cycle*, Elsevier, Amsterdam, the Netherlands, pp. 223–243.

Reay, D.S., Smith, K.A. and Edwards, A.C. (2003) Nitrous oxide emission from agricultural drainage waters. *Global Change Biology* 9, 195–203.

Robertson, L.J. and Waghorn, G.C. (2002) Dairy industry perspectives on methane emissions and production from cattle fed pasture or total mixed rations in New Zealand. *Proceedings of the New Zealand Society of Animal Production* 62, 213–218.

Rochette, P. and Janzen, H.H. (2005) Towards a revised coefficient for estimating N_2O emissions from legumes. *Nutrient Cycling in Agroecosystems* 73, 171–179.

Saggar, S., Andrew, R.M., Tate, K.R., Hedley, C.B., Rodda, N.J. and Townsend, J.A. (2004a) Modelling nitrous oxide emissions from dairy-grazed pastures. *Nutrient Cycling in Agroecosystems* 68, 243–255.

Saggar, S., Hedley, C.B., Bolan, N.S., Bhandral, R. and Luo, J. (2004b) A review of emissions of methane, ammonia, and nitrous oxide from animal excreta deposition and farm effluent application in grazed pastures. *New Zealand Journal of Agricultural Research* 47, 513–544.

Saggar, S., Giltrap, D.L., Li, C. and Tate, K.R. (2007a) Modelling nitrous oxide emissions from grazed grasslands in New Zealand. *Agriculture, Ecosystems and Environment* 119, 205–216.

Saggar, S., Tate, K.R., Giltrap, D.L. and Singh, J. (2007b) Soil-atmosphere exchange of nitrous oxide and methane in New Zealand terrestrial ecosystems and their mitigation options: a review. *Plant and Soil*. doi 10.1007/s11104-007-9421-3.

Sawamoto, T., Nakajima, Y., Kasuya, M., Tsuruta, H. and Yagi, K. (2005) Evaluation of emission factors for indirect N_2O emission due to nitrogen leaching in agro-ecosystems. *Geophysical Research Letters* 32, 1–4.

Schils, R.L.M., Verhagen, A., Aarts, H.F.M. and Sebek, L.B.J. (2005) A farm level approach to define successful mitigation strategies for GHG emissions from ruminant livestock systems. *Nutrient Cycling in Agroecosystems* 71, 163–175.

Seitzinger, S.P., Kroeze, C. and Styles, R.V. (2000) Global distribution of N_2O emissions from aquatic systems: natural emissions and anthropogenic effects. *Chemosphere – Global Change Science* 2, 267–279.

Shoun, H., Kim, D., Uchiyama, H. and Sugiyama, J. (1992) Denitrification by fungi. *FEMS Microbiology Letters* 94, 277–282.

Simpson, A.M., Webster, A.J.F., Smith, J.S. and Simpson, C.A. (1978) The efficiency of utilization of dietary energy for growth in sheep (*Ovis ovis*) and red deer (*Cervus elaphus*). *Comparative Biochemistry and Physiology* 59A, 95–99.

Skiba, U., Sheppard, L., Pitcairn, C.E.R., Leith, I., Crossley, A., van Dijk, S., Kennedy, V.H. and Fowler, D. (1998) Soil nitrous oxide and nitric oxide emissions as indicators of elevated atmospheric N deposition rates in seminatural ecosystems. *Environmental Pollution* 102, 457–461.

Skiba, U., Pitcairn, C., Sheppard, L., Kennedy, V. and Fowler, D. (2004) The influence of atmospheric N deposition on nitrous oxide and nitric oxide fluxes and soil ammonium and nitrate concentrations. *Water, Air, and Soil Pollution: Focus* 4, 37–43.

Solomon, S., Qin, D., Manning, M., Alley, R.B., Berntsen, T., Bindoff, N.L., Chen, Z., Chidthaisong, A., Gregory, J.M., Hegerl, G.C., Heimann, M., Hewitson, B., Hoskins, B.J., Joos, F., Jouzel, J., Kattsov, V., Lohmann, U., Matsuno, T., Molina, M., Nicholls, N., Overpeck, J., Raga, G., Ramaswamy, V., Ren, J., Rusticucci, M., Somerville, R., Stocker, T.F., Whetton, P., Wood, R.A. and Wratt, D. (2007) Technical summary. In: Solomon, S., Qin, D.,

Manning, M., Chen, Z., Marquis, M., Averyt, K.B., Tignor, M. and Miller, H.L. (eds) *Climate Change 2007: The Physical Science Basis. Contribution of Working Group I to the Fourth Assessment Report of the Intergovernmental Panel on Climate Change.* Cambridge University Press, Cambridge.

Soussana, J.F., Loiseau, P., Vuichard, N., Ceschia, E., Balesdent, J., Chevallier, T. and Arrouays, D. (2004) Carbon cycling and sequestration opportunities in temperate grasslands. *Soil Use and Management* 20, 219–230.

Soussana, J.F. and Lüscher, A. (2007) Temperate grasslands and global atmospheric change: a review. *Grass and Forage Science* 62, 127–134.

Steinfeld, H., Greber, P.T.W., Castel, V., Rosales, M. and de Haan, C. (2006) *Livestock's Long Shadow – Environmental Issues and Options.* FAO, Rome.

Stevens, R.J., Laughlin, R.J., Mosier, A., Abrahamsen, G., Bouwman, L., Bockman, O., Drange, H., Frolking, S., Howarth, R., Kroeze, C., Oenema, O., Smith, K. and Bleken, M.A. (1998) Measurement of nitrous oxide and di-nitrogen emissions from agricultural soils. *Nutrient Cycling in Agroecosystems* 52, 131–139.

Swainson, N.M., Hoskin, S.O., Clark, H., Pinares-Patino, C.S. and Brookes, I.M. (2007) Comparative methane production and yields from adult cattle, red deer and sheep. *Greenhouse Gas and Animal Agriculture – Conference Book,* November 2007. Christchurch, New Zealand, pp. 140–141.

Terada, F., Iwasaki, K., Tano, R., Itoh, M. and Kameoka, K. (1985) Comparison of metabolizability and metabolizable energy contents among cattle, sheep and goats fed the same diets. In: Moe, P.W., Tyrrell, H.F. and Reynolds, P.J. (eds) *Energy Metabolism of Farm Animals.* EAAP Pub. 32, Rowman & Littlefield, pp. 130–133.

Tiedje, J.M. (1988) Ecology of denitrification and dissimilatory nitrate reduction to ammonium. In: Zehnder, A.J.B. (ed.) *Biology of anaerobic microorganisms.* New York, Wiley & Sons. pp. 179–244.

Torrent, J. and Johnson, D.E. (1994) Methane production in the large intestine of sheep. In: Aguilera, J.F. (ed.) *Energy Metabolism of Farm Animals.* EAAP Publ. 76, CSIC, Servicio de Publicaciones, Mojacar, Spain, pp. 391–394.

Ulyatt, M.J. and Lassey, K.R. (2000) Methane emissions from pastoral systems: the situation in New Zealand. *Archivos Latinoamericanos de Produccion Animal* 9, 118–126.

Ulyatt, M.J., Lassey, K.R., Martin, R.J., Walker, C.F. and Shelton, I.D. (1997) Methane emission from grazing sheep and cattle. *Proceedings of the New Zealand Society of Animal Production* 57, 130–133.

Ulyatt, M.J., Lassey, K.R., Shelton, I.D. and Walker, C.F. (2002) Seasonal variation in methane emission from dairy cows and breeding ewes grazing ryegrass/white clover pasture in New Zealand. *New Zealand Journal of Agricultural Research* 45, 217–226.

UNFCCC (1997) *Kyoto Protocol to the United Nations Framework Convention on Climate Change.* United Nations, Bonn, Germany.

van den Pol-van Dasselaar, A., van Beusichem, M.L. and Oenema, O. (1999) Effects of nitrogen input and grazing on methane fluxes of extensively and intensively managed grasslands in the Netherlands. *Biology and Fertility of Soils* 29, 24–30.

van der Nagel, L.S., Waghorn, G.C. and Forgie, V.E. (2003) Methane and carbon emissions from conventional pasture and grain based total mixed rations for dairying. *Proceedings of the New Zealand Society of Animal Production* 63, 128–132.

van Dorland, H.A., Wettstein, H.-R., Leuenberger, H. and Kreuzer, M. (2007) Effect of supplementation of fresh and ensiled clovers to ryegrass on nitrogen loss and methane emission of dairy cows. *Livestock Science* 111, 57–69.

Van Groenigen, J.W., Velthof, G.L., Van der Bolt, F.J.E., Vos, A. and Kuikman, P.J. (2005) Seasonal variation in N_2O emissions from urine patches: effects of urine concentration, soil compaction and dung. *Plant and Soil* 273, 15–27.

Van Nevel, C.J. and Demeyer, D.I. (1996) Control of rumen methanogenesis. *Environmental Monitoring and Assessment* 42, 73–97.

van Spanning, R.J.M., Richardson, D.J. and Ferguson, S.J. (2007) Introduction to the biochemistry and molecular biology of denitrification. In: Bothe, H., Ferguson, S.J. and Newton, W.E. (eds) *Biology of the Nitrogen Cycle,* Elsevier, Amsterdam, the Netherlands, pp. 3–20.

Van Vugt, S.J., Waghorn, G.C., Clark, D.A. and Woodward, S.L. (2005) Impact of monensin on methane production and performance of cows fed forage diets. *Proceedings of the New Zealand Society of Animal Production* 65, 362–366.

Velthof, G.L. and Oenema, O. (1995) Nitrous oxide fluxes from grassland in The Netherlands: II. Effects of soil type, nitrogen fertiliser application and grazing. *European Journal of Soil Science* 46, 541–549.

Vlaming, J.B., Lopez-Villalobos, N., Brookes, I.M., Hoskin, S.O. and Clark, H. (2008) Within- and between-animal variance in methane emissions in non–lactating dairy cows. *Australian Journal of Experimental Agriculture* 48, 124–127.

Waghorn, G.C. (2007) Beneficial and detrimental effects of dietary condensed tannins for sustainable sheep and goat production – progress and challenges. *Animal Feed Science and Technology*. doi:10.1016/j.anifeedsci.2007.09.013.

Waghorn, G.C. and Woodward, S.L. (2006) Ruminant contributions to methane and global warming – a New Zealand perspective. In: Bhatti, J.S., Lal, R., Apps, M.J. and Price, M.A. (eds) *Climate Change and Managed Ecosystems*. CRC Press, New York, pp. 233–260.

Waghorn, G.C., Lassey, K.R., Kolver, E. and Molano, G. (2001) *Methane Emission from Dairy Cows Fed Either Pasture or Balanced Rations over One Lactation*. A report prepared for the New Zealand Dairy Board environment global programme. Hamilton, New Zealand.

Waghorn, G.C., Molano, G. and Lassey, K. (2003a) *Estimates of Whole Herd Methane Production from Cows at the Lincoln University Dairy Farm in January and March 2002*. A report prepared for Landcare Research, New Zealand. p. 26.

Waghorn, G.C., Molano, G. and Cavanagh, A. (2003b) *An Estimate of Whole Herd Methane Production from Cows at the Lincoln University Dairy Farm in October 2003*. A report prepared for Landcare Research, New Zealand. p. 23.

Waghorn, G.C., Cavanagh, A., Molano, G., Tavendale, M. and Burke, J. (2003c) *The Effect of Once Daily Supplementation with Molasses or Agri-feeds Rumax on Rumen Methane Production by Lactating Cows Grazing Pasture in October 2002*. A report prepared for Agrifeeds Limited, New Zealand. p. 25.

Waghorn, G.C., Clark, H, Taufa, V. and Cavanagh, A. (2008) Monensin controlled release capsules for mitigating methane in dairy cows fed pasture. *Australian Journal of Experimental Agriculture* 48, 65–68.

Waugh, C.D., Clark, D.A., Waghorn, G.C. and Woodward, S.L. (2005) Feeding maize silage to dairy cows: implications for methane emissions. *Proceedings of the New Zealand Society of Animal Production* 65, 356–361.

Wheeler, D.M., Ledgard, S.F., de Klein, C.A.M., Monaghan, R.M., Carey, P.L., McDowell, R.W. and Johns, K.L. (2003) OVERSEER nutrient budgets 2 – moving towards on-farm resource accounting. *Proceedings of the New Zealand Grasslands Association* 65, 191–194.

Whitehead, D.C. (1995) *Grassland Nitrogen*. CAB International, Wallingford, UK.

Williams, Y.J., Walker, G.P., Doyle, P.T., Egan, A.R. and Stockdale, C.R. (2005) Rumen fermentation characteristics of dairy cows grazing different allowances of Persian clover- or perennial ryegrass-dominant swards in spring. *Australian Journal of Experimental Agriculture* 45, 665–675.

Woodward, S.L. (2003) *Methane Emissions from Growing Dairy Calves and Mature Cows*. A report prepared for Ministry of Agriculture and Forestry, New Zealand. p. 25.

Woodward, S.L., Waghorn, G.C., Laboyrie, P.G. (2004) Condensed tannins in birdsfoot trefoil (Lotuis corniculatus) reduce methane emissions from dairy cows. *Proceedings of the New Zealand Society of Animal Production* 64, 160–164.

Woodward, S.L., Waghorn, G.C., Thomson, N.A. (2006) Supplementing dairy cows with oils to improve performance and reduce methane – does it work? *Proceedings of the New Zealand Society of Animal Production* 66, 176–181.

Wrage, N., Velthof, G.L., van Beusichem, M.L. and Oenema, O. (2001) Role of nitrifier denitrification in the production of nitrous oxide. *Soil Biology and Biochemistry* 33, 1723–1732.

Yan, T., Agnew, R.E., Gordon, F.J., Porter, M.G. (2000) Prediction of methane energy output in dairy and beef cattle offered grass silage-based diets. *Livestock Production Science* 64, 253–265.

2 Impacts of Pastoral Grazing on Soil Quality

S.P. Cuttle

Institute of Grassland and Environmental Research, Plas Gogerddan, Aberystwyth, UK

Introduction

For the purpose of this chapter, pastoral grazing systems are considered to be predominantly grass swards grazed by farmed livestock, mainly ruminants such as cattle, sheep, goats and deer. Although grasses predominate, herbaceous species, especially legumes such as white clover (*Trifolium repens*), may also be important components of the sward. Because environmental impacts are usually associated with relatively intensive managements, most attention is paid to sown and semi-natural swards, rather than to the most extensively grazed rangelands. In those parts of the world where it is necessary to house livestock during the winter months, pastures may also be cut to conserve forage as hay or silage and may receive manure from housed animals. These are integral components of the management of these pastures and must also be taken into account when considering the overall impact on the environment.

Soil is basic to the functioning of these pastoral grazing systems; it provides the physical medium in which roots develop and supplies the nutrients and water that plants need for growth. It also provides the physical surface supporting the livestock that consume this herbage. The extent to which a soil provides the conditions required for herbage growth and how it responds to the impacts of grazing and other management practices will determine its suitability for use as pasture. Many of the effects of pastoral agriculture that directly modify soil properties also act through the soil to have a wider environmental impact on water or air quality. Indeed, it is often these forms of diffuse pollution that receive the greatest attention. Mitigating these impacts requires an understanding of the soil processes involved and often needs soil-based control measures. The ways in which pastoral agriculture affects the soil and in turn affects the wider environment can be examined through the concept of soil quality.

Concepts of Soil Quality

In recent years there has been increasing concern about the widespread degradation of the world's soils. These concerns focus on the loss of productive capacity, loss of biological diversity and the wider environmental impacts of soil and nutrient loss. The need to manage soils in a more sustainable way has prompted the development of systems to more closely monitor and record the status of this vital resource, to enable land users and policy makers to determine the effects of soil management and to report on trends in the condition of the soil. Fundamental to this is the development of the concept of 'soil quality' or 'soil health'. These two terms are sometimes treated as synonymous (Doran *et al.*, 1996), but elsewhere a distinction has been made between 'soil quality', used to describe the soil's fitness for a particular use, and 'soil health', which is seen more as an inherent attribute of the soil and is independent of land use (Bloem *et al.*, 2006). Despite much published work on the origins and development of the soil quality concept (Doran *et al.*, 1996; Karlen *et al.*, 2001), there have also been criticisms that the concept is too vague or is unnecessary (Sojka *et al.*, 2003).

Definitions and assessment of soil quality

Soil quality has been defined as 'the capacity of a soil to function within ecosystem boundaries to sustain biological productivity, maintain environmental quality and promote plant and animal health' (Doran and Parkin, 1994; Stenberg, 1999). Soil quality is therefore dependent upon land use. An alternative definition describes soil quality as 'the fitness of soil for a specific use' (Larson and Pierce, 1994). In the context of pastoral grazing, soil quality primarily refers to the soil's ability to support the growth of herbaceous vegetation and the utilization of this herbage by grazing or cutting, while minimizing impacts on the wider environment. However, many grasslands are examples of multifunctional land use; at the same time as providing agricultural products, they may also function to regulate the flow of water, act as environmental buffers, preserve traditional landscapes or provide amenity benefits. It will be this combination of functions that determines the mix of soil properties required for a high-quality soil. In the case of more intensively managed pastures, high-quality soils are those most able to fulfil the land managers' objectives of supporting a profitable livestock enterprise. Their management often involves artificial inputs to increase production and, as a result, soils may be considerably modified from their original condition. These highly modified soils are considered to be of high quality because of their greater capacity to deliver a desirable mix of services (Kibblewhite, 2005). Thus, when applied to more intensively managed agricultural soils, concepts of soil quality differ fundamentally from those of air or water quality where the objective is most commonly to maintain or restore the medium to something approaching its pristine, 'natural' condition. Even so, there is recognition that soil is a finite resource that must be managed sustainably within certain environmental constraints.

Assessments of soil quality require a range of measurements that adequately describe the condition of the soil. In practice, financial and time restraints limit measurements to those that are most informative. Soil quality attributes that are most commonly measured are those that are readily determined and can be directly related to plant and animal performance and to potential environmental impacts (Larson and Pierce, 1994; Beare *et al.*, 2005); for example, soil density and porosity, nutrient content, pH, organic matter content and indicators of biological activity such as microbial biomass or potential N mineralization. Several authors have described minimum data sets required to describe soil quality for particular purposes (Gregorich *et al.*, 1994; Doran and Parkin, 1996; Paz Jimenez *et al.*, 2002). Suitable parameters should be sensitive to environmental changes and to the effects of management. They should be easily interpreted so that the significance of any change can be properly understood. It is also important that the property be sufficiently well measured and documented to provide reference or threshold values, so that values outside the normal range for that particular soil type can be highlighted as a point of concern. In the absence of reference values, temporal changes or trends in indicator values can demonstrate changes in soil quality and provide a warning that the system may be unsustainable (Karlen *et al.*, 2001). Approaches to assessing soil quality have developed independently in different countries but demonstrate considerable overlap in the parameters selected (Burns *et al.*, 2006).

Soil quality in pastures

In many respects, pasture soils are less vulnerable to adverse impacts on soil quality than arable soils. Indeed, sowing grass seed and converting land to pasture is often adopted as a means of restoring soil fertility and reversing the effects of other, more damaging managements (Logan, 1992). However, grassland soils are not sufficiently resilient to withstand all the potential impacts of pasture use, particularly those associated with the more intensive forms of management. Intensification and increased environmental damage are often associated with increases in artificial inputs of fertilizer and feed, but can also result simply from increased grazing pressure. For example, in Africa and elsewhere in the tropics, overgrazing of marginal lands has resulted in serious degradation and soil erosion (Lal, 1992). However, intensification does not inevitably result in a loss of soil quality. Kemp and Michalk (2005) suggest the 'well-adapted' perennial ryegrass-white clover (*Lolium perenne-T. repens*) pastures of New Zealand as an example of intensification with minimal environmental impact.

It is convenient in the following sections to consider the physical, chemical and biological components of soil quality separately. However, in practice, there are few properties that can be considered in isolation because of the many interactions that occur between soil properties. Recognition of this interdependence of soil properties and processes is fundamental to the sustainable management of soils. The physical and chemical properties define the environment in which the biological components of the soil have to function and are therefore considered first.

Impacts on Physical Aspects of Soil Quality

The most basic physical property of a soil is its texture, as determined by the proportions of sand, silt and clay particles making up the mineral fraction of the soil. However, in most soils other than sands, which have little structure, it is the binding of these individual particles and organic matter into aggregates that most directly determines the physical characteristics that influence soil function and its suitability as a medium for plant growth. The nature and stability of these aggregates and the pore space between them determine the movement of air and water through the soil, moisture retention and the physical environment for soil organisms and plant roots. The physical parameters that are measured in assessing soil quality include measurements of soil texture, bulk density and porosity as well as hydrological properties such as hydraulic conductivity and moisture retention capacity. Field assessments include profile depth and the depth to which roots are able to penetrate to extract nutrients and moisture.

Soil structure and the effects of grazing

Soils under pasture tend to accumulate organic matter, which favours the development of good soil structure (Tisdall and Oades, 1982; Kay, 1990; Soane, 1990; Tisdall, 1994). However, there are also particular stresses on the physical condition of pasture soils. Of particular concern is the poaching and compaction of soils caused by grazing animals (Greenwood and McKenzie, 2001).

Poaching (or pugging; Bilotta *et al.*, 2007) describes the damage to the soil surface that occurs when wet soils are trampled by stock. The treading displaces wet soil from around the animals' hooves, leaving imprints of the hoof and smearing of the soil surface. This damage restricts further infiltration of water, increases the moisture content of the soil and exacerbates the risk of further poaching. Trampling also damages the surface vegetation and root mat, leaving the soil more exposed and susceptible to erosion, particularly as the decreased infiltration in these areas makes surface runoff more likely. Poaching is most severe on poorly structured clays and silts and on soils with high organic matter concentrations (Frame, 2000). These soils have a large capacity to retain surplus water and remain wet for a greater proportion of the year. Susceptibility to poaching may also be related to the strength of the sward, which provides some protection against damage (Scholefield and Hall, 1985).

Treading by livestock also decreases soil quality by compacting the soil. Under the pressure of the animal's hooves, soil particles become more closely packed together, which produces a layer with increased bulk density and decreased pore space. The depth at which compaction occurs varies but is typically between 2 and 12 cm below the soil surface (Bryant *et al.*, 1972; Curll and Wilkins, 1983; Scholefield and Hall, 1985; Butler and Adams, 1990). The total pore space and continuity of the pores are decreased, which restricts the movement of water and air through the soil (Soane *et al.*, 1981; Greenwood and McKenzie, 2001), leading to waterlogging and decreased aeration above and within the compacted zone. Compaction also impedes root growth. Examination of the upper-soil profile is likely to show signs of restricted rooting depth and horizontal root growth and, in some cases, visible indications of waterlogging above the compacted layer (Ball *et al.*, 1997; Batey and McKenzie, 2006). Environmental impacts associated with soil compaction include an increased risk of surface runoff, leading to greater losses of nutrients and sediment, and increased denitrification losses from anaerobic sites within the soil. The effects of poaching and compaction tend to be most serious in areas where animals congregate, e.g. around water troughs and field gateways, along fencelines and in camping areas (West *et al.*, 1989; Haynes and Williams, 1999; Franzluebbers *et al.*, 2000a; White *et al.*, 2001; McDowell *et al.*, 2004; Iyyemperumal *et al.*, 2007).

Unlike poaching, compaction does not only occur in wet soils. However, the ease with which soils are compacted generally increases with increasing wetness. Accordingly, poaching is usually accompanied by a degree of compaction. Compaction is likely to be particularly severe in fields grazed by cattle, which typically exert a ground pressure double that of sheep (Betteridge *et al.*, 1999; Greenwood and McKenzie, 2001). The degree of compaction increases with increasing stocking rate. However, most of the damage appears to be caused by the first compression and only increases slightly with repeated treading (Mulholland and Fullen, 1991). Soil compaction also occurs in pastures as a result of vehicle traffic. The pressures exerted by fully laden silage trailers or slurry tankers are comparable to those exerted by grazing animals and have a similar effect.

There are natural processes in the soil that act to restore soil structure and ameliorate the effects of poaching and compaction. Wetting and drying cycles in soils with medium to high clay concentrations create structural cracks that disrupt compacted soil. In regions with sufficiently cold winters, cycles of freezing and thawing have a similar effect (Rodd *et al.*, 1999; Donkor *et al.*, 2002). The growth and decay of roots creates channels that allow movement of air and water through compacted soil (Unger and Kaspar, 1994; Yunusa and Newton, 2003). Although the initial compaction destroys earthworm burrows (Cluzeau *et al.*, 1992; Ligthart, 1997), the subsequent recovery of worm populations and renewed activity will create channels and help restore soil porosity (Aritajat *et al.*, 1977). Soils under permanent and long-term pastures provide few opportunities for cultivation to break up the compacted soil. If the natural processes that work to ameliorate compaction are insufficient to correct the damage caused by grazing or wheeled traffic, it is often necessary to change the management (Bilotta *et al.*, 2007) or resort to some form of mechanical treatment to disrupt or penetrate the compacted layer (Spoor, 2006).

Soil drainage

On fine-textured soils that are naturally slow-draining, the installation of field drains to remove surplus water and lower the water table extends the season during which pastures can be grazed without poaching the ground. Drainage improves soil aeration and changes the dominant hydrological pathway from surface to subsurface flow. Improved aeration increases microbial activity with marked increases in organic matter decomposition and N mineralization. At the same time, denitrification losses will be decreased. However, the increased supply of N from mineralization and decrease in gaseous N losses may increase the quantity of nitrate available for leaching (Jarvis, 1997). Although the increased proportion of subsurface flow will decrease the transport of pollutants in surface runoff, water moving rapidly through macropores and fissures to the drains may provide an alternative route to surface waters. Lowering the water table decreases the quantity of available moisture stored within the soil profile and may increase the risk of drought (Tyson *et al.*, 1992). Summer drought that restricts herbage growth and N uptake may leave greater concentrations of nitrate in the soil, available for leaching later in the year (Garwood and Tyson, 1977; Scholefield *et al.*, 1993; Morecroft *et al.*, 2000).

Soil erosion

Under normal circumstances, permanent pastures are less susceptible to erosion than arable fields. Pasture soils are well structured and have a continuous vegetation cover that protects the surface from the impact of raindrops and traps soil particles that might otherwise be transported in water flows across the surface. However, there is a greater risk of erosion where soils are grazed and poaching jeopardizes these protective features. The most immediate impact of erosion on soil quality is that it represents a loss of fertility, particularly as it is the most fertile topsoil that is lost first. However, the loss of sediment has a wider environmental impact if the eroded material is washed into rivers or streams where it may settle out and smother gravel beds that are often spawning and nursery areas for fish and other aquatic species. This also provides a route by which chemical pollutants, such as P and heavy metals that are associated with soil particles, may be transported into water bodies.

Extensive damage can occur under dry-land conditions as a result of overgrazing. Uncontrolled grazing or increased stock numbers destroys the vegetation that protects the soil surface against the erosive forces of wind and water. These effects have been particularly severe in Africa and other tropical countries where destruction of vegetative cover and decreased rainfall infiltration rates in areas trampled by livestock are important factors increasing the rate and volume of surface runoff into existing erosion gullies (Lal, 1992). The denuded and compacted soil surface is also susceptible to erosion by wind during the dry season. Effects are particularly severe if the overgrazing is accompanied by burning. How much overgrazing contributes to desertification in Africa, as opposed to the effects of drought, is unclear. However, the effects of livestock are clearly visible on a local scale; for example, around watering points (Reid *et al.*, 2005). Effects of overgrazing are not confined to tropical regions, though elsewhere the effects are usually less severe. There is evidence in the UK that increased stocking rates, particularly in the uplands, have increased the incidence of localized soil erosion (Evans, 1997).

Impacts on Chemical Aspects of Soil Quality

In terms of agricultural productivity, the most obvious chemical indicators of soil quality are the concentrations of nutrients that have a direct, positive effect on plant growth and the quantity of soil organic matter (SOM). SOM is important not only because of its role as a reservoir of nutrients and in the formation of soil aggregates but also because of its global significance as a repository for C that would otherwise contribute to atmospheric CO_2. Chemical indicators of soil quality therefore commonly include a measure of the concentration of organic C, together with concentrations of the major nutrients that most often limit plant growth (e.g. total N, ammonium-N, nitrate-N, plant-available P and K, Ca, Mg) and soil pH. Measurements of the concentration of these and other nutrients provide information about possible deficiencies but also indicate where contents are excessive and present an increased risk of pollution, particularly from N and P.

High-quality soils contain satisfactory concentrations of nutrients for plant growth and for satisfying the dietary requirements of livestock. If concentrations are inadequate, these limitations can usually be overcome by applications of fertilizer or other forms of nutrient input, provided that the economics of the system permit the use of such expensive inputs. Where nutrient deficiencies do occur, they are unlikely to have adverse environmental impacts except where poor plant growth leaves soil more susceptible to overgrazing and erosion or where a deficiency of one nutrient restricts the uptake of another, which may then accumulate in the soil and itself become a pollutant. Generally, however, impacts on the wider environment are usually associated with an excessive concentration of nutrients in the soil.

An important characteristic of pasture soils, affecting concentrations of nutrients and SOM, is the continuous recycling of plant material in litter and excreta. This is usually greater than in arable systems where much of the biomass is removed in the harvested crop. Although much of the growth in managed grasslands is removed by cutting or grazing, a substantial proportion is returned to the soil through the death and decay of leaves and stems, as root exudates and from the turnover of root biomass. Together, these provide a steady input of fresh organic matter and recycling of nutrients. In grazed pastures, these inputs are supplemented by the return of organic matter and nutrients in dung and urine. The quantities recycled in cut fields may approach those under grazing if the fields receive regular applications of slurry or manure.

Grassland soils as a carbon reservoir

The combination of greater organic matter inputs and an absence of disturbance from cultivation results in greater SOM concentrations in grassland than under arable land use. Because of this, grasslands are important globally as a reservoir of C that might otherwise add to levels of atmospheric CO_2 (Follett and Schuman, 2005). The C concentration of the soil at any time is determined by the balance between organic matter inputs and the rate of decomposition that ultimately releases C back to the atmosphere as CO_2. This balance is influenced by the nature of the organic residues (C/N ratio) and by stabilization processes in the soil that protect organic compounds against decomposition (Lützow *et al.*, 2006). The balance must also take account of other possible losses of C from the soil as dissolved organic matter (Ghani *et al.*, 2007) or in eroded topsoil (Schuman *et al.*, 2002). In recently established grasslands following a period of arable cropping, the organic matter input initially exceeds the decomposition rate, resulting in a net accumulation of SOM. However, as the organic matter concentration increases, there is a corresponding increase in the decomposition rate, with the result that the soil moves towards an equilibrium organic matter concentration at which the rate of decomposition is equal to the rate of addition. The increase in SOM is most rapid in the first years after pasture establishment (Tyson *et al.*, 1990; Johnston *et al.*, 1994) but may then continue at a slower rate for many years. Differences between C inputs and outputs are relatively small so that it is difficult to quantify short-term changes by direct measurement of C concentrations. Alternatively, models can be used to predict changes in soil C (Soussana *et al.*, 2004; Rees *et al.*, 2005), although there are many uncertainties about the magnitude of C fluxes in grasslands, especially under grazing (Follett and Schuman, 2005).

The amount of C stored in grassland is influenced by management practices that affect organic matter inputs and decomposition rates. In most cases, C inputs to soil will be greater in grazed swards than under cutting because of the return of organic material in litter and faeces (Franzluebbers *et al.*, 2000b) and because of the other benefits of controlled grazing on C turnover and species composition (Follett and Schuman, 2005; Rees *et al.*, 2005). Responses to management changes depend on the type of pasture and current intensity of management (Table 2.1). Moderate increases in fertilizer N and stocking rates generally increase

Table 2.1. Estimates of the effect of management options on the soil organic carbon (SOC) stock of different types of grassland in France. Values refer to the SOC stock after 20 years under the changed management. (From Soussana *et al.*, 2004.)

Management change	Initial type[a]	Final type[a]	SOC stock (t C/ha)	SOC change (t C/ha)
Decrease in N fertilizer input	A3	A2	41.4	6.4
Conversion to grass-legume mixtures	A3	B2	45.6	10.2
	A2	B2	50.3	6.3
Intensification of permanent grassland	D1	D2	53.9	3.9
Intensification of upland grassland (organic soils)	D0	D2	87.4	–22.6
	D0	C2(10/2)	91.3	–18.7
Permanent grassland to medium duration leys	D2	C2(10/2)	67.0	–3.0
Increasing duration of the ley	C2(5/2)	C2(10/2)	58.1	3.9
	C1(5/2)	C2(10/2)	50.9	9.1
Short duration leys to permanent grassland	C2(5/2)	D2	60.0	5.7
	C1(5/2)	D2	80.0	8.2

[a]Grassland types. A2, short duration grass leys (1–2 years), cut or grazed with moderate N inputs; A3, short duration grass leys (1–2 years), intensively managed, cut or grazed with high N inputs; B2, legume-based leys (3–6 years), without N fertilizer; C1, sown intensive grasslands (3–15 years), extensively managed for hay or grazing; C2, sown intensive grasslands (3–15 years), intensively managed for silage or grazing (values in parentheses are the number of years under grass/arable cropping); D0, permanent upland grass on nutrient-poor organic soils, grazed at low stocking rates; D1, permanent grasslands (>15 years), extensively managed for hay or grazing; D2, permanent grasslands (>15 years), intensively managed for silage or grazing.

C storage by increasing primary production and the quantities of litter returned to the soil. Primary production can also be increased by introducing legumes into the sward. The net reduction in greenhouse gas emissions may be greater than where fertilizers are used because N_2 fixation by legumes avoids the energy use and emissions associated with fertilizer manufacture. However, increases in soil C may be less than expected because of the rapid decomposition of low C/N legume residues (Follett and Schuman, 2005). Increased N inputs to intensively managed grasslands, already well supplied with N, are more likely to decrease C storage. This is a result of more rapid decomposition of the N-enriched residues (Loiseau and Soussana, 1999; Table 2.2) and the smaller root systems of grasses receiving high rates of N fertilizer (Whitehead, 2000). In these intensively managed grasslands, decreasing N inputs is likely to increase C storage. Increasing N inputs to upland pastures, where soils already have high SOM concentrations, is likely to bring about large decreases in C stocks.

Other management practices that increase net primary production and soil C include liming to increase P availability, P fertilization of P-deficient soils, improved grazing management, introduction of earthworms (Conant *et al.*, 2001; Follett and Schuman, 2005) and irrigation (Martens *et al.*, 2005). Where fertility limits growth, application of manure not only increases primary production but also provides an additional direct input of organic matter to the soil. The residues of warm-season grasses (C4 photosynthetic process) decompose less readily than those of temperate, cool-season species (C3 photosynthetic process). Increasing the proportion of C4 species in pastures by reseeding or controlled grazing has been shown to increase C storage (Wedin and Tilman, 1990; Frank *et al.*, 1995; Corre *et al.*, 1999; Ross *et al.*, 2002).

Grassland soils can only be an effective C sink where they are left undisturbed. If long-term swards are ploughed, the physical disruption and improved aeration of the soil causes a rapid increase in organic matter decomposition and release of CO_2 back to the atmosphere. Temporary grasslands therefore have lower concentrations of SOM than soils under permanent pasture. Increasing the duration of leys or changing to permanent grassland will increase C storage (Table 2.1). However, the C concentration of resown pasture may take many years to approach that of the soil prior to cultivation. Rates of accumulation are typically

Table 2.2. Typical C/N ratios of plant residues in pastures and of animal excreta. (From Whitehead, 2000.)

Component of pasture system	C/N ratio
Dead grass herbage, little or no fertilizer N	44:1
Dead grass herbage, high rate of fertilizer N	19:1
Dead clover herbage	18:1
Grass roots, little or no fertilizer N	46:1
Grass roots, high rate of fertilizer N	30:1
White clover roots	13:1
Faeces (cattle or sheep)	20:1
Urine (cattle or sheep)	4:1

half the rate of SOM loss following conversion of grassland to arable cropping (Soussana *et al.*, 2004; Follett and Schuman, 2005). Installation of field drains also decreases C storage. The improved aeration increases organic matter decomposition and CO_2 release but the net effect on greenhouse gas emissions may be offset to some extent by a decrease in nitrous oxide (N_2O) emissions.

Methane is also a component of global C budgets but is of greater direct importance because of its role as a potent greenhouse gas. Livestock farming is a significant contributor to CH_4 emissions but this is primarily due to the loss of enteric CH_4 from the stomachs of ruminant livestock (Johnson and Johnson, 1995) rather than to emissions from the soil. Soils are generally only a significant source when waterlogged or submerged, such as rice paddies and some upland soils (Mer and Roger, 2001). In other soils, CH_4 fluxes are dependent on the opposing processes of methanogenesis in waterlogged conditions and CH_4 oxidation in the aerated zone near the soil surface. For example, CH_4 produced in the waterlogged layers of a pasture soil in Germany was oxidized in the shallow, aerated surface layer and this prevented CH_4 fluxes from the surface (Kammann *et al.*, 2001). As a result, grasslands are more likely to be net sinks for CH_4 (Langeveld *et al.*, 1997). In the Broadbalk plots at Rothamsted in the UK, CH_4 oxidation rates were found to vary in the order woodland > grassland > arable with activities of −45, −21 and −8 nl CH_4/l/h, respectively (Willison *et al.*, 1995). Ammonium fertilizers had an irreversible inhibitory effect on methane oxidation (Hutsch *et al.*, 1994) but there are indications that this inhibition may be prevented by the rapid uptake of N in densely rooted grassland soils (Glatzel and Stahr, 2001). Compaction also decreases CH_4 oxidation (Ball *et al.*, 1999) and may increase emissions (Yamulki and Jarvis, 2002). Dung pats in grazed pastures are hot spots for CH_4 production, but the losses are small compared with those from liquid manure during storage (Holter, 1997; Amon *et al.*, 2006) or enteric fluxes (Flessa *et al.*, 1996b). Urine patches in grazed fields are not a significant source of CH_4, with adsorption equalling or exceeding emissions (Lovell and Jarvis, 1996). Applications of liquid manure to grasslands typically increase CH_4 emissions for 2–3 days, but soils then revert to being a net sink for CH_4 (Chadwick *et al.*, 2000; Sherlock *et al.*, 2002; Dittert *et al.*, 2005; Jones *et al.*, 2006; Rodhe *et al.*, 2006). Overall effects of pasture management on CH_4 fluxes are small. Measurements on intensively managed grasslands in the Netherlands showed no significant effects of grazing versus mowing, of stocking density or of withholding N fertilizer; the soil in all cases remaining a net sink for CH_4 (Pol-van Dasselaar *et al.*, 1999).

Nitrogen losses

The main reservoir of soil N is in the organic compounds that constitute the organic fraction of the soil. Pasture soils, with their relatively high concentrations of SOM, are generally well supplied with N. However, before this reservoir of organic-N can be utilized by plants, it must first be mineralized by soil organisms to ammonium and nitrate-N and this increases its susceptibility to loss. In temperate regions, the total N concentration of long-term grassland soils is typically between 5000 and 15,000 kg N/ha, with <5% of this present as inorganic N (Whitehead, 2000).

Nevertheless, this pool of mineral-N is sufficient to pose a threat to both air and water quality. Leaching from agricultural soils is a major contributor to nitrate pollution of ground and surface waters while N_2O and nitric oxide (NO) produced from ammonium and nitrate are both important gaseous pollutants. Although these losses are of inorganic forms of N and considered as chemical components of soil quality, their production is dependent on biological processes in the soil.

Nitrate leaching

Nitrate and ammonium ions both exist in the soil solution but differ in their susceptibility to leaching and loss in subsurface flow or deep drainage. The positively charged ammonium ions are attracted to, and retained by, cation exchange sites on clay mineral and organic matter surfaces. Nitrate anions, however, exist freely in solution and are readily leached whenever there is drainage of water through the soil profile. The greatest risk of leaching is associated with coarse-textured, free-draining soils containing high concentrations of nitrate. However, the significance of enriched nitrate concentrations varies with the season. Large concentrations at a time when the sward is actively growing may be a desirable soil quality factor indicating a fertile soil, but in autumn or winter are more likely to be an indicator of unacceptable losses.

Although nitrate is particularly susceptible to loss, the quantities lost from natural or semi-natural grasslands are usually small because mineral-N is continuously utilized by soil organisms and higher plants as it becomes available and concentrations in the soil remain small. Much greater losses occur where conditions lead to the accumulation of high concentrations of mineral-N in the soil. This is most likely in more intensively managed pastures where the supply of N from the mineralization of organic matter is supplemented by inputs from manures and inorganic fertilizers or by fixation of atmospheric N_2 by legumes. Whereas annual losses from extensive grassland are typically <10 kg nitrate-N/ha, losses from intensively managed pastures with high N inputs can be 100–200 kg N/ha (Scholefield *et al.*, 1993; Ledgard *et al.*, 1999; Blicher-Mathiesen and Paulsen, 2002; Eriksen *et al.*, 2004).

Large concentrations of soil nitrate occur where fertilizer or manure has been applied at rates that supply more N than the sward can utilize. Even moderate applications of fertilizer or manure with a large concentration of readily available N will increase the risk of leaching if applied late in the season when there is limited growth to utilize the added N. Similarly, applications in summer when growth is restricted by drought will leave surplus N in the soil, which may be leached when the soil re-wets. Utilizing legumes as a source of N avoids some of the factors that contribute to N leaching from N-fertilized pastures. In particular, the grass component of mixed grass/clover swards acts as a scavenger for mineral-N released from clover residues and from mineralization of SOM. The use of legumes also avoids the peaks of large concentrations of soil N that follow applications of fertilizer or N-rich manures.

Although nitrate leaching can be minimized by matching N inputs to the requirements of the sward, this is difficult to achieve in grazed pastures because of the N excreted in dung and urine. This is not spread uniformly but is concentrated into dung and urine patches, which, as a result, contain much N. Urine patches typically receive the equivalent of 400–1000 kg N/ha (Haynes and Williams, 1993; Di and Cameron, 2002); far in excess of what the sward can immediately utilize. This urine-N is rapidly hydrolysed to ammonium-N and then, over several days, converted to nitrate. The uneven distribution of urine creates a mosaic of 'hot spots' containing very enriched concentrations of nitrate, which are the source of much of the leaching from grazed pastures. The risk of loss is greatest from urine deposited in late summer or autumn when growth is slowing and there is only limited uptake of N by the sward. This leaves much of the urine-derived nitrate in the soil and at risk of being leached over the winter. Dung is less important as a source of leached nitrate. The N is less readily available than in urine and the patches affect a smaller proportion of the pasture area (Sugimoto and Ball, 1989; Stout *et al.*, 1997). For dairy cows grazing at 700 cow days/ha/year (e.g. equivalent to 7 cows/ha grazing for 100 days during the year), it has been estimated that only 6% of the pasture would be directly covered by dung, compared with 21% by urine (Whitehead, 2000). The routine determination of soil mineral-N concentrations employed as an indicator of plant-available N and of potential N losses is usually carried out on a bulked sample of soil and does not adequately reflect this heterogeneity and the very high concentrations present in urine patches.

A consequence of the high concentrations of nitrate in urine and dung-affected areas is that N losses are generally greater from grazed than from cut swards and increase with increasing stocking rate. The highest losses are likely to be from heavily fertilized, intensively stocked pastures; not only because a greater proportion of the area is affected by excreta but also because of higher concentrations of N in urine, and herbage that is already well supplied with N has less capacity to utilize the excess N in urine patches. Because urine patches are the main source of nitrate leaching from grazed pastures, losses are more closely related to stocking rate and N input than to the form of N supply. Hence, quantities of nitrate leached from grass/clover pastures are similar to those from equivalent N fertilized grass pastures with similar stocking rates or N inputs (Cuttle *et al.*, 1992; Tyson *et al.*, 1997; Ledgard, 2001). With equal stocking rates, similar proportions of pasture will be affected by urine, and because the N content of clover-rich herbage is comparable to that of N-fertilized grass, the N content of urine will be similar on both types of pasture. Relative losses may be influenced to some extent by the greater digestibility of clover compared with grass, the presence of N-deficient grass to act as a sink for mineral-N and the inhibition of N_2 fixation by the high N concentrations in urine patches (Vinther, 1998). However, any advantages of clover-based swards appear to be small and difficult to demonstrate in practice. Nitrate losses from pure clover swards grazed with high numbers of stock are comparable to those from intensively fertilized swards (Macduff *et al.*, 1990).

N uptake by the sward is important for preventing accumulations of nitrate in the soil. There is therefore a greater risk of loss from annual pastures where the plant cover does not persist throughout the year. The absence of an actively growing sward allows soil nitrate to accumulate in late summer and autumn. Similarly, cultivation of perennial swards in autumn, either for reseeding or as a part of a ley/arable rotation, interferes with the uptake of soil mineral-N. The enriched organic matter concentration of grassland soils and the surge of mineralization that follows their cultivation release large quantities of nitrate. This will be leached unless a satisfactory plant cover is established before the onset of drainage. Losses of 58–360 kg nitrate-N/ha have been reported during the winter following autumn cultivation of grasslands in the UK (Shepherd *et al.*, 2001).

Although nitrate is the main form of N lost in drainage water, N can also be leached from pastures as nitrite (Smith *et al.*, 1997). Concentrations are less than those of nitrate but are of significance because of the greater direct toxicity of nitrite to aquatic organisms and because of its role as a precursor of N_2O. Nitrite is a temporary intermediate in the transformation of ammonium to nitrate and does not normally persist in the soil. However, the conversion to nitrate is inhibited by high pH in urine patches and after application of NH_4-producing fertilizers such as urea. Nitrite is also produced during denitrification and from organic substrates (Burns *et al.*, 1996; Muller *et al.*, 2006). N is also leached as dissolved organic N and this may be of particular significance in grassland soils because of their enriched organic matter concentration (Jones *et al.*, 2004; Macdonald *et al.*, 2004; Ghani *et al.*, 2007). Annual losses of 2–18 kg N/ha as dissolved organic N have been measured from pastures on an undrained clay loam in the UK (Hawkins and Scholefield, 2000) and losses of up to 30 kg N/ha/year following cultivation of grass/clover swards on a sandy soil (Vinther *et al.*, 2006). Drainage and liming of an organic upland soil was found to markedly increase the concentrations of dissolved organic N in drainage water (Cuttle and James, 1995).

Losses of N may also occur when heavy rain falls shortly after slurry or fertilizer application and washes material from the soil in surface runoff or in preferential flow through macropores to field drains. Losses following slurry applications are mainly as ammonium and organic forms of N, whereas the form of loss from fertilizer depends on the type of fertilizer used. Mineral-N losses are greater following applications of animal slurry than from farmyard manure or compost where a greater proportion of the N content is bound as organic compounds (Chambers *et al.*, 2000).

Gaseous forms of N loss

Soil N also contributes to gaseous emissions in the form of N_2O and NO. Both are produced by the actions of soil microbes during the processes of nitrification and denitrification (for further information see Chapter 1, this volume). Nitrification describes the aerobic process by which ammonium ions are converted to nitrate. During this process, N_2O and NO are produced as by-products from the decomposition of the intermediates formed during the transformation to nitrate. Losses are increased by the use of ammonium or urea-based fertilizers that increase the supply of ammonium substrate in the soil. Similarly, applications of animal slurry with a high proportion of ammonium-N and inputs of urine during grazing both increase the potential for N_2O and NO emissions.

In contrast, denitrification occurs under anaerobic conditions. Nitrate is sequentially reduced to NO, N_2O and ultimately to dinitrogen (N_2). It is not necessary for the whole soil volume to be anaerobic before denitrification occurs – nitrification and denitrification can proceed simultaneously in soils, with nitrate being produced in aerobic zones and defusing into anaerobic microsites where it is denitrified (Abbasi and Adams, 1998). Most field measurements are unable to distinguish between N_2O produced by nitrification and by denitrification but it is generally assumed that where soils are wet, denitrification is the dominant process. Simultaneous measurements demonstrate a wide variation in the ratios of NO/N_2O fluxes (Dendooven *et al.*, 1994). Ratios >1 generally indicate aerobic soil conditions favouring nitrification, whereas in less-aerated soils where conditions favour denitrification, N_2O is the major contributor to emissions (Lipschultz *et al.*, 1981; Skiba *et al.*, 1992; del Prado *et al.*, 2006). Overall, global emissions of N_2O from grassland are estimated to be approximately double those of NO (Stehfest and Bouwman, 2006). The ratio of N_2O to N_2 also varies with soil conditions and nitrate supply. In soils without added nitrate, the product of denitrification is almost entirely N_2 but where nitrate has been applied, there may be a greater proportion of N_2O than of N_2 (Scholefield *et al.*, 1997; Bol *et al.*, 2003). The N_2O/N_2 ratio tends to decrease with increasing soil wetness.

Highest denitrification losses occur from fine-textured, poorly drained soils containing enriched concentrations of nitrate (Smith K.A. *et al.*, 1997; del Prado *et al.*, 2006) but, being heterotrophs, the denitrifier organisms also require a source of readily available C. Accordingly, denitrification rates are directly correlated with SOM contents, particularly the easily decomposable fraction (Bijay *et al.*, 1988; Paul and Clark, 1996). The enriched organic matter concentrations of grassland soils therefore favour denitrification. High concentrations of dissolved organic matter in pasture soils suggest that significant denitrification may extend to the deeper soil layers (Jarvis and Hatch, 1994), but most studies have shown only limited denitrification in subsoils, with activity restricted by C and nitrate supply (Luo *et al.*, 1998; Murray *et al.*, 2004). Unlike leaching, denitrification is temperature-dependent and is limited by low temperatures (i.e. <4°C) in winter.

High concentrations of nitrate accumulate in the soil where N inputs exceed plant uptake or occur directly from fertilizer or manure applications. The release of N from readily degradable residues in grass/clover pastures does not appear to be a significant source of N_2O (Carter and Ambus, 2006). As with nitrate leaching, urine and dung patches are a particularly important source of N_2O and result in greater emissions from grazed pastures than from mown swards. These losses are exacerbated by the physical effects of grazing, particularly the compaction and decreased aeration of soil caused by treading (Oenema *et al.*, 1997; Groenigen *et al.*, 2005; Bhandral *et al.*, 2007). In addition, the high pH in urine patches may lead to hydrolysis of SOM and increase the supply of water-soluble C to the denitrifiers (Monaghan and Barraclough, 1993). In New Zealand, where most N is supplied via legumes, excreta deposited during grazing are responsible for 90% of the N_2O emitted from pastures (Clark *et al.*, 2005). The proportion is less in those countries that use more mineral-N fertilizer; for example, it has been estimated that up to 22% of the total N_2O emission from grassland in the UK originates from excreta (Yamulki *et al.*, 1998).

N_2O losses from urine are usually several times greater than those from dung and represent a greater proportion of the total N concentration (Allen *et al.*, 1996; Flessa *et al.*, 1996a; Oenema *et al.*, 1997; Yamulki *et al.*, 1998). Typically, <2–16% of the total N in urine patches is lost as N_2O (Fowler *et al.*, 1997; Vermoesen *et al.*, 1997). Although nitrification may contribute to the N_2O emitted from urine patches (Koops *et al.*, 1997; Carter, 2007), the relationship between soil moisture and N_2O fluxes suggests that in most circumstances denitrification is the dominant process

(Monaghan and Barraclough, 1993; Stevens and Laughlin, 2001). Because of the seasonal variation in soil moisture content and effect on denitrification rates, the timing of grazing has a marked influence on the total quantities of N_2O emitted during the year (Anger *et al.*, 2003). Fluxes from cattle dung are influenced by the possible effects of dung sealing the soil surface and restricting diffusion of O_2 and denitrification products into and out of the soil (Allen *et al.*, 1996).

N_2O fluxes also increase following applications of mineral-N fertilizers and animal manures. Both increase the nitrate concentration of the soil but manures are also a source of readily biodegradable C (Chadwick *et al.*, 2000) and may contribute to the creation of anaerobic conditions in the soil, further stimulating N_2O production. This may be less important in grassland than in arable soils where available C is more likely to be limiting (McTaggart *et al.*, 1997). The proportion of total N lost as N_2O varies widely, with much of the variation explained by soil conditions and the type of fertilizer or manure applied. Typically 2–4% of the N in ammonium fertilizers and cattle slurry is emitted as N_2O following applications to wet ground but a greater proportion is lost from nitrate fertilizers (5–12%) and pig slurries (4–17%; Christensen, 1983; Egginton and Smith, 1986; Stevens and Laughlin, 1997; Velthof *et al.*, 1997, 2003). Applying slurry and inorganic fertilizer together has been shown to increase N_2O fluxes compared with separate applications (McTaggart *et al.*, 1997; Stevens and Laughlin, 2001; Dittert *et al.*, 2005), and emissions are also greater following slurry injection than from conventional surface applications (Ellis *et al.*, 1998; Velthof *et al.*, 2003; Rodhe *et al.*, 2006). Overall, the total N_2O emissions from slurry applications are considered to be less than those from N excreted during grazing (Oenema *et al.*, 1997).

Grassland is also a source of ammonia emissions to the atmosphere (Sommer and Hutchings, 1997). This volatilization is of concern because much of the ammonia lost from agricultural land is deposited within a relatively short distance and may cause acidification and N enrichment of nearby, more sensitive ecosystems (Krupa, 2003). Ammonia emissions from pastures are primarily short-term, surface responses to additions to the soil such as urine deposited during grazing and applications of animal manures or N fertilizer, rather than representing a loss of N from the soil itself. Nevertheless, losses from these applications can be considerable. Ammonia volatilization accounts for 5–25% of the N in urine patches and 20–80% of the total N in slurry following surface applications to grassland (van der Putten and Ketelaars, 1997; Whitehead, 2000). Typically, 10–25% of the N in urea fertilizer can be lost as ammonia, with smaller losses (1–3%) from ammonium nitrate and calcium ammonium nitrate fertilizers (Whitehead and Raistrick, 1990; Harrison and Webb, 2001). Although ammonia emissions are largely determined by external inputs and predominantly from the soil surface, soil properties are important in determining how much of the potentially volatilizable ammonia is lost following application (Stevens and Laughlin, 1997). The permeability of the soil determines how rapidly urine and the liquid portion of slurry infiltrate into the soil where it is less likely to be volatilized. Soil pH influences the equilibrium between dissolved NH_4^+ ions and gaseous ammonia, with high pH favouring ammonia and volatilization (see also Chapters 1 and 3, this volume). Less ammonia is lost from acid soils and from clay soils with high cation exchange capacities and H^+ buffering that are better able to resist the pH increases induced by urine, slurry and fertilizer applications (Whitehead and Raistrick, 1990; Sommer and Ersboll, 1996; van der Putten and Ketelaars, 1997). The rate of hydrolysis of urea to ammonium-N is determined by urease activity in the soil, which varies with cation exchange capacity and C content (Bremner and Mulvaney, 1978).

Phosphorus

Transfers of P from agricultural land are a widespread cause of diffuse pollution of surface waters. Annual P losses from grasslands are typically in the range <0.1–3.0 kg P/ha (Haygarth and Jarvis, 1999; McDowell *et al.*, 2001). Although these losses are small compared with those of N, P concentrations in unpolluted surface waters are very low and even small increases in P loadings can be sufficient to cause eutrophication.

There is wide variation in the relative proportions of organic and inorganic P in soils, but in most agricultural soils, inorganic P accounts for more than half of the total and this proportion increases with depth (Whitehead, 2000). The proportion of inorganic P also tends to be greater in soils enriched with P (Dougherty *et al.*, 2006). Phosphate ions are strongly adsorbed on to the surfaces of soil particles through reactions with hydrous oxides of Fe and Al or, in neutral or alkaline soils, precipitated as insoluble Ca and Mg phosphates. The limited solubility of these compounds normally maintains P at very low equilibrium concentrations in the soil solution, mainly as $H_2PO_4^-$ and HPO_4^{2-} ions (Frossard *et al.*, 2000), and only small quantities are available for loss in water flow. Soil P is predominantly associated with finer soil fractions, which have relatively larger adsorption surfaces. Other inorganic forms include P adsorbed by kaolinitic clays and P in unweathered soil minerals. Organic forms include P in the cellular components of plant litter and in microbial biomass (Brookes *et al.*, 1984), together with complex humus polymers and small molecular compounds such as inositol phosphates (Celi and Barberis, 2007). Other sources of mobile P include inorganic and organic P released from the mineralization of crop residues (Sharpley and Smith, 1989). Dung is an important source of potentially mobile P in grazed pastures. Livestock are inefficient at utilizing the P they consume and the surplus is excreted, almost entirely in the faeces. Dairy cows typically convert about 36% of the dietary P into milk but the conversion efficiency can be as low as 10% in stock kept for meat and fibre production (Haynes and Williams, 1993; Whitehead, 2000).

Forms of P in soil water include inorganic phosphates and organic P compounds in true solution and as colloidal and coarser particles, though in practice, it is difficult to distinguish between dissolved P and P associated with the finer colloidal fractions. The loss of P from soil first requires a source of potentially mobile P and then a mechanism, usually water flow, to transport the mobilized P from the soil – together constituting a source–mobilization–transfer continuum (Haygarth and Jarvis, 1999). P can be mobilized by solubilization and by detachment.

P solubilization

Solubilization describes the processes that control the concentration of dissolved P in the soil solution. In the absence of excessive P inputs to the soil, sorption and precipitation ensure that concentrations of dissolved P remain low and little P is lost in solution. However, if P inputs to the soil exceed the quantities removed by grazing and in harvested herbage, there will be a gradual accumulation of P on sorption sites and a progressive decrease in the soil's capacity to adsorb further P. This is accompanied by increasing concentrations of P in the soil solution and a corresponding increased risk of loss in surface or subsurface flow (Smith *et al.*, 1995; Jordan *et al.*, 2000; McDowell and Condron, 2004). Losses will start to increase well before the sorption capacity is fully saturated (Holford *et al.*, 1997). Soils with the lowest sorption capacities are typically sands and peat soils that have few sorption sites (Daly *et al.*, 2001). Concentrations of P in surface and subsurface flow increase with increasing concentrations of soil P, as determined by various chemical extractants (e.g. Olsen-P; Olsen and Sommers, 1982) used to determine levels of plant-available P in the soil (Haygarth and Jarvis, 1999). Measurements indicate that leaching losses increase with increasing concentrations of Olsen-P, especially above a point termed the 'threshold' or 'change-point', beyond which, losses increase more sharply (Heckrath *et al.*, 1995). There is also close agreement between P losses and the percentage saturation of the P sorption capacity, as determined from P sorption isotherms or the concentration of oxalate-extractable Al and Fe in the soil (Bolland *et al.*, 1996; Leinweber *et al.*, 1997; Hooda *et al.*, 2000; Pautler and Sims, 2000; Burkitt *et al.*, 2006). The use of percentage saturation as an indicator of P loss appears to be less dependent on soil type than indicators based on extractable P contents (Sharpley, 1995). P sorption has been found to be correlated with clay content and organic C in some studies (Saini and MacLean, 1965; Samadi and Gilkes, 1999; Dodor and Oya, 2000) but not in others (Harter, 1969; Singh and Gilkes, 1991; Burkitt *et al.*, 2002).

Enrichment of soil P is common in grasslands that receive regular applications of P fertilizer and particularly in those that also receive animal manures. Purchased feed can provide a significant input of P to the farm; much of which will be excreted and transferred to the soil in manure, effluent or dung. Excessive inputs are often due, in part, to a failure to decrease inorganic fertilizer inputs sufficiently to compensate for the additional P supplied in manure. Because P offtakes under grazing are less than those where herbage is cut and removed, P is most likely to accumulate in grazed pastures, especially in regions with high stock densities (Leinweber *et al.*, 1997). Regular soil testing to determine contents of extractable P as part of a balanced programme of fertilizer and manure use can help avoid excessive accumulations. However, care is needed when using data from routine soil fertility tests as an indicator of potential leaching from pastures. In the absence of regular cultivation, P accumulates in the upper centimetres of grassland soils and the deeper sampling depth (typically to 7.5 or 10 cm) employed for fertility testing may not be representative of the P content of the uppermost soil (Haygarth *et al.*, 1998; Dougherty *et al.*, 2006), although there are likely to be correlations between the different depths (Daly and Casey, 2005).

High P concentrations in the surface soil will not necessarily result in increased leaching if the mobile P is adsorbed further down the soil profile (Sinaj *et al.*, 2002). Hence, the installation of field drains decreases P losses by increasing the proportion of subsurface flow and increasing the contact between drainage water and the soil matrix (Haygarth *et al.*, 1998). Adsorption elsewhere in the profile is also less likely where P is transported as dissolved organic compounds that are protected against sorption, e.g. from dung or following applications of animal manures (Eghball *et al.*, 1996; Chardon *et al.*, 1997; McDowell *et al.*, 2001; Toor *et al.*, 2004, 2005). The organic matter in manures and effluent may also increase P mobility by complexing with Fe and Al and decreasing P sorption (Leytem and Maguire, 2007). Inositol phosphates and other organic P compounds with high affinities for soil surfaces may also increase leaching by competing with inorganic P for sorption sites (Iyamuremye *et al.*, 1996). Their preferential association with the colloidal soil fraction also increases their potential mobility within the soil (Celi and Barberis, 2007).

Mobilization of P by detachment

This form of P loss is caused by the physical detachment of soil particles and organic matter by the impact of raindrops, followed by transport of the suspended material and associated P in surface runoff (Sharpley and Smith, 1990). Similarly, in clay soils, particles may be lost in subsurface flow where there is rapid water movement through macropores and field drains (Haygarth and Jarvis, 1999). Detachment and transport of particulate matter is responsible for 75–90% of the P loss from arable land but is less important in grasslands where most of the loss is as dissolved P (McDowell *et al.*, 2001). However, there will be a temporarily increased risk of particulate loss where pastures are cultivated for reseeding or where heavy treading damage occurs. In addition to the soil's susceptibility to detachment, the amount of P lost is determined by the P concentration of the soil, and thus of the suspended sediment, and by the frequency and intensity of runoff events. The greater accumulation of P in the upper few centimetres of uncultivated grassland soils increases the amount of P lost as it is this layer that is most likely to be eroded. The P concentration of the eroded material may be further enriched relative to the surface soil by the selective erosion of the finer soil particles and by sorption of P from solution during transport (Haygarth and Jarvis, 1999).

Grazing increases losses of particulate P. Soil compaction caused by treading increases the volume of overland flow while damage to soil structure and the vegetation cover in poached areas makes the soil surface more susceptible to detachment and erosion of soil particles (Haygarth and Jarvis, 1997; Nguyen *et al.*, 1998; McDowell *et al.*, 2003a). Dung also provides an additional source of particulate organic matter that is rich in P (Haynes and Williams, 1993; Whitehead, 2000).

Incidental losses

In addition to these losses of soil P arising from solubilization and detachment, there are also P losses from non-soil sources. These incidental losses occur when heavy rain falls soon after the application of fertilizer or manure/effluent. If the rainfall is of sufficient intensity to initiate runoff, this will transport a portion of the applied material and associated P in surface flow or in subsurface flow through fissures and drains (Smith *et al.*, 1998). Incidental losses can involve the transport of both dissolved and particulate P. Manufactured P fertilizers, such as superphosphate, supply P in a soluble form but this need not contribute to P losses, provided there is sufficient opportunity for the P to be adsorbed on soil particles. However, if surface runoff occurs shortly after the application, fertilizer P may be dissolved and washed off the surface with only limited soil contact and little chance of adsorption. Measurements on pastures in New Zealand indicated that concentrations of dissolved P in overland flow were increased for about 60 days following application of superphosphate. The resulting loss was much greater than where P was supplied in a less-soluble form as reactive phosphate rock (McDowell *et al.*, 2003b). The same processes can also result in high losses of P from freshly applied animal manures (Heathwaite *et al.*, 1998; Smith *et al.*, 1998). Although incidental losses occur relatively infrequently, a single high intensity event can account for much of the total P loss during the year (Haygarth and Jarvis, 1997).

The soil quality attributes most useful as indicators of potential P losses are those that provide a measure of the content of extractable P in the soil, particularly when accompanied by a measurement of the P sorption capacity and percentage saturation. Because P is lost by both surface and subsurface pathways, the physical properties describing the hydrological behaviour of the soil may be less useful as indicators of potential loss. However, this information is essential for understanding the processes involved and for implementing measures to control P losses.

Other nutrients

Although environmentally damaging losses are usually a result of excessive accumulations of nutrients in the soil, nutrient deficiencies can also have an impact through the indirect effect of one nutrient on the uptake of another. For example, deficiencies of S or K in the soil may limit herbage growth and restrict the uptake of N fertilizer. N that is not utilized will add to the pool of soil nitrate that is available for loss by leaching or denitrification. Deficiencies of S and K are more likely to occur in cut than in grazed pastures because of the greater offtake of nutrients in the harvested grass. A study in the UK showed that the addition of S to a cut grass sward receiving 450kg fertilizer-N/ha significantly increased herbage dry matter yields and N offtakes and decreased nitrate leaching in the following two winters by 72% and 58% (Brown *et al.*, 2000).

Sewage sludge and heavy metals

Heavy metals such as Cd, Cu, Hg, Ni, Pb, Zn and Cr are toxic to soil organisms and plants if present in sufficiently high concentrations (Brookes and McGrath, 1984; Barkay *et al.*, 1985; McGrath *et al.*, 1988; Giller *et al.*, 1998; Khan and Scullion, 2002). They can be toxic to grazing animals that consume herbage grown on contaminated sites and also by direct soil ingestion (Hillman *et al.*, 2003). Contamination of soils by heavy metals is not a particular problem of pastures, but can occur where land receives repeated applications of sewage sludge (Percival, 2003). Sewage sludges (or biosolids) from wastewater treatment works contain useful quantities of plant nutrients and their organic matter content may benefit soil physical properties (Aggelides and Londra, 2000; Debosz *et al.*, 2002). There are therefore benefits of recycling sludges on farmland, particularly following bans on dumping at sea and because of the environmental drawbacks of other forms of disposal such as landfill or incineration. Grass swards are particularly suitable for sludge applications because they are available at times of the year when access to arable land is restricted by growing crops (Hillman *et al.*, 2003). However, sewage sludges also frequently contain large concentrations of heavy metals. Once applied, these metals are strongly bound on soil mineral surfaces and as complexes with SOM and retained within the soil (McBride *et al.*, 1997). Repeated applications of sludge can lead to the accumulation of high concentrations of heavy metals unless statutory controls are imposed to limit the quantities that can be applied. Uptakes of heavy metals by grasses are relatively low (O'Riordan *et al.*, 1994) and they may not exhibit symptoms of toxicity until high concentrations have accumulated in the soil. Clovers are more sensitive than grasses. This appears to be due to impaired activity of *Rhizobia* and decreased N_2 fixation rather than to phytotoxicity directly affecting clover growth (McGrath *et al.*, 1988; Lakzian *et al.*, 2002). Studies in the UK on grassland that had received repeated applications of sewage sludge indicated that though concentrations in herbage were increased, accumulations of potentially toxic elements in the edible body tissues of sheep that had grazed the pastures were generally low (Wilkinson *et al.*, 2001). Heavy metal contamination in sludge-treated soils has been shown to decrease microbial biomass and N mineralization, increase biomass-specific respiration rates and cause changes in the microbial community (Fliebach *et al.*, 1994; Chander *et al.*, 1995; Abaye *et al.*, 2005).

The solubility and plant availability of these metals is greatest at low soil pH and their toxicity increases if the soil pH is allowed to fall (Bolan *et al.*, 2003). Legislation controlling maximum permitted concentrations of heavy metals in sludge-amended soils therefore stipulates different maxima for different ranges of soil pH. Because heavy metals are of limited mobility in the soil, surface applications of sewage sludge to permanent grassland result in higher concentrations of these elements (particularly Pb, Cd and Cu) in the surface soil (Davis *et al.*, 1988; Wilkinson *et al.*, 2001; Hillman *et al.*, 2003). This may increase intakes of heavy metals by stock through soil ingestion. The environmental impact of heavy metal contamination is generally confined to the sludge-treated area, though where soil erosion occurs, surface runoff may transport metals adsorbed on to soil particles and extend the contamination to adjoining surface waters. There is also evidence that significant downward movement through the soil profile can occur as a result of the complexation of these metals with soluble organics and transport in preferential flow (Richards *et al.*, 1998).

Cadmium also occurs as a contaminant in the parent rock from which phosphatic fertilizers are manufactured and may accumulate in soils that have received regular fertilizer applications (Rothbaum *et al.*, 1986; Loganathan *et al.*, 1997; Gray *et al.*, 1999). Uptake by herbage and soil ingestion may then lead to accumulations of Cd in the tissues of grazing stock (Lee *et al.*, 1994). The mobility of Cd in soils is limited by adsorption on SOM (Loganathan and Hedley, 1997). Its availability is greatest in acid soils and is decreased by liming (Rothbaum *et al.*, 1986; Nicholson and Jones, 1994). Small amounts of Cd may be leached from pastures (Gray *et al.*, 2003). Phosphatic fertilizers are also a source of fluoride and accumulations of Cd may be accompanied by increased fluoride concentrations in soils and forages (Loganathan *et al.*, 2001).

Soil acidification

Acidification is a natural process that occurs wherever rainfall is sufficient to cause leaching through the soil profile; however, the rate of acidification can be accelerated by management practices that increase the production of H^+ ions or the leaching of basic cations. Soil pH is an important component of soil quality, in particular, controlling nutrient availability, microbial activity and the concentrations of potentially toxic Al and Mn ions. Differing tolerance to soil acidity is important in determining the range of pasture species that can be grown on acid soils. Soil acidity decreases *Rhizobium* populations and N_2 fixation (Slattery *et al.*, 2001). Earthworm numbers are also decreased in acid soils (Ma *et al.*, 1990). In northern Europe, the acidification of managed grasslands is remedied by regular liming of pastures to maintain soil pH within a satisfactory range. In other parts of the world that rely on more extensive long-term pastures of low profitability (e.g. Australia), lime use is not economically viable. It is in these areas that acidification is of greatest concern (Scott *et al.*, 2000).

The main processes involved in acidification of soils under managed grassland are (Oenema, 1990; Ridley *et al.*, 1990; Bolan *et al.*, 1991):

- excretion of H^+ to balance the excess of cation over anion uptake in plants either fixing atmospheric N_2 or utilizing NH_4^+ ions as the major source of N;
- net nitrification of the N derived from fixation or from NH_4^+- or urea-based fertilizers;
- removal of plant and animal products containing N derived from these sources;
- leaching of nitrate where the N input is by fixation or as NH_4^+ or urea fertilizer;
- C cycle acidification through export of organic anions in agricultural products and accumulation of SOM;
- acidic atmospheric deposition in regions affected by industrial emissions.

The uptake of excess cations over anions resulting in the acidification of the rhizosphere can be balanced by the release of OH^- ions during subsequent plant decomposition. However, nitrification of fixed N_2 or N fertilizers and loss of N from the soil in products or by nitrate leaching lead to a net acidification of the system. The greater nitrate leaching under grazing is responsible for the accelerated rate of acidification in grazed compared with ungrazed grasslands (Carran and Theobald, 1995). Where nitrate anions are leached, it is necessary for them to be accompanied by a cation (usually Ca or Mg) to maintain electrical neutrality (Haynes and Williams, 1992). The loss of basic cations therefore increases with increasing stocking rate and nitrate leaching. Increasing productivity by application of P fertilizer to grass/clover pastures or by application of N fertilizer also increases the output of products that further accelerates acidification. Acidification is most rapid in coarse-textured soils with low cation exchange capacities that are poorly buffered against pH change (Crocker and Holford, 1991).

There is only limited evidence of off-site impacts of acidification but it is likely that soil acidification could increase losses of N, P and sediment in surface waters as a result of poor plant growth and less-effective ground cover

(Scott *et al.*, 2000). In the UK, the main cause of stream water acidification has been the input of acid rain coupled with the widespread afforestation of upland areas. In these circumstances, the leaching of bases from limed pastures has had a beneficial effect of ameliorating the acidification originating from elsewhere in the catchment (Hornung *et al.*, 1995).

Salinity

Primary salinization is a natural process affecting soils in many parts of the world. In contrast, secondary salinity is a product of man-made changes to the hydrological cycle (Nash and Haygarth, 2005). It is an increasing problem in dry-land regions such as southern Australia and California where large areas of pasture and arable land have been lost to agriculture because of the development of secondary salinity. This dry land, or seepage salinity, occurs where shallow, salt-rich groundwater rises into the root zone or to the soil surface where the high salt concentration inhibits pasture growth. Where the water reaches the surface, evaporation can produce bare ground with encrustations of dry salt on the surface. These shallow water tables are found on foot slopes and valley floors that receive drainage from further upslope. The rising water tables are a result of the replacement of native perennial species with shallower-rooting annual crops and pasture species that use less water than the native species. This allows more water to reach the subsoil and causes groundwater levels to rise. Grazing of the understorey of partly cleared woodland accelerates the replacement of the remaining perennial grasses and herbs by shallow-rooting annual species (Eberbach, 2003). This is a catchment-scale problem, with the source often being deforestation in the upper parts of the catchment rather than directly in the lower regions that are most affected by the salinity. There is also a risk of more widespread effects on water quality because of increased leakage of saline water from the affected area.

Concentrations of Na^+ ions in affected soils may be sufficient to completely degrade the physical structure of the soil. Saturation of the cation exchange complex by Na^+ ions in these sodic soils causes the deflocculation of clay particles and collapse of soil aggregates. The resulting compaction decreases permeability and porosity, restricting water storage but at the same time slowing internal drainage and increasing waterlogging. Additional secondary stresses restricting plant growth in sodic soils include crusting, acidity or alkalinity, nutrient deficiencies and toxicities (Rengasamy *et al.*, 2003). Soil quality indicators of salinity are high concentrations of exchangeable Na^+ and high electrical conductivity.

Remedies for seepage salinity involve the replanting of deep-rooted trees in agroforestry systems to lower the water table. However, this may require planting of 70–80% of the catchment to be effective. Pastures may also be resown with perennial species that are able to develop deep roots and limit water movement to the water table. The properties of lucerne (*Medicago sativa*) make it potentially useful for this purpose (Ward *et al.*, 2003). However, this species was ineffective in other studies where its growth increased concentrations of salts in the rooting zone (Rengasamy *et al.*, 2003). The structural limitations of sodic soils can be remedied by applications of lime or gypsum, to replace Na^+ with Ca^{2+} ions, combined with deep cultivation and increasing SOM. This is less effective at overcoming the constraints imposed by subsoil salinity. An alternative where the salinity is not too severe may be to sow pastures with more salt-tolerant grasses.

Impacts on Biological Aspects of Soil Quality

Biological activity is important because of the key role that microbes and other organisms play in organic matter breakdown, nutrient release and developing soil structure (Tisdall, 1994; Stenberg, 1999; Bloem and Breure, 2003). Hence, it is also a valuable indicator of soil quality. Measurement of the activities of soil organisms provides a measure of general soil health that integrates the combined effects of the wide range of physical and chemical properties that define the soil environment in which these organisms live. Microorganisms are particularly useful as indicators because of their sensitivity to changes in their environment. Soil respiration, microbial biomass, potential N mineralization and various enzyme activities are commonly included as part of minimum data sets for assessing the biological component of soil quality (Bloem *et al.*, 2006). Although these properties are readily determined, assessments for broader ecological monitoring exercises and for detecting changes in biodiversity require more detailed measurements of the main functional groups of the soil food web (Mulder *et al.*, 2005).

The different trophic groups of invertebrates, protozoa, fungi and bacteria that make up the complex food web that exists in the soil require a regular input of readily available organic substrate. The relatively high inputs of organic matter in pasture soils therefore favour high rates of biological activity, and properties such as respiration rate and microbial biomass are generally greater than in equivalent arable soils (Lynch and Panting, 1980). However, the aspects of pastoral agriculture that have an adverse effect on the physical and chemical properties of the soil also impact on biological activity. For example, mineralization and nitrification are decreased and denitrification increased in soils that have been compacted by treading. Many biological processes are inhibited by acidification and salinity.

Synthetic inhibitors

Although high rates of biological activity are seen as desirable soil qualities, these processes also contribute to the environmentally damaging losses of nutrients that occur from soils. In particular, the nitrification process is responsible for much of the N that is lost because it converts the relatively immobile NH_4^+ ion to NO_3^-, which is much more readily lost by leaching and by denitrification to NO and N_2O. These gases are also produced as by-products of the nitrification process itself. This has led to the use of artificial nitrification inhibitors to block the action of the *Nitrosomonas* bacteria that are responsible for the first stage of the nitrification process, preventing the conversion of NH_4^+ to NO_3^- and retaining it in a form that is less susceptible to loss. Inhibitors such as dicyandiamide (DCD) have been shown to decrease leaching and denitrification from urea- and ammonium-based fertilizers, from dairy slurry and from urine patches in grazed pastures (Smith K.A. *et al.*, 1997; Merino *et al.*, 2002; Dalal *et al.*, 2003; Edmeades, 2004; Di *et al.*, 2007). Although the use of artificial inhibitors to disrupt natural soil process may appear to be contrary to the wider ecological concepts of soil quality and the preservation of biological diversity, the inhibition is only temporary (Hauser and Haselwandter, 1990) and there is no evidence of wider environmental impacts arising from the use of these products (Edmeades, 2004). Within the context of productive agricultural soils, where nitrification inhibitors are likely to be employed, the resulting increase in N efficiency and decreased losses can be seen as representing an improvement in soil quality. Other inhibitors of biological processes that have been used to decrease N losses include urease inhibitors added to urea fertilizer to control NH_3 volatilization, by preventing the hydrolysis of urea to NH_3 and similarly to control NH_3 emissions from beef feedlots and pig units (Varel, 1997; Watson *et al.*, 1998).

Agrochemicals and veterinary residues

Biological processes may also be disturbed by other chemical inputs that have unintentional effects on soil organisms. As described above, heavy metals added in sewage sludge may inhibit *Rhizobia* and decrease N_2 fixation in clover-based pastures. Other biologically active inputs include herbicides and fungicides, though when applied at normal field rates, levels of herbicides reaching the soil are usually too low to affect soil biological processes. Fungicides are more likely to have an impact (Paul and Clark, 1996), but are applied to pastures less frequently than to arable crops. Of greater concern are the residues of veterinary medicines used to treat farm livestock. Residues of these drugs are excreted in the faeces and this provides a route by which these chemicals may reach the soil and impact on the normal functioning of soil organisms.

One concern is that these residues may be toxic to soil-dwelling insects and earthworms. The most visible effect of this would be to delay the rate of incorporation of dung pats into the soil. Decreases in invertebrate numbers would also have secondary effects on insectivorous birds that rely on dung pats as a source of prey. Avermectins are a family of antiparasitic agents that are commonly used to treat farm livestock and a number of studies have concluded that their residues have detrimental effects on several species of dung-dwelling diptera, coleoptera and nematodes (Wall and Strong, 1987; Madsen *et al.*, 1990; Barth *et al.*, 1993; Strong, 1993; Wratten *et al.*, 1993; Wardhaugh and Mahon, 1998; Suarez *et al.*, 2003; Webb *et al.*, 2007). There have also been concerns about fenbendazole, an anthelmintic that is effective against endoparasitic nematodes. Although it belongs to the benzimidazole group that includes several fungicides that are toxic to earthworms, it has not been shown to have harmful effects on dung-related insects. Studies in Denmark indicated that neither ivermectin (one of the avermectin family) nor fenbendazole had serious negative long-term effects on populations of *Lumbricus terrestris* following normal use in cattle (Svendsen *et al.*, 2003, 2005). Observed difference in dung incorporation rates were attributed to an effect of ivermectin on soil insects.

A related area of concern is the widespread use of antibiotics in agriculture (Kumar *et al.*, 2005). A large proportion of the antibiotic administered to livestock is excreted in dung and urine and can then be transferred to the soil in excreta and manures. The medicinal use of antibiotics is necessary for treating infection but particular concerns are centred on their routine use as feed supplements to promote growth. Although this non-clinical use is largely limited to confined stock, the excreted antibiotics may reach the soil in manure. The use of antibiotics as growth promoters is no longer permitted in the countries of the European Union and is likely to decline elsewhere as a result of consumer concerns and voluntary initiatives. The excessive use of antibiotics is of concern because it increases the likelihood of antibiotic resistance developing in livestock and humans. There is also a risk that antibiotics may alter soil microbial systems and be toxic to some plants and soil organisms. In addition to their possible direct effects within pastures, there is also a wider risk of leaching into groundwater. Even where they are firmly bound to soil particles, there is still a risk that antibiotics may be lost with sediment in surface runoff. Survival times for antibiotics in soil, manure and water vary depending on their photostability, binding to soil solids, biodegradability and water solubility. Their survival is greater in cold weather; for example, after autumn applications of manure (Kumar *et al.*, 2005). In contrast to the concerns about antibiotic use, there are also claims that adding antibiotics to ruminant diets may provide environmental benefits through decreases in N excretion and methane production (Tedeschi *et al.*, 2003).

Pathogens

Excretion by grazing animals and spreading of animal manures on grassland also create a risk of contamination of soil and water bodies with faecal pathogens. Pathogenic organisms that normally inhabit the gut are transferred to the soil in faeces and manures, where they may transmit infection to other stock and to humans. Pathogens from livestock farming are an important cause of gastrointestinal illness in the human population. Although direct human exposure is limited within the pasture itself, there is a much greater risk of infection where pathogens are transported from the soil in surface runoff or macropore flow (Oliver *et al.*, 2005). The transfer of pathogens from soils to surface waters may contaminate sources of drinking water or waters used for recreational purposes. There are similar risks following the application of sewage sludges (biosolids) to farmland. Spreading these materials on pasture creates a risk of exposure for livestock within the treated area and of wider contamination of groundwater and surface waters (Pepper *et al.*, 2006).

Bacterial pathogens commonly found in livestock manures include some strains of *Escherichia coli* (notably *E. coli* O157), *Salmonella* spp., *Campylobacter* spp. and *Listeria* spp.; protozoan parasites include *Cryptosporidium parvum* and *Giardia intestinalis*. Manures may also contain helminths. Viruses may be present in excreta but few zoonotic viruses infect cattle and the risk of viral infection is less than from domestic sewage sludge. The incidence of pathogens varies greatly for different organisms and between herds and countries. There is also considerable temporal variation (Oliver *et al.*, 2005). Human pathogens routinely found in sewage sludge similarly include bacteria, protozoan parasites, viruses and helminths.

These organisms are not a threat to the soil biota; indeed, the soil's ability to degrade them is seen as a benefit of land-spreading as a means of disposal. However, survival times vary greatly for different forms of pathogen and are further affected by environmental conditions and soil type. The length of time for which pathogens are able to survive in the soil is an important factor determining their risk of transmission and transport to water bodies. Conditions affecting survival include temperature, moisture, sunlight (UV radiation), presence of indigenous microorganisms, soil pH, organic matter content and the availability of nutrients (Table 2.3). Lower temperatures favour the survival of most organisms, provided that the temperature remains above freezing. Warm temperatures created in manure heaps and during composting play an important role in decreasing the numbers of viable organisms in these materials before they are spread on land. Survival is decreased in drier soils and by extremes of soil pH. There is a relatively narrow pH optimum for bacteria of around 6–7. Predation by soil protozoa decreases bacterial numbers but this is less effective in clay-rich soils where microbes can achieve some protection against predation by forming complexes with clay particles or by sheltering in pores that are too small to allow the entry of protozoa. Bacteria also require a supply of nutrients and of utilizable C; the high solids concentration of slurries and manures aids their survival. In addition, the formation of a dry crust on the surface of dung pats increases pathogen survival by affording protection against sunlight and maintaining favourable conditions within the pat. Survival times for total and faecal coliform bacteria and faecal streptococci in soils range from weeks to months, depending on temperature and moisture, whereas helminth eggs can survive for several years.

Soil properties affect not only survival times and the numbers of pathogens in the soil but also determine the potential for the transport of these organisms to water bodies. Pathogens are transported in surface runoff and in water moving through the soil profile. In more permeable soils, pathogens can be transported down the profile in percolating water but this movement may be slowed by adsorption on to clay or organic matter surfaces. Sorption is greatest in clay-rich soils. Filtering of microorganisms as water passes through finer pores and the blocking of pores by microbial cells impedes their movement through the soil matrix. Movement of pathogens through the bulk soil is also influenced by water

Table 2.3. Effect of environmental factors on the survival of pathogenic microbes. (From Pepper *et al.*, 2006.)

	Effect on survival time		
Parameter	Viruses	Bacteria	Protozoa
Increasing temperature	–	–	–
Decreasing soil moisture	–	–	–
Increasing rate of desiccation	–	–	–
Increasing clay content	+	+	Not known
pH in range 6–8	+	+	+

– denotes decreased survival time; + denotes increased survival time.

flow rates, predation, physiological state of the cells and by their intrinsic mobility. The diameters of bacteria and viruses fall into the size range of colloids that are particularly mobile in soil water (Kretzschmar *et al.*, 1999). Significant movement of pathogens only occurs where there are sufficient water-filled pores, but where conditions are suitable, this provides a pathway by which contaminants may reach the groundwater.

In less-permeable soils, the primary route of pathogen transport is in surface runoff following storm events or where such events initiate preferential flow through soil macropores to field drains. Organisms can be transported as freely suspended cells or attached to soil or manure particles. Transport is rapid because there is little contact between the suspended material and the soil, and therefore little opportunity for interactions with soil surfaces or for the filtering processes that restrict movement through the soil matrix. Earthworms facilitate the movement of pathogens by creating channels for macropore flow and by the burial of contaminated plant residues (Thorpe *et al.*, 1997). Transport by these routes is primarily to surface waters. Where farm animals have access to unfenced streams or rivers, there is also the likelihood of contamination from direct defaecation into the water body.

Conclusions

The main effects of pastoral agriculture on soil quality and their potential environmental impact are summarized in Table 2.4. Although grassland does offer environmental benefits over arable cropping and some other land uses, these advantages tend to be lost with increasing intensity of management. Many of these adverse environmental impacts are a direct result of grazing. The supply of nutrients and pathogens to the soil in dung and urine, coupled with relatively low offtakes of nutrients, provide a source of pollutants, while poaching and compaction caused by treading increases their mobilization and transport. The innate behaviour of livestock and marked spatial heterogeneity of dung and urine returns that contribute to the large losses from pastures also add to the difficulties of controlling these losses. Animal manures also contribute to the pool of potential pollutants in the soil. Even where there is a benefit of grazing and manure applications through increased organic matter inputs and C storage, the net benefit is decreased by the increased emissions of non-CO_2 greenhouse gases arising from the presence of livestock. Grazing also contributes to increased rates of acidification in pasture soils through the increased leaching of bases that accompanies the loss of nitrate from urine patches.

Not all environmental impacts originating from grassland soils are a direct result of grazing. Other characteristics of grasslands that contribute to environmental damage include the absence of regular cultivation that would otherwise disrupt compacted soil layers and allow less-mobile pollutants to accumulate in the topsoil, and the high organic matter content of these soils that may lead to large losses of N and CO_2 if they are cultivated.

There are positive and negative interactions between the various impacts of pasture management on soil quality. Inverse relationships between losses of

Table 2.4. Summary of the main effects of pastoral agriculture on soil quality and their potential environmental impact.

Process	Effect on soil quality	Potential environmental impact
1. Treading by livestock	Loss of soil structure Increased bulk density Decreased pore space and pore continuity Decreased aeration and hydraulic conductivity	Increased surface runoff and loss of sediment and associated P Increased loss of N and P in surface runoff Increased denitrification (N_2O loss) Loss of nutrients decreases productivity
2. Deposition of excreta (and application of manures)	Increased N contents (particularly in urine and dung patches)	Increased nitrate leaching Increased NO and N_2O loss by nitrification and denitrification Volatilization of NH_3 Increased rate of soil acidification Loss of nutrients decreases productivity
	Accumulation of soil P and increasing saturation of sorption capacity	Increased loss of dissolved and particulate P in surface runoff and subsurface flow
	Increased organic matter content	See (3) below
	Introduction of pathogenic organisms	Direct infection and transfer to surface water
	Introduction of veterinary residues	Decreased activity of soil invertebrates and rate of dung incorporation
3. Organic matter inputs and turnover	Increased organic matter content	Provides a sink for atmospheric CO_2 Risk of large C and N losses if management changes (especially if cultivated)
	Improved soil structure	Decreased loss of sediment, P and N in surface runoff Decreased denitrification (N_2O loss)
	Increased biological activity	Increased nutrient turnover but may increase losses when high rates of N mineralization occur at times of limited N uptake
4. Acidification	Decreased soil pH Increased solubility of Al and Mn Decreased availability of plant nutrients Decreased biological activity	Decreased productivity of pasture Decreased N fixation Production of acidic drainage water
5. Salinization	High Na content Loss of soil structure	Decreased productivity of pasture Production of saline drainage water
6. Application of biosolids	Increased heavy metal content Introduction of pathogenic organisms	Toxicity to crops and stock Effects on soil organisms and decreased N fixation Loss of heavy metals in runoff Direct infection and transfer to surface water

different forms of the same pollutant arise because a loss by one process decreases the quantity available for loss by other routes; also, soil conditions that favour leaching are less likely to promote surface runoff or denitrification and vice versa. Positive interactions occur because of direct associations between pollutants, e.g. between sediment and P adsorbed on soil particles, and more generally where management affects the source strength of a number of pollutants concurrently. Hence, intensification of pasture management, which increases all inputs to the soil, will tend to have a similar effect of increasing all forms of nutrient and pathogen loss. However, because of the non-linearity of individual responses, the relative increases may differ.

Soil quality is determined by the productive capacity of the soil and by its interactions with the wider environment. Many of the features of pastoral agriculture that result in environmental damage, particularly those associated with nutrient loss, also have detrimental effects on productivity and resource-use efficiency. Thus, there are opportunities for developing more sustainable pasture systems that simultaneously address the environmental and production aspects of soil quality.

References

Abaye, D.A., Lawlor, K., Hirsch, P.R. and Brookes, P.C. (2005) Changes in the microbial community of an arable soil caused by long-term metal contamination. *European Journal of Soil Science* 56, 93–102.

Abbasi, M.K. and Adams, W.A. (1998) Loss of nitrogen in compacted grassland soil by simultaneous nitrification and denitrification. *Plant and Soil* 200, 265–277.

Aggelides, S.M. and Londra, P.A. (2000) Effects of compost produced from town wastes and sewage sludge on the physical properties of a loamy and a clay soil. *Bioresource Technology* 71, 253–259.

Allen, A.G., Jarvis, S.C. and Headon, D.M. (1996) Nitrous oxide emissions from soils due to inputs of nitrogen from excreta return by livestock on grazed grassland in the UK. *Soil Biology and Biochemistry* 28, 597–607.

Amon, B., Kryvoruchko, V., Amon, T. and Zechmeister-Boltenstern, S. (2006) Methane, nitrous oxide and ammonia emissions during storage and after application of dairy cattle slurry and influence of slurry treatment. *Agriculture, Ecosystems and Environment* 112, 153–162.

Anger, M., Hoffmann, C. and Kuhbauch, W. (2003) Nitrous oxide emissions from artificial urine patches applied to different N-fertilized swards and estimated annual N_2O emissions for differently fertilized pastures in an upland location in Germany. *Soil Use and Management* 19, 104–111.

Aritajat, U., Madge, D.S. and Gooderham, P.T. (1977) The effects of compaction of agricultural soils on soil fauna. I. Field investigations. *Pedobiologia* 17, 262–282.

Ball, B.C., Campbell, D.J., Douglas, J.T., Henshall, J.K. and O'Sullivan, M.F. (1997) Soil structural quality, compaction and land management. *European Journal of Soil Science* 48, 593–601.

Ball, B.C., Scott, A. and Parker, J.P. (1999) Field N_2O, CO_2 and CH_4 fluxes in relation to tillage, compaction and soil quality in Scotland. *Soil and Tillage Research* 53, 29–39.

Barkay, T., Tripp, S.C. and Olson, B.H. (1985) Effect of metal-rich sewage sludge application on the bacterial communities of grasslands. *Applied and Environmental Microbiology* 49, 333–337.

Barth, D., Heinzemutz, E.M., Roncalli, R.A., Schluter, D. and Gross, S.J. (1993) The degradation of dung produced by cattle treated with an ivermectin slow-release bolus. *Veterinary Parasitology* 48, 215–227.

Batey, T. and McKenzie, D.C. (2006) Soil compaction: identification directly in the field. *Soil Use and Management* 22, 123–131.

Beare, M.H., Curtin, D., Thomas, S., Fraser, P.M. and Francis, G.S. (2005) Chemical components and effects on soil quality in temperate grazed pasture systems. In: Jarvis, S.C., Murray, P.J. and Roker, J.A. (eds) *Optimisation of Nutrient Cycling and Soil Quality for Sustainable Grasslands. Proceedings of a Satellite Workshop of the 20th International Grassland Congress, Oxford, UK, July 2005*. Wageningen Academic Publishers, Wageningen, the Netherlands, pp. 25–36.

Betteridge, K., Mackay, A.D., Shepherd, T.G., Barker, D.J., Budding, P.J., Devantier, B.P. and Costall, D.A. (1999) Effect of cattle and sheep treading on surface configuration of a sedimentary hill soil. *Australian Journal of Soil Research* 37, 743–760.

Bhandral, R., Saggar, S., Bolan, N.S. and Hedley, M.J. (2007) Transformation of nitrogen and nitrous oxide emission from grassland soils as affected by compaction. *Soil and Tillage Research* 94, 482–492.

Bijay, S., Ryden, J.C. and Whitehead, D.C. (1988) Some relationships between denitrification potential and fractions of organic carbon in air-dried and field-moist soils. *Soil Biology and Biochemistry* 20, 737–741.

Bilotta, G.S., Brazier, R.E. and Haygarth, P.M. (2007) The impacts of grazing animals on the quality of soils, vegetation, and surface waters in intensively managed grasslands. *Advances in Agronomy* 94, 237–280.

Blicher-Mathiesen, G. and Paulsen, I. (2002) Nitrate leaching from temporary grasslands at dairy farms. *Multi-function Grasslands: Quality Forages, Animal Products and Landscapes. Proceedings of the 19th General Meeting of the European Grassland Federation, La Rochelle, France, May 2002*. pp. 662–663.

Bloem, J. and Breure, A.M. (2003) Microbial indicators. In: Markert, B., Breure, A.M. and Zechmeister, H. (eds) *Bioindicators/Biomonitors – Principles, Assessment, Concepts*. Elsevier, Amsterdam, the Netherlands, pp. 259–282.

Bloem, J., Schouten, A.J., Sørensen, S.J., Rutgers, M., Van der Werf, A. and Breure, A.M. (2006) Monitoring and evaluating soil quality. In: Bloem, J., Hopkins, D.W. and Benedetti, A. (eds) *Microbiological Methods for Assessing Soil Quality*. CAB International, Wallingford, UK, pp. 23–49.

Bol, R., Toyoda, S., Yamulki, S., Hawkins, J.M.B., Cardenas, L.M. and Yoshida, N. (2003) Dual isotope and isotopomer ratios of N2O emitted from a temperate grassland soil after fertilizer application. *Rapid Communications in Mass Spectrometry* 17, 2550–2556.

Bolan, N.S., Hedley, M.J. and White, R.E. (1991) Processes of soil acidification during nitrogen cycling with emphasis on legume-based pastures. *Plant and Soil* 134, 53–63.

Bolan, N.S., Adriano, D.C. and Curtin, D. (2003) Soil acidification and liming interactions with nutrient and heavy metal transformations and bioavailability. *Advances in Agronomy* 78, 215–272.

Bolland, M.D.A., Gilkes, R.J., Brennan, R.F. and Allen, D.G. (1996) Comparison of seven phosphorus sorption indices. *Australian Journal of Soil Research* 34, 81–89.

Bremner, J.M. and Mulvaney, R.L. (1978) Urease activity in soils. In: Burns, R.G. (ed.) *Soil Enzymes*. Academic Press, London, pp. 149–196.

Brookes, P.C. and McGrath, S.P. (1984) Effects of metal toxicity on the size of the soil microbial biomass. *Journal of Soil Science* 35, 341–346.

Brookes, P.C., Powlson, D.S. and Jenkinson, D.S. (1984) Phosphorus in the soil microbial biomass. *Soil Biology and Biochemistry* 16, 169–175.

Brown, L., Scholefield, D., Jewkes, E.C., Preedy, N., Wadge, K.J. and Butler, M.R. (2000) The effect of sulphur application on the efficiency of nitrogen use in two contrasting grassland soils. *Journal of Agricultural Science* 135, 131–138.

Bryant, H.T., Blaser, R.E. and Peterson, J.R. (1972) Effect of trampling by cattle on bluegrass yield and soil compaction of a Meadowville loam. *Agronomy Journal* 64, 331–334.

Burkitt, L.L., Moody, P.W., Gourley, C.J.P. and Hannah, M.C. (2002) A simple phosphorus buffering index for Australian soils. *Australian Journal of Soil Research* 40, 1–18.

Burkitt, L.L., Gourley, C.J.P., Hannah, M.C. and Sale, P.W.G. (2006) Assessing alternative approaches to predicting soil phosphorus sorption. *Soil Use and Management* 22, 325–333.

Burns, L.C., Stevens, R.J. and Laughlin, R.J. (1996) Production of nitrite in soil by simultaneous nitrification and denitrification. *Soil Biology and Biochemistry* 28, 609–616.

Burns, R.G., Nannipieri, P., Benedetti, A. and Hopkins, D.W. (2006) Defining soil quality. In: Bloem, J., Hopkins, D.W. and Benedetti, A. (eds) *Microbiological Methods for Assessing Soil Quality*. CAB International, Wallingford, UK, pp. 15–22.

Butler, A.R. and Adams, W.A. (1990) Denitrification characteristics of a compacted pasture soil. *Mitteilungen der Deutschen Bodenkundlichen Gesellschaft* 60, 45–50.

Carran, R.A. and Theobald, P.W. (1995) Nitrogen cycle processes and acidification of soils in grazed pastures receiving or not receiving excreta for 23 years. *Australian Journal of Soil Research* 33, 525–534.

Carter, M.S. (2007) Contribution of nitrification and denitrification to N_2O emissions from urine patches. *Soil Biology and Biochemistry* 39, 2091–2102.

Carter, M.S. and Ambus, P. (2006) Biologically fixed N_2 as a source for N_2O production in a grass-clover mixture, measured by $^{15}N_2$. *Nutrient Cycling in Agroecosystems* 74, 13–36.

Celi, L. and Barberis, E. (2007) Abiotic reactions of inositol phosphates in soil. In: Turner, B.L., Richardson, A.E. and Mullaney, E.J. (eds) *Inositol Phosphates: Linking Agriculture and the Environment*. CAB International, Wallingford, UK, pp. 207–220.

Chadwick, D.R., Pain, B.F. and Brookman, S.K.E. (2000) Nitrous oxide and methane emissions following application of animal manures to grassland. *Journal of Environmental Quality* 29, 277–287.

Chambers, B.J., Smith, K.A. and Pain, B.F. (2000) Strategies to encourage better use of nitrogen in animal manures. *Soil Use and Management* 16, 157–166.

Chander, K., Brookes, P.C. and Harding, S.A. (1995) Microbial biomass dynamics following addition of metal-enriched sewage sludges to a sandy loam. *Soil Biology and Biochemistry* 27, 1409–1421.

Chardon, W.J., Oenema, O., delCastilho, P., Vriesema, R., Japenga, J. and Blaauw, D. (1997) Organic phosphorus in solutions and leachates from soils treated with animal slurries. *Journal of Environmental Quality* 26, 372–378.

Christensen, S. (1983) Nitrous oxide emission from a soil under permanent grass: seasonal and diurnal fluctuations as influenced by manuring and fertilization. *Soil Biology and Biochemistry* 15, 531–536.

Clark, H., Pinares-Patiño, C. and deKlein, C. (2005) Methane and nitrous oxide emissions from grazed grasslands. *Grassland: A Global Resource*. Wageningen Academic Publishers, Wageningen, the Netherlands, pp. 279–293.

Cluzeau, D., Binet, F., Vertes, F., Simon, J.C., Riviere, J.M. and Trehen, P. (1992) Effects of intensive cattle trampling on soil-plant-earthworms system in two grassland types. *Soil Biology and Biochemistry* 24, 1661–1665.

Conant, R.T., Paustian, K. and Elliott, E.T. (2001) Grassland management and conversion into grassland: effects on soil carbon. *Ecological Applications* 11, 343–355.

Corre, M.D., Schnabel, R.R. and Shaffer, J.A. (1999) Evaluation of soil organic carbon under forests, cool-season and warm-season grasses in the north-eastern US. *Soil Biology and Biochemistry* 31, 1531–1539.

Crocker, G.J. and Holford, I.C.R. (1991) Effects of pasture improvement with superphosphate on soil pH, nitrogen and carbon in a summer rainfall environment. *Australian Journal of Experimental Agriculture* 31, 221–224.

Curll, M.L. and Wilkins, R.J. (1983) The comparative effects of defoliation, treading and excreta on a *Lolium perenne – Trifolium repens* pasture grazed by sheep. *Journal of Agricultural Science* 100, 451–460.

Cuttle, S.P. and James, A.R. (1995) Leaching of lime and fertilizers from a reseeded upland pasture on a stagnogley soil in mid-Wales. *Agricultural Water Management* 28, 95–112.

Cuttle, S.P., Hallard, M., Daniel, G. and Scurlock, R.V. (1992) Nitrate leaching from sheep grazed grass-clover and fertilized grass pastures. *Journal of Agricultural Science* 119, 335–343.

Dalal, R.C., Wang, W., Robertson, G.P. and Parton, W.J. (2003) Nitrous oxide emission from Australian agricultural lands and mitigation options: a review. *Australian Journal of Soil Research* 41, 165–195.

Daly, K. and Casey, A. (2005) Environmental aspects of soil phosphorus testing. *Irish Journal of Agricultural and Food Research* 44, 261–279.

Daly, K., Jeffrey, D. and Tunney, H. (2001) The effect of soil type on phosphorus sorption capacity and desorption dynamics in Irish grassland soils. *Soil Use and Management* 17, 12–20.

Davis, R.D., Carlton-Smith, C.H., Stark, J.H. and Campbell, J.A. (1988) Distribution of metals in grassland soils following surface applications of sewage sludge. *Environmental Pollution* 49, 99–115.

Debosz, K., Petersen, S.O., Kure, L.K. and Ambus, P. (2002) Evaluating effects of sewage sludge and household compost on soil physical, chemical and microbiological properties. *Applied Soil Ecology* 19, 237–248.

del Prado, A., Merino, P., Estavillo, J.M., Pinto, M. and González-Murua, C. (2006) N_2O and NO emissions from different N sources and under a range of soil water contents. *Nutrient Cycling in Agroecosystems* 74, 229–243.

Dendooven, L., Splatt, P., Anderson, J.M. and Scholefield, D. (1994) Kinetics of the denitrification process in a soil under permanent pasture. *Soil Biology and Biochemistry* 26, 361–370.

Di, H.J. and Cameron, K.C. (2002) The use of a nitrification inhibitor, dicyandiamide (DCD), to decrease nitrate leaching and nitrous oxide emissions in a simulated grazed and irrigated grassland. *Soil Use and Management* 18, 395–403.

Di, H.J., Cameron, K.C. and Sherlock, R.R. (2007) Comparison of the effectiveness of a nitrification inhibitor, dicyandiamide, in reducing nitrous oxide emissions in four different soils under different climatic and management conditions. *Soil Use and Management* 23, 1–9.

Dittert, K., Lampe, C., Gasche, R., Butterbach-Bahl, K., Wachendorf, M., Papen, H., Sattelmacher, B. and Taube, F. (2005) Short-term effects of single or combined application of mineral-N fertilizer and cattle slurry on the fluxes of radiatively active trace gases from grassland soil. *Soil Biology and Biochemistry* 37, 1665–1674.

Dodor, D.E. and Oya, K. (2000) Phosphate sorption characteristics of major soils in Okinawa, Japan. *Communications in Soil Science and Plant Analysis* 31, 227–288.

Donkor, N.T., Gedir, J.V., Hudson, R.J., Bork, E.W., Chanasyk, D.S. and Naeth, M.A. (2002) Impacts of grazing systems on soil compaction and pasture production in Alberta. *Canadian Journal of Soil Science* 82, 1–8.

Doran, J.W. and Parkin, T.B. (1994) Defining and assessing soil quality. In: Doran, J.W., Coleman, D.C., Bezdicek, D.E. and Stewart, B.A. (eds) *Defining Soil Quality for a Sustainable Environment, Special Publication 35*. American Society of Agronomy, Madison, Wisconsin, pp. 3–21.

Doran, J.W. and Parkin, T.B. (1996) Quantitative indicators of soil quality: a minimum data set. In: Doran, J.W. and Jones, A.J. (eds) *Methods for Assessing Soil Quality. SSSA Special Publication No. 49*. Soil Science Society of America, Madison, Wisconsin, pp. 25–37.

Doran, J.W., Sarrantonio, M. and Liebig, M.A. (1996) Soil health and sustainability. *Advances in Agronomy* 56, 2–54.

Dougherty, W.J., Nash, D.M., Chittleborough, D.J., Cox, J.W. and Fleming, N.K. (2006) Stratification, forms, and mobility of phosphorus in the topsoil of a Chromosol used for dairying. *Australian Journal of Soil Research* 44, 277–284.

Eberbach, P.L. (2003) The eco-hydrology of partly cleared, native ecosystems in southern Australia: a review. *Plant and Soil* 257, 357–369.

Edmeades, D.C. (2004) Nitrification and urease inhibitors: a review of the national and international literature on their effects on nitrate leaching, greenhouse gas emissions and ammonia volatilization from temperate legume-based pastoral systems. *Environment Waikato Technical Report*. Environment Waikato, Hamilton East, New Zealand.

Egginton, G.M. and Smith, K.A. (1986) Nitrous oxide emission from a grassland soil fertilized with slurry and calcium nitrate. *Journal of Soil Science* 7, 59–67.

Eghball, B., Binford, G.D. and Baltensperger, D.D. (1996) Phosphorus movement and adsorption in a soil receiving long-term manure and fertilizer application. *Journal of Environmental Quality* 25, 1339–1343.

Ellis, S., Yamulki, S., Dixon, E., Harrison, R. and Jarvis, S.C. (1998) Denitrification and N_2O emissions from a UK pasture soil following the early spring application of cattle slurry and mineral fertilizer. *Plant and Soil* 202, 15–25.

Eriksen, J., Vinther, F.P. and Soegaard, K. (2004) Nitrate leaching and N_2-fixation in grasslands of different composition, age and management. *Journal of Agricultural Science* 142, 141–151.

Evans, R. (1997) Soil erosion in the UK initiated by grazing animals – a need for a national survey. *Applied Geography* 17, 127–141.

Flessa, H., Dörsch, P. and Beese, F. (1996a) N_2O emissions from cattle excrements in pasture land – a global inventory. *Transactions of the 9th Nitrogen Workshop, Braunschweig, September 1996*. Technical University of Braunschweig, Braunschweig, Germany, pp. 141–144.

Flessa, H., Dorsch, P., Beese, F., Konig, H. and Bouwman, A.F. (1996b) Influence of cattle wastes on nitrous oxide and methane fluxes in pasture land. *Journal of Environmental Quality* 25, 1366–1370.

Fliebach, A., Martens, R. and Reber, H.H. (1994) Soil microbial biomass and microbial activity in soils treated with heavy metal contaminated sewage sludge. *Soil Biology and Biochemistry* 26, 1201–1205.

Follett, R.F. and Schuman, G.E. (2005) Grazing land contributions to carbon sequestration. In: McGilloway, D.A. (ed.) *Grassland: A Global Resource*. Wageningen Academic Publishers, Wageningen, the Netherlands, pp. 265–277.

Fowler, D., Skiba, U. and Hargreaves, K.J. (1997) Emissions of nitrous oxide from grasslands. In: Jarvis, S.C. and Pain, B.F. (eds) *Gaseous Nitrogen Emissions from Grasslands*. CAB International, Wallingford, UK, pp. 147–164.

Frame, J. (2000) *Improved Grassland Management*, Farming Press, Tonbridge, UK.

Frank, A.B., Tanaka, D.L., Hofmann, L. and Follett, R.F. (1995) Soil carbon and nitrogen of Northern great plains grasslands as influenced by long-term grazing. *Journal of Range Management* 48, 470–474.

Franzluebbers, A.J., Stuedemann, J.A. and Schomberg, H.H. (2000a) Spatial distribution of soil carbon and nitrogen pools under grazed tall fescue. *Soil Science Society of America Journal* 64, 635–639.

Franzluebbers, A.J., Stuedemann, J.A., Schomberg, H.H. and Wilkinson, S.R. (2000b) Soil organic C and N pools under long-term pasture management in the Southern Piedmont USA. *Soil Biology and Biochemistry* 32, 469–478.

Frossard, E., Condron, L.M., Oberson, A., Sinaj, S. and Fardeau, J.C. (2000) Processes governing phosphorus availability in temperate soils. *Journal of Environmental Quality* 29, 15–23.

Garwood, E.A. and Tyson, K.C. (1977) High loss of nitrogen in drainage from soil under grass following a prolonged period of low rainfall. *Journal of Agricultural Science, Cambridge* 89, 767–768.

Ghani, A., Dexter, M., Carran, R.A. and Theobald, P.W. (2007) Dissolved organic nitrogen and carbon in pastoral soils: the New Zealand experience. *European Journal of Soil Science* 58, 832–843.

Giller, K.E., Witter, E. and McGrath, S.P. (1998) Toxicity of heavy metals to microorganisms and microbial processes in agricultural soils: a review. *Soil Biology and Biochemistry* 30, 1389–1414.

Glatzel, S. and Stahr, K. (2001) Methane and nitrous oxide exchange in differently fertilized grassland in southern Germany. *Plant and Soil* 231, 21–35.

Gray, C.W., McLaren, R.G., Roberts, A.H.C. and Condron, L.M. (1999) The effect of long-term phosphatic fertilizer applications on the amounts and forms of cadmium in soils under pasture in New Zealand. *Nutrient Cycling in Agroecosystems* 54, 267–277.

Gray, C.W., McLaren, R.G. and Roberts, A.H.C. (2003) Cadmium leaching from some New Zealand pasture soils. *European Journal of Soil Science* 54, 159–166.

Greenwood, K.L. and McKenzie, B.M. (2001) Grazing effects on soil physical properties and consequences for pastures: a review. *Australian Journal of Experimental Agriculture* 41, 1231–1250.

Gregorich, E.G., Carter, M.R., Angers, D.A., Monreal, C.M. and Ellert, B.H. (1994) Towards a minimum data set to assess soil organic matter quality in agricultural soils. *Canadian Journal of Soil Science* 74, 367–385.

Groenigen, J.W.v., Velthof, G.L., Bolt, F.J.E.v.d., Vos, A. and Kuikman, P.J. (2005) Seasonal variation in N_2O emissions from urine patches: effects of urine concentration, soil compaction and dung. *Plant and Soil* 273, 15–27.

Harrison, R. and Webb, J. (2001) A review of the effect of N fertilizer type on gaseous emissions. *Advances in Agronomy* 73, 67–108.

Harter, R.D. (1969) Phosphorus adsorption sites in soils. *Soil Science Society of America Proceedings* 33, 630–631.

Hauser, M. and Haselwandter, K. (1990) Degradation of dicyandiamide by soil bacteria. *Soil Biology and Biochemistry* 22, 113–114.

Hawkins, J. and Scholefield, D. (2000) Leaching of dissolved organic N from grass-white clover pasture in SW England. In: Søegaard, K., Ohlsson, C., Sehested, J., Hutchings, N.J. and Kristensen, T. (eds) *Grassland Farming – Balancing Environmental and Economic Demands. Grassland Science in Europe. Volume* 5. European Grassland Federation (Danish Institute of Agricultural Sciences) Tjele, Denmark, pp. 378–380.

Haygarth, P.M. and Jarvis, S.C. (1997) Soil derived phosphorus in surface runoff from grazed grassland lysimeters. *Water Research (Oxford)* 31, 140–148.

Haygarth, P.M. and Jarvis, S.C. (1999) Transfer of phosphorus from agricultural soils. *Advances in Agronomy* 66, 195–249.

Haygarth, P.M., Hepworth, L. and Jarvis, S.C. (1998) Forms of phosphorus transfer in hydrological pathways from soil under grazed grassland. *European Journal of Soil Science* 49, 65–72.

Haynes, R.J. and Williams, P.H. (1992) Changes in soil solution composition and pH in urine-affected areas of pasture. *Journal of Soil Science* 43, 323–334.

Haynes, R.J. and Williams, P.H. (1993) Nutrient cycling and soil fertility in the grazed pasture ecosystem. *Advances in Agronomy* 49, 119–199.

Haynes, R.J. and Williams, P.H. (1999) Influence of stock camping behaviour on the soil microbiological and biochemical properties of grazed pastoral soils. *Biology and Fertility of Soils* V28, 253–258.

Heathwaite, A.L., Griffiths, P. and Parkinson, R.J. (1998) Nitrogen and phosphorus in runoff from grassland with buffer strips following application of fertilizers and manures. *Soil Use and Management* 14, 142–148.

Heckrath, G., Brookes, P.C., Poulton, P.R. and Goulding, K.W.T. (1995) Phosphorus leaching from soils containing different phosphorus concentrations in the Broadbalk Experiment. *Journal of Environmental Quality* 24, 904–910.

Hillman, J.P., Hill, J., Morgan, J.E. and Wilkinson, J.M. (2003) Recycling of sewage sludge to grassland: a review of the legislation to control of the localization and accumulation of potential toxic metals in grazing systems. *Grass and Forage Science* 58, 101–111.

Holford, I.C.R., Hird, C. and Lawrie, R. (1997) Effects of animal effluents on the phosphorus sorption characteristics of soils. *Australian Journal of Soil Research* 35, 365–373.

Holter, P. (1997) Methane emissions from Danish cattle dung pats in the field. *Soil Biology and Biochemistry* 29, 31–37.

Hooda, P.S., Rendell, A.R., Edwards, A.C., Withers, P.J.A., Aitken, M.N. and Truesdale, V.W. (2000) Relating soil phosphorus indices to potential phosphorus release to water. *Journal of Environmental Quality* 29, 1166–1171.

Hornung, M., Bull, K.R., Cresser, M., Ullyett, J., Hall, J.R., Langan, S., Loveland, P.J. and Wilson, M.J. (1995) The sensitivity of surface waters of Great Britain to acidification predicted from catchment characteristics. *Environmental Pollution* 87, 207–214.

Hutsch, B.W., Webster, C.P. and Powlson, D.S. (1994) Methane oxidation in soil as affected by land use, soil pH and N fertilization. *Soil Biology and Biochemistry* 26, 1613–1622.

Iyamuremye, F., Dick, R.P. and Baham, J. (1996) Organic amendments and phosphorus dynamics: I. Phosphorus chemistry and sorption. *Soil Science* 161, 426–435.

Iyyemperumal, K., Israel, D.W. and Shi, W. (2007) Soil microbial biomass, activity and potential nitrogen mineralization in a pasture: impact of stock camping activity. *Soil Biology and Biochemistry* 39, 149–157.

Jarvis, S.C. (1997) Emission processes and their interactions in grassland soils. In: Jarvis, S.C. and Pain, B.F. (eds) *Gaseous Nitrogen Emissions from Grasslands.* CAB International, Wallingford, UK, pp. 1–17.

Jarvis, S.C. and Hatch, D.J. (1994) Potential for denitrification at depth below long-term grass swards. *Soil Biology and Biochemistry* 26, 1629–1636.

Johnson, K.A. and Johnson, D.E. (1995) Methane emissions from cattle. *Journal of Animal Science* 73, 2483–2492.

Johnston, A.E., McEwan, J., Lane, P.W., Hewitt, M.V., Poulton, P.R. and Yeoman, D.P. (1994) Effects of one to six year old grass-clover leys on soil nitrogen and on subsequent yields and fertilizer nitrogen requirements of the arable sequence winter wheat, potatoes, winter wheat, winter beans (*Vicia faba*) grown on a sandy loam soil. *Journal of Agricultural Science* 122, 73–89.

Jones, D.L., Farrar, J.F. and Willett, V.B. (2004) Dissolved organic nitrogen concentration in two grassland soils. In: Hatch, D.J., Chadwick, D.R., Jarvis, S.C. and Roker, J.A. (eds) *Proceedings of the 12th Nitrogen Workshop: Controlling Nitrogen Flows and Losses* Wageningen Academic Publishers, Wageningen, the Netherlands, pp. 441–443.

Jones, S.K., Rees, R.M., Kosmas, D., Ball, B.C. and Skiba, U.M. (2006) Carbon sequestration in a temperate grassland; management and climatic controls. *Soil Use and Management* 22, 132–142.

Jordan, C., McGuckin, S.O. and Smith, R.V. (2000) Increased predicted losses of phosphorus to surface waters from soils with high Olsen-P concentrations. *Soil Use and Management* 16, 27–35.

Kammann, C., Grunhage, L., Jager, H.J. and Wachinger, G. (2001) Methane fluxes from differentially managed grassland study plots: the important role of CH_4 oxidation in grassland with a high potential for CH_4 production. *Environmental Pollution* 115, 261–273.

Karlen, D.L., Andrews, S.S. and Doran, J.W. (2001) Soil quality: current concepts and applications. *Advances in Agronomy* 74, 2–40.

Kay, B.D. (1990) Rates of change of soil structure under different cropping systems. *Advances in Soil Science* 12, 1–52.

Kemp, D.R. and Michalk, D.L. (2005) Grasslands for production and the environment. In: McGilloway, D.A. (ed.) *Grassland: A Global Resource*. Wageningen Academic Publishers, Wageningen, the Netherlands, pp. 193–208.

Khan, M. and Scullion, J. (2002) Effects of metal (Cd, Cu, Ni, Pb or Zn) enrichment of sewage-sludge on soil microorganisms and their activities. *Applied Soil Ecology* 20, 145–155.

Kibblewhite, M.G. (2005) Soil quality assessment and management. In: McGilloway, D.A. (ed.) *Grassland: A Global Resource. Plenary and Invited Papers from the 20th International Grassland Congress, Dublin, Ireland, June/July, 2005*. Wageningen Academic Publishers, Wageningen, the Netherlands, pp. 219–226.

Koops, J.G., Beusichem, M.L.v. and Oenema, O. (1997) Nitrous oxide production, its source and distribution in urine patches on grassland on peat soil. *Plant and Soil* 191, 57–65.

Kretzschmar, R., Borkovec, M., Grolimund, D. and Elimelech, M. (1999) Mobile subsurface colloids and their role in contaminant transport. *Advances in Agronomy* 66, 121–193.

Krupa, S.V. (2003) Effects of atmospheric ammonia (NH_3) on terrestrial vegetation: a review. *Environmental Pollution* 124, 179–221.

Kumar, K., Gupta, S.C., Chandler, Y. and Singh, A.K. (2005) Antibiotic use in agriculture and its impact on the terrestrial environment. *Advances in Agronomy* 87, 1–54.

Lakzian, A., Murphy, P., Turner, A., Beynon, J.L. and Giller, K.E. (2002) Rhizobium leguminosarum bv. viciae populations in soils with increasing heavy metal contamination: abundance, plasmid profiles, diversity and metal tolerance. *Soil Biology and Biochemistry* 34, 519–529.

Lal, R. (1992) Restoring tropical land degraded by gully erosion. *Advances in Soil Science* 17, 123–152.

Langeveld, C.A., Segers, R., Dirks, B.O.M., Pol-van Dasselaar, A.v.d., Velthof, G.L. and Hensen, A. (1997) Emissions of CO_2, CH_4 and N_2O from pasture on drained peat soils in the Netherlands. *European Journal of Agronomy* 7, 35–42.

Larson, W.E. and Pierce, F.J. (1994) The dynamics of soil quality as a measure of sustainable management. In: Doran, J.W., Coleman, D.C., Bezedick, D.F. and Stewart, B.A. (eds) *Defining Soil Quality for a Sustainable Environment. Special Pub. No. 35*. American Society of Agronomy, Madison, Wisconsin, pp. 37–51.

Ledgard, S.F. (2001) Nitrogen cycling in low input legume-based agriculture, with emphasis on legume/grass pastures. *Plant and Soil* 228, 43–59.

Ledgard, S.F., Penno, J.W. and Sprosen, M.S. (1999) Nitrogen inputs and losses from clover/grass pastures grazed by dairy cows, as affected by nitrogen fertilizer application. *Journal of Agricultural Science* 132, 215–225.

Lee, J., Grace, N.D. and Rounce, J.R. (1994) Cadmium accumulation in liver and kidney of sheep grazing ryegrass/white clover pastures. *Proceedings of the New Zealand Society of Animal Production* 54, 31–34.

Leinweber, P., Lunsmann, F. and Eckhardt, K.U. (1997) Phosphorus sorption capacities and saturation of soils in two regions with different livestock densities in north-west Germany. *Soil Use and Management* 13, 82–89.

Leytem, A.B. and Maguire, R.O. (2007) Environmental implications of inositol phosphates in animal manures. In: Turner, B.L., Richardson, A.E. and Mullaney, E.J. (eds) *Inositol Phosphates: Linking Agriculture and the Environment*. CAB International, Wallingford, UK, pp. 150–168.

Ligthart, T.N. (1997) Thin section analysis of earthworm burrow disintegration in a permanent pasture. *Geoderma* 75, 135–148.

Lipschultz, F., Zafiriou, O.C., Wofsy, S.C., McElroy, M.B., Valois, F.W. and Watson, S.W. (1981) Production of NO and N_2O by soil nitrifying bacteria. *Nature* 294, 641–643.

Logan, T.J. (1992) Reclamation of chemically degraded soils. *Advances in Soil Science* 17, 13–35.

Loganathan, P. and Hedley, M.J. (1997) Downward movement of cadmium and phosphorus from phosphatic fertilizers in a pasture soil in New Zealand. *Environmental Pollution* 95, 319–324.

Loganathan, P., Hedley, M.J., Gregg, P.E.H. and Currie, L.D. (1997) Effect of phosphate fertilizer type on the accumulation and plant availability of cadmium in grassland soils. *Nutrient Cycling in Agroecosystems* 46, 169–178.

Loganathan, P., Hedley, M.J., Wallace, G.C. and Roberts, A.H.C. (2001) Fluoride accumulation in pasture forages and soils following long-term applications of phosphorus fertilizers. *Environmental Pollution* 115, 275–282.

Loiseau, P. and Soussana, J.F. (1999) Elevated CO_2, temperature increase and N supply effects on the turnover of below-ground carbon in a temperate grassland ecosystem. *Plant and Soil* 210, 233–247.

Lovell, R.D. and Jarvis, S.C. (1996) Effects of urine on soil microbial biomass, methanogenesis, nitrification and denitrification in grassland soils. *Plant and Soil* 186, 265–273.

Luo, J., Tillman, R.W., White, R.E. and Ball, P.R. (1998) Variation in denitrification activity with soil depth under pasture. *Soil Biology and Biochemistry* 30, 897–903.

Lützow, M.v., Kögel-Knabner, I., Ekschmitt, K., Matzner, E., Guggenberger, G., Marschner, B. and Flessa, H. (2006) Stabilization of organic matter in temperate soils: mechanisms and their relevance under different soil conditions – a review. *European Journal of Soil Science* 57, 426–445.

Lynch, J.M. and Panting, L.M. (1980) Variations in the size of the soil biomass. *Soil Biology and Biochemistry* 12, 547–550.

Ma, W.C., Brussaard, L. and Ridder, J.A.d. (1990) Long-term effects of nitrogenous fertilizers on grassland earthworms (Oligochaeta; Lumbricidae): their relation to soil acidification. *Agriculture, Ecosystems and Environment* 30, 71–80.

Macdonald, A.J., Francis, S.M.J., Stockdale, E.A., Goulding, K.W.T., Willett, V.B. and Jones, D.L. (2004) Effects of land use and management on soluble organic N in UK soils. In: Hatch, D.J., Chadwick, D.R., Jarvis, S.C. and Roker, J.A. (eds) *Proceedings of the 12th Nitrogen Workshop: Controlling Nitrogen Flows and Losses*. Wageningen Academic Publishers, Wageningen, the Netherlands, pp. 600–605.

Macduff, J.H., Jarvis, S.C. and Roberts, D.H. (1990) Nitrates: leaching from grazed grassland systems. In: Calvet, R. (ed.) *Nitrates-Agriculture-Water. International Symposium, Paris-La Defense 7–8 November 1990*. INRA, Paris, pp. 405–410.

Madsen, M., Nielsen, B.O., Holter, P., Pedersen, O.C., Jespersen, J.B., Jensen, K.M.V., Nansen, P. and Gronvold, J. (1990) Treating cattle with ivermectin – effects on the fauna and decomposition of dung pats. *Journal of Applied Ecology* 27, 1–15.

Martens, D.A., Emmerich, W., McLain, J.E.T. and Johnsen, T.N. (2005) Atmospheric carbon mitigation potential of agricultural management in the south-western USA. *Soil and Tillage Research* 83, 95–119.

McBride, M.B., Richards, B.K., Steenhuis, T., Russo, J.J. and Sauvé, S. (1997) Mobility and solubility of toxic metals and nutrients in soil fifteen years after sludge application. *Soil Science* 162, 487–500.

McDowell, R.W. and Condron, L.M. (2004) Estimating phosphorus loss from New Zealand grassland soils. *New Zealand Journal of Agricultural Research* 47, 137–145.

McDowell, R.W., Sharpley, A.N., Condron, L.M., Haygarth, P.M. and Brookes, P.C. (2001) Processes controlling soil phosphorus release to runoff and implications for agricultural management. *Nutrient Cycling in Agroecosystems* 59, 269–284.

McDowell, R.W., Drewry, J.J., Paton, R.J., Carey, P.L., Monaghan, R.M. and Condron, L.M. (2003a) Influence of soil treading on sediment and phosphorus losses in overland flow. *Australian Journal of Soil Research* 41, 949–961.

McDowell, R.W., Monaghan, R.M. and Carey, P.L. (2003b) Potential phosphorus losses in overland flow from pastoral soils receiving long-term applications of either superphosphate or reactive phosphate rock. *New Zealand Journal of Agricultural Research* 46, 329–337.

McDowell, R.W., Drewry, J.J. and Paton, R.J. (2004) Effects of deer grazing and fence-line pacing on water and soil quality. *Soil Use and Management* 20, 302–307.

McGrath, S.P., Brookes, P.C. and Giller, K.E. (1988) Effects of potentially toxic metals in soil derived from past applications of sewage sludge on nitrogen fixation by *Trifolium repens* L. *Soil Biology and Biochemistry* 20, 415–424.

McTaggart, I.P., Douglas, J.T., Clayton, H. and Smith, K.A. (1997) Nitrous oxide emission from slurry and mineral nitrogen fertilizer applied to grassland. In: Jarvis, S.C. and Pain, B.F. (eds) *Gaseous Nitrogen Emissions from Grasslands*. CAB International, Wallingford, UK, pp. 201–209.

Mer, J.I. and Roger, P. (2001) Production, oxidation, emission and consumption of methane by soils: a review. *European Journal of Soil Biology* 37, 25–50.

Merino, P., Estavillo, J.M., Graciolli, L.A., Pinto, M., Lacuesta, M., Munoz-Rueda, A. and Gonzalez-Murua, C. (2002) Mitigation of N_2O emissions from grassland by nitrification inhibitor and Actilith F2 applied with fertilizer and cattle slurry. *Soil Use and Management* 18, 135–141.

Monaghan, R.M. and Barraclough, D. (1993) Nitrous oxide and dinitrogen emissions from urine-affected soil under controlled conditions. *Plant and Soil* 151, 127–138.

Morecroft, M.D., Burt, T.P., Taylor, M.E. and Rowland, A.P. (2000) Effects of the 1995–1997 drought on nitrate leaching in lowland England. *Soil Use and Management* 16, 117–123.

Mulder, C., Cohen, J.E., Setälä, H., Bloem, J. and Breure, A.M. (2005) Bacterial traits, organism mass, and numerical abundance in the detrital soil food web of Dutch agricultural grasslands. *Ecology Letters* 8, 80–90.

Mulholland, B. and Fullen, M.A. (1991) Cattle trampling and soil compaction on loamy sands. *Soil Use and Management* 7, 189–193.

Muller, C., Stevens, R.J. and Laughlin, R.J. (2006) Sources of nitrite in a permanent grassland soil. *European Journal of Soil Science* 57, 337–343.

Murray, P.J., Hatch, D.J., Dixon, E.R., Stevens, R.J., Laughlin, R.J. and Jarvis, S.C. (2004) Denitrification potential in a grassland subsoil: effect of carbon substrates. *Soil Biology and Biochemistry* 36, 545–547.

Nash, D.M. and Haygarth, P.M. (2005) Grassland productivity and water quality: a 21st Century issue. In: McGilloway, D.A. (ed.) *Grassland: A Global Resource*. Wageningen Academic Publishers, Wageningen, the Netherlands, pp. 239–249.

Nguyen, M.L., Sheath, G.W., Smith, C.M. and Cooper, A.B. (1998) Impact of cattle treading on hill land. 2. Soil physical properties and contaminated runoff. *New Zealand Journal of Agricultural Research* 41, 279–290.

Nicholson, F.A. and Jones, K.C. (1994) Effect of phosphate fertilizers and atmospheric deposition on long-term changes in the cadmium content of soils and crops. *Environmental Science and Technology* 28, 2170–2175.

Oenema, O. (1990) Calculated rates of soil acidification of intensively used grassland in the Netherlands. *Fertilizer Research* 26, 217–228.

Oenema, O., Velthof, G.L., Yamulki, S. and Jarvis, S.C. (1997) Nitrous oxide emissions from grazed grassland. *Soil Use and Management* 13, 288–295.

Oliver, D.M., Clegg, C.D., Haygarth, P.M. and Heathwaite, A.L. (2005) Assessing the potential for pathogen transfer from grassland soils to surface waters. *Advances in Agronomy* 85, 125–180.

Olsen, S.R. and Sommers, L.E. (1982) Phosphorus. In: Page, A.L., Miller, R.H. and Keeney, D.R. (eds) *Methods of Soil Analysis. Part 2. Chemical and Microbiological Properties.* Soil Science Society of America, Madison, Wisconsin, pp. 403–430.

O'Riordan, E.G., Dodd, V.A., Fleming, G.A. and Tunney, H. (1994) Repeated application of a metal-rich sewage sludge to grassland. 2. Effects on herbage metal levels. *Irish Journal of Agricultural and Food Research* 33, 53–60.

Paul, E.A. and Clark, F.E. (1996) *Soil Microbiology and Biochemistry,* 2nd edn. Academic Press, San Diego, California.

Pautler, M.C. and Sims, J.T. (2000) Relationships between soil test phosphorus, soluble phosphorus, and phosphorus saturation in Delaware soils. *Soil Science Society of America Journal* 64, 765–773.

Paz Jimenez, M.d.l., Horra, A.M.d.l., Pruzzo, L. and Palma, R.M. (2002) Soil quality: a new index based on microbiological and biochemical parameters. *Biology and Fertility of Soils* 35, 302–306.

Pepper, I.L., Brooks, J.P. and Gerba, C.P. (2006) Pathogens in biosolids. *Advances in Agronomy* 90, 1–41.

Percival, H.J. (2003) Soil and soil solution chemistry of a New Zealand pasture soil amended with heavy metal-containing sewage sludge. *Australian Journal of Soil Research* 41, 1–17.

Pol-van Dasselaar, A.v.d., Beusichem, M.L.v. and Oenema, O. (1999) Effects of nitrogen input and grazing on methane fluxes of extensively and intensively managed grasslands in the Netherlands. *Biology and Fertility of Soils* 29, 24–30.

Rees, R.M., Bingham, I.J., Baddeley, J.A. and Watson, C.A. (2005) The role of plants and land management in sequestering soil carbon in temperate arable and grassland ecosystems. *Geoderma* 128, 130–154.

Reid, R.S., Serneels, S., Nyabenge, M. and Hanson, J. (2005) The changing face of pastoral systems in grass-dominated ecosystems of eastern Africa. In: Suttie, J.M., Reynolds, S.G. and Batello, C. (eds) *Grasslands of the World. Plant Production and Protection Series No. 34.* FAO, Rome, Italy, pp. 19–76.

Rengasamy, P., Chittleborough, D. and Helyar, K. (2003) Root-zone constraints and plant-based solutions for dryland salinity. *Plant and Soil* 257, 249–260.

Richards, B.K., Steenhuis, T.S., Peverly, J.H. and McBride, M.B. (1998) Metal mobility at an old, heavily loaded sludge application site. *Environmental Pollution* 99, 365–377.

Ridley, A.M., Slattery, W.J., Helyar, K.R. and Cowling, A. (1990) Acidification under grazed annual and perennial grass based pastures. *Australian Journal of Experimental Agriculture* 30, 539–544.

Rodd, A.V., Papadopoulos, Y.A., Laflamme, L.F., McRae, K.B., Fillmore, S.A.E. and Wilson, R.W. (1999) Effect of rotational grazing on selected physical properties of a gleyed brunisolic gray Luvisol loam in Nova Scotia. *Canadian Journal of Soil Science* 79, 117–125.

Rodhe, L., Pell, M. and Yamulki, S. (2006) Nitrous oxide, methane and ammonia emissions following slurry spreading on grassland. *Soil Use and Management* 22, 229–237.

Ross, D.J., Tate, K.R., Newton, P.C.D. and Clark, H. (2002) Decomposability of C3 and C4 grass litter sampled under different concentrations of atmospheric carbon dioxide at a natural CO2 spring. *Plant and Soil* 240, 275–286.

Rothbaum, H.P., Goguel, R.L., Johnston, A.E. and Mattingly, G.E.G. (1986) Cadmium accumulation in soils from long-continued applications of superphosphate. *Journal of Soil Science* 37, 99–107.

Saini, G.r. and MacLean, A.A. (1965) Phosphorus retention capacities of some New Brunswick soils and their relationship with soil properties. *Canadian Journal of Soil Science* 45, 15–18.

Samadi, A. and Gilkes, R.J. (1999) Phosphorus transformations and their relationships with calcareous soil properties in southern Western Australia. *Soil Science Society of America Journal* 63, 809–815.

Scholefield, D. and Hall, D.M. (1985) A method to measure the susceptibility of pasture soils to poaching by cattle. *Soil Use and Management* 1, 134–138.

Scholefield, D., Tyson, K.C., Garwood, E.A., Armstrong, A.C., Hawkins, J. and Stone, A.C. (1993) Nitrate leaching from grazed grassland lysimeters: effects of fertilizer input, field drainage, age of sward and patterns of weather. *Journal of Soil Science* 44, 601–613.

Scholefield, D., Hawkins, J.M.B. and Jackson, S.M. (1997) Use of a flowing helium atmosphere incubation technique to measure the effects of denitrification controls applied to intact cores of a clay soil. *Soil Biology and Biochemistry* 29, 1337–1344.

Schuman, G.E., Janzen, H.H. and Herrick, J.E. (2002) Soil carbon dynamics and potential carbon sequestration by rangelands. *Environmental Pollution* 116, 391–396.

Scott, B.J., Ridley, A.M. and Conyers, M.K. (2000) Management of soil acidity in long-term pastures of south-eastern Australia: a review. *Australian Journal of Experimental Agriculture* 40, 1173–1198.

Sharpley, A.N. (1995) Dependence of runoff phosphorus on extractable soil phosphorus. *Journal of Environmental Quality* 24, 920–926.

Sharpley, A.N. and Smith, S.J. (1989) Mineralization and leaching of phosphorus from soil incubated with surface-applied and incorporated crop residue. *Journal of Environmental Quality* 18, 101–105.

Sharpley, A.N. and Smith, S.J. (1990) Phosphorus transport in agricultural runoff: the role of soil erosion. In: Boardman, J., Foster, L.D.L. and Dearing, J.A. (eds) *Soil Erosion on Agricultural Land.* Wiley, Chichester, UK, pp. 351–366.

Shepherd, M.A., Hatch, D.J., Jarvis, S.C. and Bhogal, A. (2001) Nitrate leaching from reseeded pasture. *Soil Use and Management* 17, 97–105.

Sherlock, R.R., Sommer, S.G., Khan, R.Z., Wood, C.W., Guertal, E.A., Freney, J.R., Dawson, C.O. and Cameron, K.C. (2002) Ammonia, methane, and nitrous oxide emission from pig slurry applied to a pasture in New Zealand. *Journal of Environmental Quality* 31, 1491–1501.

Sinaj, S., Stamm, C., Toor, G.S., Condron, L.M., Hendry, T., Di, H.J., Cameron, K.C. and Frossard, E. (2002) Phosphorus exchangeability and leaching losses from two grassland soils. *Journal of Environmental Quality* 31, 319–330.

Singh, B. and Gilkes, R.J. (1991) Phosphorus sorption in relation to soil properties for the major soil types of south-western Australia. *Australian Journal of Soil Research* 29, 603–618.

Skiba, U., Hargreaves, K.J., Fowler, D. and Smith, K.A. (1992) Fluxes of nitric and nitrous oxides from agricultural soils in a cool temperate climate. *Atmospheric Environment. Part A, General Topics* 26, 2477–2488.

Slattery, J.F., Coventry, D.R. and Slattery, W.J. (2001) Rhizobial ecology as affected by the soil environment. *Australian Journal of Experimental Agriculture* 41, 289–298.

Smith, K.A., McTaggart, I.P. and Tsuruta, H. (1997) Emissions of N_2O and NO associated with nitrogen fertilization in intensive agriculture, and the potential for mitigation. *Soil Use and Management* 13, 296–304.

Smith, K.A., Chalmers, A.G., Chambers, B.J. and Christie, P. (1998) Organic manure accumulation, mobility and management. *Soil Use and Management* 14, 154–159.

Smith, R.V., Lennox, S.D., Jordan, C., Foy, R.H. and McHale, E. (1995) Increase in soluble phosphorus transported in drainflow from a grassland catchment in response to soil phosphorus accumulation. *Soil Use and Management* 11, 204–209.

Smith, R.V., Burns, L.C., Doyle, R.M., Lennox, S.D., Kelso, B.H.L., Foy, R.H. and Stevens, R.J. (1997) Free ammonia inhibition of nitrification in river sediments leading to nitrite accumulation. *Journal of Environmental Quality* 26, 1049–1055.

Soane, B.D. (1990) The role of organic matter in soil compactability: a review of some practical aspects. *Soil and Tillage Research* 16, 179–201.

Soane, B.D., Blackwell, P.S. and Dickson, J.W. (1981) Compaction by agricultural vehicles: a review. 1. Soil and wheel characteristics. *Soil and Tillage Research* 1, 207–237.

Sojka, R.E., Upchurch, D.R. and Borlaug, N.E. (2003) Quality soil management or soil quality management: performance versus semantics. *Advances in Agronomy* 79, 1–68.

Sommer, S.G. and Ersboll, A.K. (1996) Effect of air flow rate, lime amendments, and chemical soil properties on the volatilization of ammonia from fertilizers applied to sandy soils. *Biology and Fertility of Soils* 21, 53–60.

Sommer, S.G. and Hutchings, N.J. (1997) Components of ammonia volatilization from cattle and sheep production. In: Jarvis, S.C. and Pain, B.F. (eds) *Gaseous Nitrogen Emissions from Grasslands.* CAB International, Wallingford, UK, pp. 79–93.

Soussana, J.F., Loiseau, P., Vuichard, N., Ceschia, E., Balesdent, J., Chevallier, T. and Arrouays, D. (2004) Carbon cycling and sequestration opportunities in temperate grasslands. *Soil Use and Management* 20, 219–230.

Spoor, G. (2006) Alleviation of soil compaction: requirements, equipment and techniques. *Soil Use and Management* 22, 113–122.

Stehfest, E. and Bouwman, L. (2006) N2O and NO emission from agricultural fields and soils under natural vegetation: summarizing available measurement data and modeling of global annual emissions. *Nutrient Cycling in Agroecosystems* 74, 207–228.

Stenberg, B. (1999) Monitoring soil quality of agricultural land: microbiological indicators. *Acta Agriculturae Scandinavica, Section B, Soil and Plant Science* 49, 1–24.

Stevens, R.J. and Laughlin, R.J. (1997) The impact of cattle slurries and their management on ammonia and nitrous oxide emissions from grassland. In: Jarvis, S.C. and Pain, B.F. (eds) *Gaseous Nitrogen Emissions from Grasslands.* CAB International, Wallingford, UK, pp. 233–256.

Stevens, R.J. and Laughlin, R.J. (2001) Cattle slurry affects nitrous oxide and dinitrogen emissions from fertilizer nitrate. *Soil Science Society of America Journal* 65, 1307–1314.

Stout, W.L., Fales, S.A., Muller, L.D., Schnabel, R.R., Priddy, W.E. and Elwinger, G.F. (1997) Nitrate leaching from cattle urine and feces in North-east USA. *Soil Science Society of America Journal* 61, 1787–1794.

Strong, L. (1993) Overview – the impact of avermectins on pastureland ecology. *Veterinary Parasitology* 48, 3–17.

Suarez, V.H., Lifschitz, A.L., Sallovitz, J.M. and Lanusse, C.E. (2003) Effects of ivermectin and doramectin faecal residues on the invertebrate colonization of cattle dung. *Journal of Applied Entomology* 127, 481–488.

Sugimoto, Y. and Ball, P.R. (1989) Nitrogen losses from cattle dung. In: Desroches, R. (ed.) *Proceedings of the 16th International Grassland Congress, Nice, France.* Association Française pour la Production Fourragère, Centre National de Recherche Agronomique, Versailles, France, pp. 153–154.

Svendsen, T., Gronvold, J., Holter, P. and Sommer, C. (2003) Field effects of ivermectin and fenbendazole on earthworm populations and the disappearance of dung pats from bolus-treated cattle. *Applied Soil Ecology* 24, 207–218.

Svendsen, T., Hansen, P., Sommer, C., Martinussen, T., Gronvold, J. and Holter, P. (2005) Life history characteristics of *Lumbricus terrestris* and effects of the veterinary antiparasitic compounds ivermectin and fenbendazole. *Soil Biology and Biochemistry* 37, 927–936.

Tedeschi, L.O., Fox, D.G. and Tylutki, T.P. (2003) Potential environmental benefits of ionophores in ruminant diets. *Journal of Environmental Quality* 32, 1591–1602.

Thorpe, I.S., Prosser, J.I., Glover, L.A. and Killham, K. (1997) The role of the earthworm *Lumbricus terrestris* in the transport of bacterial inocula through soil. *Biology and Fertility of Soils* 23, 132–139.

Tisdall, J.M. (1994) Possible role of soil microorganisms in aggregation in soils. *Plant and Soil* 159, 115–121.

Tisdall, J.M. and Oades, J.M. (1982) Organic matter and water-stable aggregates in soils. *Journal of Soil Science* 33, 141–163.

Toor, G.S., Condron, L.M., Di, H.J., Cameron, K.C. and Sims, J.T. (2004) Impact of farm-dairy effluent application on the amounts and forms of phosphorus loss by leaching from irrigated grassland. *New Zealand Journal of Agricultural Research* 47, 479–490.

Toor, G.S., Condron, L.M., Cade-Menun, B.J., Di, H.J. and Cameron, K.C. (2005) Preferential phosphorus leaching from an irrigated grassland soil. *European Journal of Soil Science* 56, 155–167.

Tyson, K.C., Roberts, D.H., Clement, C.R. and Garwood, E.A. (1990) Comparison of crop yields and soil conditions during 30 years under annual tillage or grazed pasture. *Journal of Agricultural Science* 115, 29–40.

Tyson, K.C., Garwood, E.A., Armstrong, A.C. and Scholefield, D. (1992) Effects of field drainage on the growth of herbage and the live weight gain of grazing beef cattle. *Grass and Forage Science* 47, 290–301.

Tyson, K.C., Scholefield, D., Jarvis, S.C. and Stone, A.C. (1997) A comparison of animal output and nitrogen leaching losses recorded from drained fertilized grass and grass/clover pasture. *Journal of Agricultural Science* 129, 315–323.

Unger, P.W. and Kaspar, T.C. (1994) Soil compaction and root growth: a review. *Agronomy Journal* 86, 759–766.

van der Putten, A.H.J. and Ketelaars, J.J.M.H. (1997) Ammonia volatilization from dairy farms: experiments and model. In: Jarvis, S.C. and Pain, B.F. (eds) *Gaseous Nitrogen Emissions from Grasslands*. CAB International, Wallingford, UK, pp. 123–130.

Varel, V.H. (1997) Use of urease inhibitors to control nitrogen loss from livestock waste. *Bioresource Technology* 62, 11–17.

Velthof, G.L., Oenema, O., Postma, R. and Beusichem, M.L.v. (1997) Effects of type and amount of applied nitrogen fertilizer on nitrous oxide fluxes from intensively managed grassland. *Nutrient Cycling in Agroecosystems* 46, 257–267.

Velthof, G.L., Kuikman, P.J. and Oenema, O. (2003) Nitrous oxide emission from animal manures applied to soil under controlled conditions. *Biology and Fertility of Soils* 37, 221–230.

Vermoesen, A., Van Cleemput, O. and Hofman, G. (1997) Contribution of urine patches to the emission of nitrous oxide. In: Jarvis, S.C. and Pain, B.F. (eds) *Gaseous Nitrogen Emissions from Grasslands*. CAB International, Wallingford, UK, pp. 189–194.

Vinther, F.P. (1998) Biological nitrogen fixation in grass-clover affected by animal excreta. *Plant and Soil* 203, 207–215.

Vinther, F.P., Hansen, E.M. and Eriksen, J. (2006) Leaching of soil organic carbon and nitrogen in sandy soils after cultivating grass-clover swards. *Biology and Fertility of Soils* 43, 12–19.

Wall, R. and Strong, L. (1987) Environmental consequences of treating cattle with the antiparasitic drug ivermectin. *Nature* 327, 418–421.

Ward, P.R., Fillery, I.R.P., Maharaj, E.A. and Dunin, F.X. (2003) Water budgets and nutrients in a native *Banksia* woodland and an adjacent *Medicago sativa* pasture. *Plant and Soil* 257, 305–319.

Wardhaugh, K.G. and Mahon, R.J. (1998) Comparative effects of abamectin and two formulations of ivermectin on the survival of larvae of a dung-breeding fly. *Australian Veterinary Journal* 76, 270–272.

Watson, C.J., Poland, P. and Allen, M.B.D. (1998) The efficacy of repeated applications of the urease inhibitor N-(n-butyl) thiophosphoric triamide for improving the efficiency of urea fertilizer utilization on temperate grassland. *Grass and Forage Science* 53, 137–145.

Webb, L., Beaumont, D.J., Nager, R.G. and McCracken, D.I. (2007) Effects of avermectin residues in cattle dung on yellow dung fly *Scathophaga stercoraria* (Diptera: Scathophagidae) populations in grazed pastures. *Bulletin of Entomological Research* 97, 129–138.

Wedin, D.A. and Tilman, D. (1990) Species effects on nitrogen cycling: a test with perennial grasses. *Oecologia* 84, 433–441.

West, C.P., Mallarino, A.P., Wedin, W.F. and Marx, D.B. (1989) Spatial variability of soil chemical properties in grazed pastures. *Soil Science Society of America Journal* 53, 784–789.

White, S.L., Sheffield, R.E., Washburn, S.P., King, L.D. and Green, J.T, Jr. (2001) Spatial and time distribution of dairy cattle excreta in an intensive pasture system. *Journal of Environmental Quality* 30, 2180–2187.

Whitehead, D.C. (2000) *Nutrient Elements in Grassland: Soil-Plant-Animal Relationships*, CAB International, Wallingford, UK.

Whitehead, D.C. and Raistrick, N. (1990) Ammonia volatilization from five nitrogen compounds used as fertilizers following surface application to soils. *Journal of Soil Science* 41, 387–394.

Wilkinson, J.M., Hill, J. and Livesey, C.T. (2001) Accumulation of potentially toxic elements in the body tissues of sheep grazed on grassland given repeated applications of sewage sludge. *Animal Science* 72, 179–190.

Willison, T.W., Webster, C.P., Goulding, K.W.T. and Powlson, D.S. (1995) Methane oxidation in temperate soils: effects of land use and the chemical form of nitrogen fertilizer. *Chemosphere* 30, 539–546.

Wratten, S.D., Meadbriggs, M., Gettinby, G., Ericsson, G. and Baggott, D.G. (1993) An evaluation of the potentia effects of ivermectin on the decomposition of cattle dung pats. *Veterinary Record* 133, 365–371.

Yamulki, S. and Jarvis, S.C. (2002) Short-term effects of tillage and compaction on nitrous oxide, nitric oxide, nitrogen dioxide, methane and carbon dioxide fluxes from grassland. *Biology and Fertility of Soils* 36, 224–231.

Yamulki, S., Jarvis, S.C. and Owen, P. (1998) Nitrous oxide emissions from excreta applied in a simulated grazing pattern. *Soil Biology and Biochemistry* 30, 491–500.

Yunusa, I.A.M. and Newton, P.J. (2003) Plants for amelioration of subsoil constraints and hydrological control: the primer-plant concept. *Plant and Soil* 257, 261–281.

3 Land–Water Interactions: Impacts on the Aquatic Environment

R.J. WILCOCK

National Institute of Water and Atmospheric Research Ltd, PO Box 11–115, Hamilton, New Zealand

Introduction

Land use affects water resources across a range of spatial scales, from microhabitats, where localized disturbances have their greatest effect, to the watershed (catchment) scale, where channel form and hydrological regime are important (Allan *et al.*, 1997). For example, inputs of coarse particulate organic matter depend on local, streamside vegetation, whereas hydrologic regime affects sediment delivery and channel conditions and is the product of regional climate, geology and vegetation (Allan and Johnson, 1997). Agricultural practices may extend over an area that is similar to, or even larger than, many river catchments so that impacts may occur throughout the stream length. Many small streams in agricultural catchments receive inputs of nitrogen (N), phosphorus (P), pathogens, agrichemicals and sediment from a multitude of surface and subsurface sources, from the headwaters to the catchment outlet at the confluence with a larger river or lake, or an estuary. Similarly, when pasture grasses or crops extend to the stream bank with little riparian shading provided by taller trees and shrubs, stream habitat is likely to be affected along much of the stream length with little respite for aquatic life from resulting high temperatures. Habitat loss and water quality degradation are both linked with land-use intensity in such streams, and riparian zone management provides a buffer against farming activities causing such impacts. Ways in which pastoral agriculture modifies stream habitats and water quality are summarized in Table 3.1. In order to understand how land uses in general, and agriculture in particular, impact waterways it is first necessary to consider the relationships between land use, hydrology and runoff and then examine the pathways and mechanisms connecting land with water, and the nature and impacts of the pollutants and habitat degradation.

Table 3.1. Changes to physical habitat and water quality characteristics of streams affected by pastoral agriculture. (From Parkyn and Wilcock, 2004.)

Response	Mechanisms	Impacts to stream and downstream environments
Physical habitat and channel morphology		
Riparian vegetation cover and diversity decreased	• Deforestation • Livestock grazing, browsing, trampling of remnant vegetation • Exposure to wind and sun-drying	• Decreased shade, increased water and air temperature • Loss of cool-water organisms • Increased growth of nuisance plants and algae with increased light • Cumulative increases in water temperature downstream • Decreased channel stability • Decreased food supply to the stream • Decreased habitat cover for fish
Soil condition degraded	• Compaction and decreased water infiltration	• Greater surface runoff • Increased delivery of contaminants
Channel stability decreased	• Trampling by livestock • Tree removal	• Bed siltation, local widening • Decreased in-stream habitat quality • Decreased visual appeal and amenity values
Channel width decreased	• Pasture grasses armour against fluvial erosion and trap sediments • Soil creep from hillslopes into channels • Channel width may locally increase at livestock crossings	• Decreased benthic habitat • Decreased quality of benthic habitat
Bed sediment texture decreased	• Siltation of stream bed by fines	• Decreased interstitial water exchange • Decreased epilithic food quality • Decreased benthic habitat quality
Contaminants and water quality		
Suspended sediment load and turbidity increased	• Trampling and grazing leading to bank erosion and sediment suspension • Hillslope instability • Decreased entrapment in riparian vegetation • Stock crossings	• Altered habitat and food selection for sighted animals • May affect gills of fish, crayfish and other invertebrates • Siltation of estuaries and lakes • Decreased visual appeal
Nutrients increased (N and P)	• Stock defaecation in stream channel	• Proliferation of nuisance plants and algae in streams

continued

Table 3.1. Continued

Response	Mechanisms	Impacts to stream and downstream environments
	• Surface runoff from dung on hillslopes • Leaching of urine (N) • Decreased entrapment in riparian soils and vegetation	• Eutrophication of downstream lakes and estuaries
Agrichemicals (herbicides, insecticides, fungicides, etc.)	• Poorly managed application of agrichemicals • Poorly managed sheep dip operations	• Potential for fish and invertebrate mortalities
Faecal microbes increased	• Defaecation in stream channel • Stock crossings • Runoff from farm tracks and raceways • Decreased entrapment in riparian vegetation • Farm dairy effluent and manure spreading	• Health risk to human water supply • Unsafe recreation and swimming • Health risk to domestic livestock • Contamination of shellfish in downstream estuaries

Hydrology and Runoff

Hydrological pathways

Runoff is the generic term for the movement of water from where it falls as precipitation (rainfall, snow melt or irrigation water) to where it reaches a stream or river channel (Fig. 3.1). Overland flow, or surface runoff, is water that flows over the ground and is a result either of: (i) infiltration excess; or (ii) saturation overland flow. Infiltration excess overland flow is that which occurs when the rainfall rate exceeds the infiltration rate of the surface soil, and is sometimes called Hortonian overland flow (Horton, 1933). Saturation overland flow is that which occurs through the flux of water reaching the saturated zone causing the groundwater table to rise until it reaches the surface, so that overland flow is a combination of rainfall falling on to an already saturated soil and water returning from a saturated soil matrix (Hewlett and Hibbert, 1967; Davie, 2004).

Contaminants entering waterways from agriculture favour particular hydrological pathways and as a consequence are modified and mitigated differently according to the interactions they have with soil, plants and microbes. Particulate substances (e.g. fine sediment and associated nutrients like P, and pathogenic microbes in animal dung) are mainly transported in overland flow or via coarse macropores to shallow groundwater, whereas dissolved contaminants

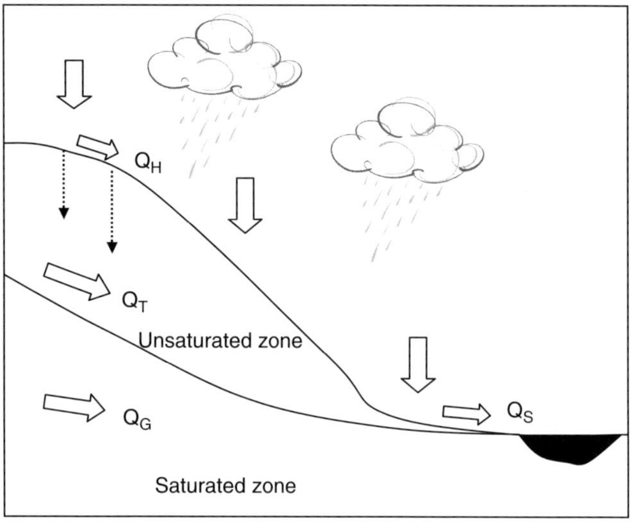

Fig. 3.1. Hillslope runoff processes. Q is flow and the subscripts refer to Hortonian overland flow (H), saturation overland flow (S), throughflow (T) and groundwater flow (G). (Modified from Davie, 2004.)

(e.g. nitrate and dissolved (<0.45 µm) reactive P) are transported via surface and subsurface pathways. Surface and subsurface drains collect shallow surface waters at depths within 1 m of the surface and transport it rapidly to nearby streams, often with very little attenuation (Monaghan *et al.*, 2007). Drain waters pose special problems for surface waters because they often discharge directly into streams without the benefits of attenuation by riparian processes. Riparian grasses are often able to filter particulate materials from overland flow, and wetlands can remove nitrate (a potential pollutant) via denitrification, a microbially mediated process whereby nitrate is decreased to nitrogen gas (N_2) (see Chapter 1, this volume).

Contaminant sources

It is convenient to divide land-based sources of aquatic pollution into two classes: point sources (PSs) and diffuse (or non-point) sources (NPSs). PSs are clearly identifiable, have specific locations and are typically pipes and drains discharging wastes from industry and municipal waste disposal networks. Non-point (or diffuse) pollution has been defined as: 'Pollution arising from land-use activities (urban and rural) that are dispersed across a catchment or subcatchments.' (Novotny, 2003). NPS pollution occurs when water flows over land or through the soil, mobilizes pollutants, and deposits them in surface water or groundwater. Pollutants often tend to enter waterways predominantly from either diffuse or PSs (Table 3.2) due to the nature of their sources and the main pathways for transport to waterways.

Table 3.2. Point sources and diffuse (non-point) sources of agricultural pollution and key contaminants.

Pollution source	Pollutant type	Contaminant
Point source		
Surface and subsurface drains	Farm wastes, irrigation water, dairy pond effluent, silage leachate	N, P, SS, faecal microbes BOD
Industrial discharge	Processing wastes (e.g. abattoir, dairy factory)	BOD, toxic organics, faecal microbes, heat (warm water)
Non-point (diffuse) source		
Surface runoff from agriculture	Particulate pollutants[a]	TP, TN, SS, faecal microbes
Subsurface runoff	Dissolved pollutants[b]	DIN, DRP
Riparian grazing by livestock (including livestock in channels)	Animal wastes, sediment, decreased stream bank stability	Faecal microbes, SS, N, P
Spray drift	Farm operations	Pesticides, fertilizer

[a]Surface drains often collect drainage from subsurface drains and hence collect dissolved and particulate pollutants.
[b]Subsurface drains can convey particulates if there are soil macropores (e.g. soil cracks).

The distinction between PSs and NPSs is important when addressing ways of treating or mitigating agricultural pollution of waterways. Wastewater discharges from pipes are mostly treated at source and often have to meet stringent discharge conditions for receiving waters that take into account the composition and flow rates of the wastewater and the receiving water. Agricultural wastes in this category include discharges from piggeries, effluent from dairy shed oxidation ponds and other treated farm wastes (Vanderholm, 1985). PS yields of nutrients to lakes and rivers in most developed countries are considerably lower than NPSs. For example, in New Zealand, the total P NPS load is roughly 50 t/day, and is considerably greater than the 6.1 t/day PS load entering inland waterways (Elliott and Sorrell, 2002). The total N NPS load is about 400 t/day, compared with the PS estimate of 29 t/day. A similar NPS/PS ratio was reported for N inputs to waterways in the USA (Gianessi and Peskin, 1984).

Base flow and storm flow

Base flow in a river derives from seepage of groundwater into the channel, or from the outflows of lakes and reservoirs and is characterized by a slow rate of change. Flood flows are produced by direct precipitation into the channel, from overland flow down surfaces sloping into the channel, from water that moves laterally through the upper layers of soil above the water table until it reaches a stream channel (interflow), and runoff from wet areas near channels (contributing source areas; Duncan and Woods, 2004). Flood events mobilize large amounts of materials from the catchment, as well as from within stream channels, and concentrations of contaminants entering streams tend to rise and fall with the stream flow depending upon the proportions of material being washed into the stream compared with those that are mobilized within the channel. Loads produced in flood events can be much greater than those in base flow. A recent study has shown that total numbers of the faecal bacterium *Escherichia coli* (*E. coli*) mobilized in a dairy catchment stream during several large storm events were each greater than the total number of *E. coli* transported in a year of base flow: 95% of the total yield occurred during storm flows (Davies-Colley *et al.*, 2007). As a rule of thumb, flood flows are important for the loads they carry and the downstream effects these have on sensitive waters, whereas with base flows it is the concentrations of contaminants that are important to on-site or proximal water use. Sensitive waters affected by loads from agricultural catchments include lakes (eutrophication caused by N and P) and coastal waters used for shellfish farming (affected by faecal microbial loads, especially pathogens, from grazing livestock). Concentrations of N and P may cause unwanted periphyton blooms during base flows, and high concentrations of *E. coli* indicate faecal contamination and an enhanced risk of infection from waterborne disease.

A summary of catchment experiments throughout the world showed that replacement of forest with other vegetation types (most notably pasture) increases water yield (Bosch and Hewlett, 1982). Rainfall interception by pasture is generally lower than for scrub or forest, for any given event, so that specific water yields (flow per unit of catchment area, $l/s/km^2$) and peak storm flows (Fig. 3.2) are commonly greater in pasture catchments than in forest

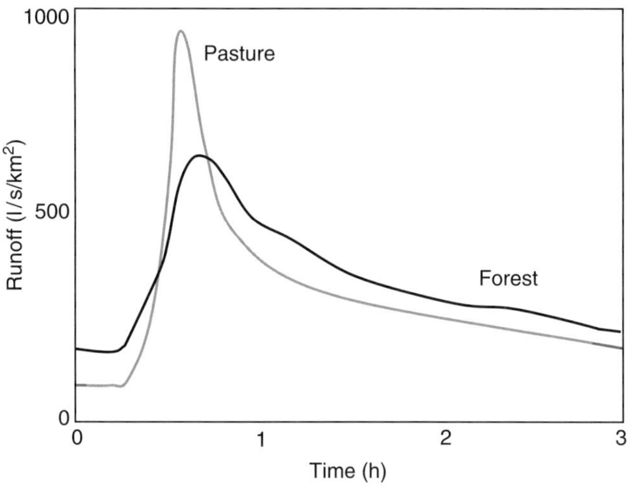

Fig. 3.2. Schematic comparison of hydrographs in a pasture catchment (grey) and in the pre-existing forest (black) during the same rainstorm. Note that peakflow is increased and the duration of storm flow is decreased such that there is a crossover of hydrographs on the receding limb. (From R.J. Davies-Colley, New Zealand, 2007, personal communication.)

catchments with similar topography, climate and rainfall (Duncan and Woods, 2004). Thus, contaminant transport from pastoral runoff is generally greater, too. Depending on farming practices, sediment loss in pastoral runoff may be so large that it causes degraded water clarity and excessive sedimentation and degradation of the stream channel habitat (Loehr, 1979; Schaller and Bailey, 1983; Wilcock, 1986).

Pastoral Agriculture and the Aquatic Environment

Stream water quality is determined by the prevailing lithology, climate, soil types and land uses within the catchment. The constituents of runoff from agriculture can influence stream morphology and water quality, and aquatic biota. Furthermore, downstream environments, such as lakes and estuaries, can also be affected by what happens on the land many kilometres away. Thus, the impacts of agricultural inputs to streams can be localized, but will also have cumulative effects downstream (Parkyn and Wilcock, 2004). These impacts include: changes in stream habitat caused by decreased riparian shading, siltation and channel modifications (e.g. straightening clearance to improve drainage), and changes in the physical, chemical and microbial attributes of water quality.

Habitat impacts

Agricultural practices alter the habitat of streams and hence, ecosystem structure, particularly by increasing exposure to direct sunlight. Clearance of riparian (stream bank) shading vegetation allows more sunlight on streams so that they tend to be warmer than otherwise, which can be stressful for aquatic organisms (Rutherford *et al.*, 1999). The combination of increased solar radiation and enriched concentrations of nutrients, notably N and P, stimulates plant growth through photosynthesis and fertilization of stream water and sediments. Streams that are cool and weed-free when riparian shading is present become warmer and often choked with algae and aquatic weeds (macrophytes) in open and unshaded habitats that are typical for many streams in agricultural catchments (Wilcock *et al.*, 1999; Wilcock and Nagels, 2001).

Increased sunlight and nutrients stimulates plant growth and photosynthetic production, often resulting in marked diel changes in dissolved oxygen (DO) and pH (Wilcock and Chapra, 2005). By contrast, shaded forest streams exhibit little diel variation in DO and pH. Respiration associated with decaying vegetation may cause minimum DO values to be lowered near dawn to the extent that streams can no longer support aquatic life, while photosynthetic production may cause maximum pH values in the afternoon to be high enough to cause ammonia toxicity (Fig. 3.3). Ammonia, a common agricultural pollutant that is a major component of animal waste products, exists in two forms in water: the non-toxic ionized ammonium form (NH_4^+), and the unionized ammonia (NH_3) form, which is very toxic to many aquatic species at low concentrations (commonly, <1 mg N/l). The

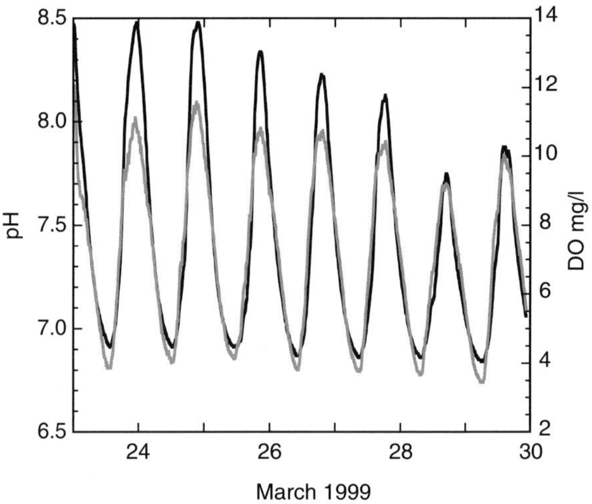

Fig. 3.3. Diel changes in dissolved oxygen (DO, light grey line) and pH (black line) in a lowland stream affected by agriculture. The water average temperature was 20°C. (From Wilcock and Chapra, 2005.)

ratio of toxic NH_3 ammonia to NH_4^+ depends on the pH, and the dissociation constant, *K*a, for the following reaction (Eqn 3.1):

$$NH_4^+ \rightleftharpoons NH_3 + H^+ \tag{3.1}$$

The value of *p*Ka ($-\log_{10}Ka$) is a function of temperature, *T* (°C; Erickson, 1985).

$$pKa = 0.09018 + 2729.92/(273.2 + T) \tag{3.2}$$

The ammonia and ammonium concentrations are approximately equal at a temperature of 20°C and a pH of 9.4 (Eqn 3.3).

$$\frac{[NH_3]}{[NH_4^+]} = 10^{(pH - pKa)} \tag{3.3}$$

In the following example (Fig. 3.3), pH in a photosynthetically productive stream reaches 8.4 at an average water temperature of 20°C; $[NH_3]$ comprises about 9% of total ammonia $[NH_4]$. Streams in pasture catchments can have $[NH_4]$ concentrations of 1–2 mg/l (Wilcock *et al.*, 1999, 2007), and under these conditions unionized ammonia concentrations are toxic to stream life, especially invertebrates (Hickey and Vickers, 1994).

Diel changes in pH may also affect the toxicity of heavy metals and organic acids in streams via acid-base equilibria, adsorption–desorption processes and solubility phenomena (Wilcock and Chapra, 2005). In this way, agriculture may affect the physical chemistry of a stream by modifying the pH regime, and thereby alter the toxicity of sediment-bound metals and arsenic deriving from industrial or mining activities.

Sediments

Fine-grained sediments transported in suspension can be a major contaminant in flowing waters because of the damage caused to aquatic habitats and biotic communities. Waters become more turbid (less visually clear) because of light scattering by suspended sediment particles (Davies-Colley and Smith, 2001), thereby reducing the ability of fish to detect prey, and suspended sediment may damage the respiratory structures of aquatic animals (Ryan, 1991; Waters, 1995). Normally hard-bottom streams with substrates comprising fine sands and cobbles can become soft-bottom streams dominated by silts and, as a consequence, have totally different ecosystems. As well as the intrinsic effects of unwanted fine sediment in stream channels there are sediment-associated contaminants, like particulate P and faecal matter that may contain pathogenic microorganisms. Agriculture on slopes of greater than 3% increases the risk of soil erosion, notably on tilled soils in the south-eastern USA (Wischmeier and Smith, 1978) that can lead to increases in nutrient and sediment loadings to surface waters. A survey of reaches on 11 streams showed that plantation forest and pasture streams had threefold greater suspended solids and fine sediment stored in the streambed than streams in native forest catchments (Quinn *et al.*, 1997). A comparison of sediment yields from studies of large catchments with hilly topography showed that grazed pasture is a major source of sediment by comparison with other major land uses, such as disturbed (by logging) and undisturbed exotic forest, and native forest (Table 3.3). Sediment yields from pasture on flat-to-gently rolling land (2–12%) are much lower than from hilly land (Table 3.3) and are commonly 30–200 kg/ha/year (McDowell and Wilcock, 2004; Wilcock *et al.*, 2007). Stream bank collapse caused by cattle grazing in riparian zones is a major source of sediment in pasture catchment streams and because of this it is likely that sediment yields from forest catchments will be much lower than from intensive pasture, in corresponding geoclimatic regions (Laubel *et al.*, 1999; McDowell and Wilcock, 2007).

Agricultural practices influence sedimentation rates by changing the runoff characteristics of catchments and by changing stream bank stability. Treading by grazing animals on stream banks destabilizes soils causing slumping and loss of soil into stream channels. This effect is exacerbated by shallow-rooted pasture species that do not stabilize soil as well as the roots of larger trees and shrubs. Natural stream channels in forested catchments tend to be wider and more stable than channels in grassy agricultural catchments (Davies-Colley, 1997). This

Table 3.3. Specific yields (kg/ha/year) for different land uses in New Zealand.

Land use	SS	TN	TP	Reference
Intensive flat-land dairy	142	35	1.16	Wilcock *et al.* (1999)
Average grazed pasture	600–2000	4–14	0.3–1.7	Wilcock (1986)
Hill-land pasture	1000–3000	10–20	1.5–3.2	Quinn and Stroud (2002)
Exotic forest – disturbed	300–2000	0.06–0.8	0.4–8	Wilcock (1986)
– undisturbed	500	0.07–0.2	0.15	Wilcock (1986)
Native forest	27–300	2–7	0.04–0.68	Wilcock (1986)

is because sediment is temporarily stored by pasture grasses, resulting in narrow and incised channels that release sediment during flood events.

Sediment tracing techniques using ^{137}Cs isotopes have been used to determine the key agricultural sources of sediment to streams. Subsurface drains have been identified by this technique as a key pathway for sediment transport to waterways in a UK study, with 30–60% coming from drains compared with 6–10% from channel bank erosion (Walling *et al.*, 2002). A similar result was found for sediment-associated P entering a stream in an intensively farmed dairying catchment in New Zealand (McDowell and Wilcock, 2004). Sediment inputs to streams in agricultural catchments without extensive drainage networks may be dominated by sheet and rill erosion from tilled systems (Wischmeier and Smith, 1978), and by bank erosion in grazed pastures. A Danish study, for example, found that bank erosion was approximately 11 mm/year and contributed more than half the annual catchment export. Bank erosion in grassland areas used for cattle grazing was greater than in forest (Laubel *et al.*, 1999).

Chemical contaminants

Agricultural runoff carries a range of chemical and microbial contaminants, which may be exacerbated by overgrazing, causing soil infiltration rates to decrease so that surface runoff occurs more frequently, resulting in a higher than usual loading of contaminants to waterways. Key chemical contaminants entering waterways from agriculture are shown below (Table 3.4).

Table 3.4. Chemical contaminants entering streams in agricultural catchments and their effects on aquatic ecosystems.

Abbreviation	Description	Effect
Major contaminants and macronutrients		
BOD_5	Biochemical oxygen demand (5 day) caused by inputs of oxidizable matter (C, N)	Low DO stress
NH_4^+, NH_3	Ammonia (ionized/unionized)	Toxicity, eutrophication
NO_3^-, NO_2^-	Nitrate, nitrite	Eutrophication
Organic-N	Plant and animal waste products	Eutrophication
DRP	Dissolved reactive P	Eutrophication
TP	Total P, some of which is bioavailable	Eutrophication
Minor and trace contaminants		
Pesticides	A wide range of insecticides, herbicides and fungicides	Toxicity, low DO
Monofluoracetate, CN^-	Poisons for noxious animals	Acute toxicity
DDT, chlordane, etc.	Persistent organochlorine pesticides no longer used but still present	Bioaccumulating toxins

continued

Table 3.4. Continued

Abbreviation	Description	Effect
Organophosphates, pyrethroids	Pesticides applied externally for control of ectoparasites	Acute toxicity
Helminthicides, e.g. levamisole	Drenches for controlling internal parasites in cattle and sheep	May be persistent and toxic
Zn, Cu and Cd	Pharmaceuticals (zinc and copper); cadmium in superphosphate fertilizer	Acute and chronic toxicity
Se, B and Mo	Trace elements added as soil amendments	Acute and chronic toxicity

C, N and P in aquatic ecosystems

Carbon (C), N and P are three essential macronutrients for stream life. Other macronutrients are sulfur, potassium, magnesium and calcium, but they are usually plentiful in streams (as major ionic species; Odum, 1971). Eyster (1964) listed ten micronutrients that are essential for plants: Fe, Mn, Cu, Zn, B, Si, Mo, Cl, V and Co. Most of these are essential for animals too, and a few others, such as iodine, are essential for vertebrates (Davies-Colley and Wilcock, 2004). Some, notably Cu and Zn, are toxic at greater (than optimum) concentrations.

The C cycle describes the circulation of C through ecosystems. Carbon dioxide (CO_2) is incorporated into organic compounds in green plants during photosynthesis. These compounds are eventually oxidized during respiration by plants, or by herbivores, carnivores and saprophytes, releasing CO_2 back to the atmosphere. Photosynthesis and respiration by plants are the key processes transferring C between the atmosphere and biosphere. They can be characterized by the following chemical equations for algal photosynthesis that show the relative importance of C, N and P forms (Stumm and Morgan, 1996). In each case the forward reaction (left to right) describes photosynthesis, while the reverse reaction (right to left) describes respiration, and the quantity in braces {} represents biomass (actually, the average composition of algae). It is important to note that for both equations the C/N and N/P *mole* ratios are 106:16 (or about 7:1) and 16:1, respectively. For limnologists using *masses* instead of molar values, the C/N and N/P ratios are approximately 6:1 and 7:1, respectively.

$$106\,CO_2 + 16\,NO_3^- + HPO_4^{2-} + 122\,H_2O + 18\,H^+ \rightleftharpoons \{C_{106}H_{263}O_{110}N_{16}P\} + 138\,O_2 \quad (3.4)$$

$$106\,CO_2 + 16\,NH_4^+ + HPO_4^{2-} + 106\,H_2O \rightleftharpoons \{C_{106}H_{263}O_{110}N_{16}P\} + 106\,O_2 + 14\,H^+ \quad (3.5)$$

C supplied to rivers is composed of allochthonous and autochthonous inputs. Allochthonous inputs are mainly terrestrial litter, such as leaves, tree stems, branches and flowers, that fall or are washed into streams. Autochthonous inputs are generated within stream channels, photosynthetically by algae, bryophytes, and large aquatic plants (macrophytes). Thus, conversion of land from forest to pasture (with concomitant changes in lighting of the stream and therefore aquatic plant production) changes the source of terrestrial C as well as altering flow paths, thereby affecting the release of C from vegetation and litter (Findlay *et al.*, 2001).

N is a key element within biological systems and is a constituent of amino acids that are the building blocks of all proteins (Hamilton *et al.*, 2004). It can form more than 10% of the dry weight of organisms and occurs in freshwaters in the following forms: nitrogen gas (N_2), ammoniacal-N ($NH_4 = NH_3 + NH_4^+$), N_2O, NO_2^-, NO_3^- and in various organic forms (Hamilton *et al.*, 2004). In streams, N enters mainly in dissolved inorganic ($NH_4 + NO_2^- + NO_3^-$), and organic forms. NH_4 removal is due to uptake by primary producers, bacteria and fungi plus direct nitrification. Indirect nitrification is the conversion of NH_4 mineralized from organic matter to NO_3^-. Removal of NO_3^- (and NO_2^-) from the water is primarily via assimilation by biota and denitrification on the channel bottom. Regeneration is the release of NH_4 and NO_3^- from the stream bottom back to the water column and is the net result of several interacting processes, including mineralization, indirect nitrification, denitrification and reuptake by organisms. Nitrate and NH_4 remaining in the water are exported downstream. Denitrification (see also Chapters 1 and 2, this volume) is the process occurring under anaerobic conditions in which NO_3^- (and NO_2^-) is returned in gaseous form to the atmosphere as N_2 and N_2O (Peterson *et al.*, 2001). Pasture inputs of N to waterways include direct inputs of fertilizer and animal wastes (dung and urine) as well as indirect inputs in runoff. Pastoral agriculture is a major source of N to waterways (Table 3.3).

P is an essential constituent of DNA and RNA, is involved in energy transfer processes in living cells (ADP-ATP processes) and is a component of fats of cell membranes. P moves slowly from deposits on land and in sediments to living organisms, and even more slowly back into soil and sediment. In aquatic ecosystems N and P are most often the elements in shortest supply for algal growth (Eqns 3.4 and 3.5) with P most often the 'limiting nutrient' (Hecky and Kilham, 1988). Research on the North American Great Lakes in the 1960s established that phytoplankton growth was driven by inputs of P, rather than N or other elements, and this also been shown to be true for many European lakes (Vollenweider, 1968). However, phytoplankton growth in some other lakes (e.g. a few in New Zealand) has been shown to be limited by N availability, for a variety of biological and geochemical reasons (Shallenberg, 2004). The influence of P on plant growth in pasture-fed rivers and streams is more complex because: (i) shade and available light generally have a much greater influence than nutrient concentrations; and (ii) macrophytes tend to take up nutrients from sediments that have large stores of N and P (Dawson and Haslam, 1983; Chambers *et al.*, 1989). Ratios of N/P in rivers are expressed in terms of dissolved inorganic N ($NH_4 + NO_3^- + NO_2^-$) and DRP (Biggs, 2000). An example of how this ratio varies seasonally in an agricultural stream is shown in Fig. 3.4, whereby P is 'limiting' for most of the year except during late summer (March–April in the southern hemisphere), when N 'limitation' is evident, reflecting the seasonal inputs of N to the stream (Monaghan *et al.*, 2007). Nitrate accumulates in the topsoil during dry periods as a result of mineralization of soil organic matter and plant

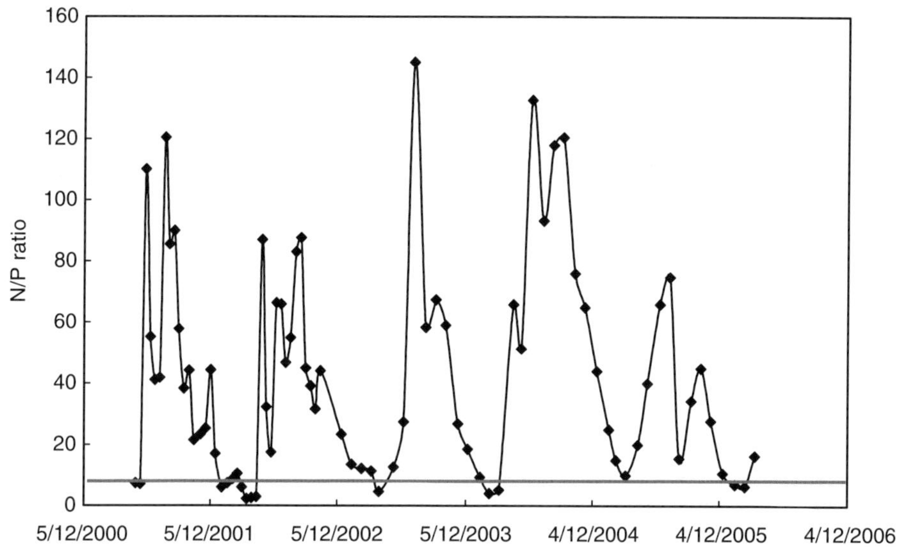

Fig. 3.4. Time series of N/P mass ratios (dissolved inorganic N and dissolved reactive P, respectively) in Bog Burn, a small stream in a predominantly pastoral dairy farming catchment in southern New Zealand. The horizontal line indicates the N/P requirement of aquatic plants (~7; Eqns 3.4 and 3.5). (From Monaghan *et al.*, 2007.)

material, and the input of dung and urine. With evapotranspiration exceeding rainfall in summer there is little downward movement of NO_3^- in drainage water. During this time of vigorous stream plant growth, demand for N may exceed supply, causing NO_3^- concentrations to be negligible. As soil moisture increases NO_3^- is flushed from the soil causing stream concentrations to rise to a maximum value in mid-winter (Wilcock *et al.*, 1999).

Specific yields (also called catchment export coefficients) are the annual load per unit area of land and have the units kg/ha/year. They vary according to land use, climate, topography and soil type and are used to characterize differences between land uses in a given region where the other features are similar. Examples of specific yields for suspended solids/sediment (SS), total N (TN) and total P (TP) are given in Table 3.3.

The production of NO_3^- by grazing animals leads not only to high concentrations in waterways, with a potential for eutrophication, but also to increased production of the greenhouse gas nitrous oxide (N_2O). Emissions of N_2O from pasture derive mainly from fertilizer and animal waste and are designated as being 'indirect' if they originate from nitrification and denitrification of N in runoff losses to waterways (Groffman *et al.*, 2002). Using a global land-use model, Seitzinger and Kroeze (1998) estimate that rivers could contribute more than 30% of the anthropogenic production of N_2O on land. Recent studies show that N_2O fluxes from small streams in intensively grazed catchments are generally much less than those from intensively farmed pasture having high N loadings, but are within a factor of 10 of annual emissions from a wide range of rural land uses (including many grazed pasture systems; Hlaváčová *et al.*, 2006; Wilcock and Sorrell, 2008).

Agrichemicals

'Agrichemicals' are synthetic chemical products manufactured and sold for use in agriculture, and include fertilizers and pesticides. In the past, persistent organochlorine pesticides like dichloro-diphenyl-trichloroethane (DDT) and lindane were used to control insect pests, such as pasture insect larvae, with the result that residues of these compounds and their biologically active metabolites (e.g. 1, 1-dichloro-2, 2-bis(p-dichlorodiphenyl) ethylene (DDE) and 1, 1-dichloro-2, 2-bis (p-chlorophenyl) ethane (DDD) are still present at nuisance levels more than 30 years after their use was prohibited (Boul *et al.*, 1994). Greater awareness of the problems created by persistent and bioaccumulating pesticides like DDT (Carson, 1962) has led to the development and widespread use of pesticides that have much shorter half-lives in soil than the 'legacy' pesticides. Herbicides generally have a low toxicity to aquatic life with LC50s (the concentration that is lethal to 50% of test animals within a specified time; commonly 96 h) often greater than 10 mg/l. Insecticides such as organophosphates are usually much more toxic (<1 mg/l) and synthetic pyrethroids are extremely toxic to fish and aquatic invertebrates (e.g. the LC50 for cypermethrin is <1 μg/l for rainbow trout; Morgan and Brunson, 2002).

Pesticide toxicity to aquatic organisms is deemed to be either acute or chronic, depending upon the dose, mode of toxic action and the exposure period. Acute toxicity manifests itself after a relatively short exposure time (commonly a few days), whereas chronic toxicity occurs over a longer period of exposure (weeks to months, or longer). Many pesticides exert toxicity by interfering with membrane processes (blocking of ion channels), enzymatic reactions (acetyl-choline esterase inhibitors) or by acting as hormone mimics. Chronic toxicity is commonly observed at doses ten times lower than acute toxicity. Many modern pesticides exert high acute toxicity, but are rapidly metabolized (broken down) so that they do not accumulate within organisms or in the aquatic environment. First generation pesticides, such as DDT and many other organochlorines, are only metabolized very slowly or not at all. As a consequence, they may persist in the aquatic environments for a long time (months to years), increasing their chance of bioaccumulation by organisms (resulting in contaminant tissue concentrations that are several times greater than in the external environment), and increasing the likelihood of chronic toxicity. The degree to which a pesticide accumulates in organism tissues is not only determined by its rate of breakdown but also by its fat solubility: pesticides with higher fat solubility accumulate to a greater extent and are consequently more toxic than less fat-soluble compounds. Chronic toxicity may also be the result of biomagnification (increasing tissue concentrations through the food chain due to minimal breakdown at lower level in the food chain). Chronic toxicity may manifest itself through sublethal effects (changes in metabolism or physiological function that do not lead to immediate death) that can affect ecological fitness, such as reproduction and growth. Chronic toxicity can also be the result of permanent genetic alterations (mutagenicity) or cancers induced through lengthy exposure to chemicals (carcinogenicity).

Endocrine disruptors act by disrupting the physiologic function of hormones. Because endocrine disruptors may give rise to adverse effects in aquatic and terrestrial animals, there are concerns that low-level exposure might cause similar effects in human beings as well. This is especially so where water is used multiple times (Petrovic *et al.*, 2002). Agricultural endocrine disruptors include the pesticides DDT, methoxychlor, linuron and vinclozolin, and their various metabolites, as well as natural gonadal hormones released in animal waste products, namely oestrogens, androgens and progestins (Lintelmann *et al.*, 2003).

Some short-lived but acutely toxic agricultural pesticides pose additional threats to the aquatic environment. Of particular concern is the use of organophosphates and pyrethroids, known to be especially toxic to stream invertebrates, for external control of ectoparasites on sheep and cattle (Wilcock *et al.*, 1994; Virtue and Clayton, 1997). Helminthicides (such as levamisole) applied orally for controlling worms in livestock may be of concern where dosing is carried out near waterways and pesticides are spilt or excreted directly into stream channels. There is anecdotal evidence (but few, if any, publications) suggesting that avermectins released into streams may be harmful to sediment-dwelling invertebrates.

Microbial contaminants

Dung and manure (i.e. dung that has had anthropogenic influence such as storage) from ruminant livestock contains microbes that may include pathogenic zoonoses (infectious diseases that may be transmitted from other animals, both wild and domestic, to humans) such as *Giardia*, *Salmonella*, *Cryptosporidium* and *Campylobacter* (Grau, 1988; Stanley *et al.*, 1998; Donnison and Ross, 1999; Collins *et al.*, 2007). Most microbiological agents of disease (pathogens) are derived from the faeces of warm-blooded animals and people. The presence in waters of pathogens is sporadic, only occurring when waters are polluted by faecal matter from sick individuals or 'carriers'. Animal dung contains large numbers of benign

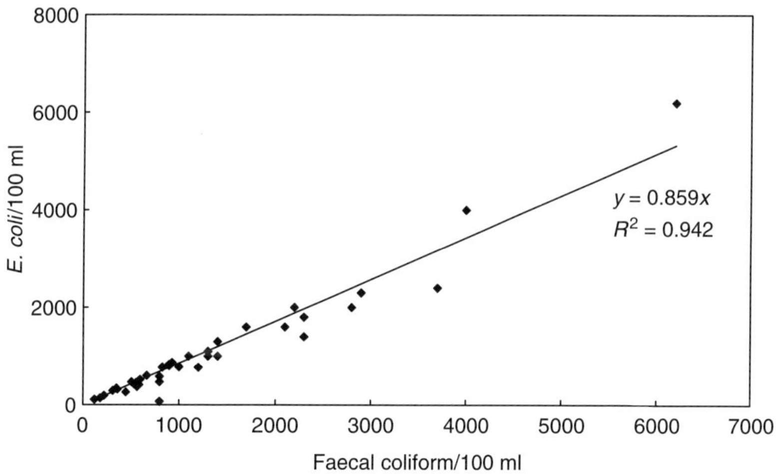

Fig. 3.5. Relationship between concentrations of *Escherichia coli* (*E. coli*) and faecal coliform in Bog Burn, a stream in a dairy farming catchment in New Zealand. (Data from Monaghan *et al.*, 2007.)

bacteria, most notably *E. coli* that is now routinely determined in water monitoring programmes to indicate faecal pollution levels. *Escherichia coli* comprise 85–90% of faecal coliforms in natural waters over several orders of magnitude (Fig. 3.5). Thus, it is relatively straightforward to compare *E. coli* data with faecal (thermotolerant) coliform data as used in monitoring of some shellfish harvesting waters.

Faecal pollution of waterways from agriculture derives from diffuse sources, such as runoff from grazed pasture and feed-pads, and PSs such as purpose-built waste treatment systems (e.g. oxidation ponds for dairy shed effluent). Other diffuse sources include runoff farm stock tracks, livestock accessing unfenced streams, and cattle crossings of streams (Collins *et al.*, 2007). Thus, it is important that key land uses within the catchment of rivers being used for water supply are known so that implications for water treatment are understood. A recent analysis of *E. coli* loadings to waterways in the Waikato region (an intensively farmed area of New Zealand) showed that surface runoff was the major source of faecal pollution from agriculture, but that inputs from dairy herds crossing streams and from drains were almost equally important (Fig. 3.6; Davies-Colley *et al.*, 2004; Wilcock, 2006). Median *E. coli* concentrations of streams in five catchments in which the dominant land use is pastoral dairy farming (Fig. 3.7) are 2–10 times the guideline value (126 *E. coli*/100 ml) for contact recreation in New Zealand (ANZECC, 2000; Wilcock *et al.*, 2007).

Faecal contamination of streams can be very high during floods owing to mobilization of contaminated sediments and wash-in from contributing pasture areas of catchments. *Escherichia coli* concentrations of 41,000 most probable number (MPN) /100 ml were measured in a single flood event in an agricultural stream, compared to a pre-flood level of about 100 MPN/100 ml (Nagels *et al.*, 2002).

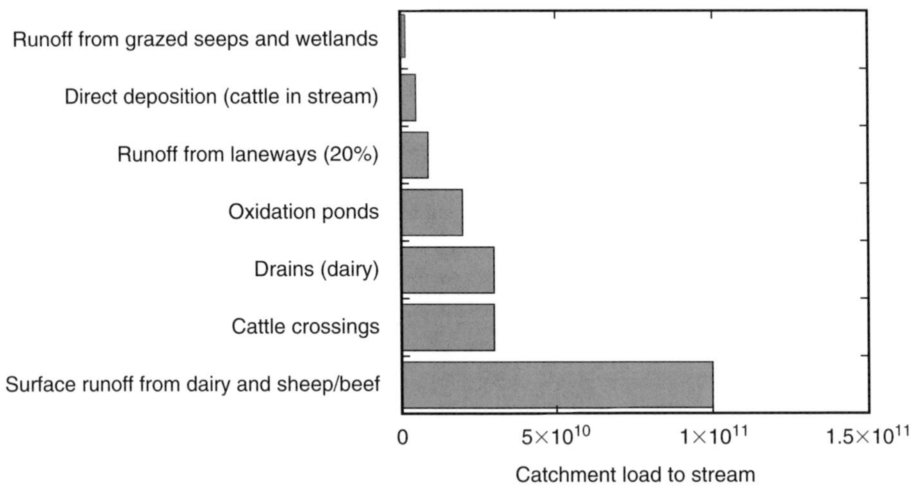

Fig. 3.6. Waterway loadings (*Escherichia coli*/ha-pasture/year) for major sources of faecal matter in the Waikato region, New Zealand. Loadings from farm laneways (tracks) were made assuming that 20% of deposited faecal matter is washed by rainfall into waterways. (From Wilcock, 2006.)

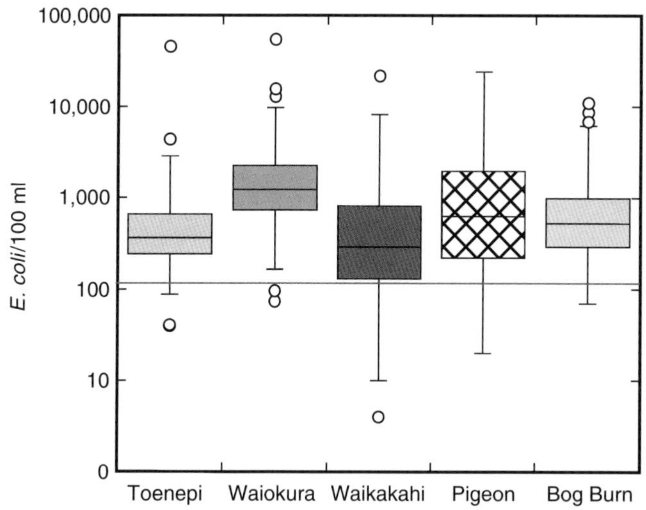

Fig. 3.7. Box plots of *Escherichia coli* concentrations in five dairy catchments (Wilcock *et al.*, 2007). Each box encloses 50% of the data with the median value of the variable displayed as a line. The top and bottom of the box mark the limits of ±25% of the variable population. The lines extending from the top and bottom of each box mark the minimum and maximum values within the data set that fall within an acceptable range. Any value outside of this range, called an outlier, is displayed as an individual point. The horizontal line is the median contact recreation guideline in New Zealand. (From ANZECC, 2000.)

Baseflow concentrations are important when considering health risk (from pathogens) to downstream water users, including bathers and other recreational users, and drinking water for livestock. Storm-flow loads are particularly important to water users well downstream, including aquaculture (e.g. Davies-Colley *et al.*, 2007). Filter-feeding shellfish (e.g. mussels, oysters and clams) can accumulate microorganisms when contaminated by runoff from agriculture during storm events and thereby present a potential health risk to consumers (Graczyk *et al.*, 2006). For this reason, shellfish farms are commonly restricted from harvesting during and after floods. Median concentrations of *E. coli* in New Zealand streams and rivers are much higher in pasture than in forest catchments (Table 3.5).

Modelling Agricultural Impacts on Water Resources

Models have been used to predict the effects of agriculture on water quality and to examine different land-use options ('scenarios'). One such model is Chemicals, Runoff and Erosion from Agricultural Management Systems (CREAMS) developed by the US Department of Agriculture. The CREAMS model was designed to predict runoff, soil erosion and the transport of nutrients and pesticides from agricultural land in the USA on a field scale, and has had wide application. Other models have been developed from CREAMS, including Groundwater Loading Effects of

Table 3.5. Faecal contamination in a range of New Zealand streams and rivers expressed by median concentrations of *Escherichia coli*.

Slope and elevation	Land use	*E. coli*/100 ml	References
Hill-country (Waikato)	Sheep/beef	398	Donnison *et al.* (2004)
Native forest		100	Donnison *et al.* (2004)
Pine forest		83	Donnison *et al.* (2004)
Lowland (Waikato)	Dairy	370	Wilcock *et al.* (2007)
Lowland (Taranaki)	Dairy	1250	Wilcock *et al.* (2007)
Lowland (Southland)	Dairy	530	Wilcock *et al.* (2007)
Lowland – high rainfall	Dairy	640	Wilcock *et al.* (2007)
	Native forest	4	Davies-Colley and Nagels (2002)

Agricultural Management Systems (GLEAMS), and the catchment-scale models Basins New Zealand (BNZ) and Watershed Assessment Model (WAMview; Novotny, 1986; Gillingham and Thorrold, 2000). These models have been used to evaluate loads and specific yields of N, P and sediment for different land uses and intensities. The Spatially Referenced Regression on Watershed Attributes (SPARROW) surface water quality model has been used to calculate annual nutrient fluxes in non-tidal streams throughout the USA and New Zealand, on the basis of N sources, landscape characteristics and stream properties. The model has been useful for assessment of water quality in stream networks (Smith *et al.*, 1997; Elliott *et al.*, 2005).

A range of modelling approaches is needed to solve different issues relating to agricultural impacts on water quality. Contaminant generation and transformations often occur at small spatial scales (e.g. raindrop erosion, leaching from urine patches). Some ecosystem processes must be understood at small spatial and temporal scales while others require understanding at large scales. Management decisions often require information at large spatial and temporal scales (e.g. decadal 'end of catchment' loads). One approach is to use different models for different purposes. Farm-scale models are required to help design and build mitigation measures. Catchment-scale models are required to estimate the combined effect of spatially distributed management practices and to examine downstream impacts (Rutherford *et al.*, 2006).

Conclusions and Future Trends

Increasing agricultural intensification and demand for limited water resources will force us to think more carefully about impacts of farming on water quality and ecology, and the effects on downstream water use. Climate changes brought about by greenhouse gas emissions will affect water resource management and place additional responsibility on agriculturalists to use resources wisely and sustainably. Water quantity and quality and aquatic and riparian ecology are all inextricably connected so that an impact on one affects the others. There is increasing demand for a whole-systems approach to farming and resource management that may best be achieved through the use of soundly based models. There is no doubt that agriculture degrades surface water quality and that there are limits to what can be done to decrease loadings on to land or to intercept contaminants in run-off pathways, through the use of best management practices. Agricultural landscapes are inherently 'leaky' and there are limits to land-use intensification that will be tolerated without breaching receiving water quality standards.

Public awareness of the negative effects of agriculture on water quality and increasing reluctance to put up with it are key drivers for a systems-based approach. Past reliance on voluntary actions by individual farmers and sector groups has had limited success and regulatory enforcement of rules for land and water management are an increasing likelihood. The European Union Water Framework Directive (WFD) requires the restoration of water bodies to good ecological status within a prescribed timetable (Johnes, 2007). The WFD requires each Member State to assess the ecological status of all inland and coastal waters, defined as a deviation from an undamaged (reference) state with timetables for waters of 'moderate, poor or bad' ecological status to be restored to 'good' ecological status. Activities that degrade the ecological structure and function of water bodies will be controlled, including modification of hydrochemistry through direct and indirect discharges of polluting substances. Of these, nutrient enrichment through discharge of N and P from agriculture is recognized as being widespread in Europe and a target for tighter control measures (Johnes, 2007). Land uses that cannot meet these targets will come under the new Environmental Liability Directive, which specifically implements the 'polluter pays principle'. Its fundamental aim is to hold operators whose activities have caused environmental damage financially liable for remedying this damage.

Future pastoral agricultural practices will have to be carried out with greater consideration of local water resource sensitivity and scarcity and in a way that is more harmonious with community aspirations and the need for a clean and healthy environment.

References

Allan, J.D. and Johnson, L.B. (1997) Catchment-scale analysis of aquatic ecosystems. *Freshwater Biology* 37, 107–111.

Allan, J.D., Erikson, D.L. and Fay, J. (1997) The influence of catchment land use on stream integrity across multiple spatial scales. *Freshwater Biology* 37, 149–161.

ANZECC (2000) *National Water Quality Management Strategy: Australian and New Zealand Guidelines for Fresh and Marine Water Quality*. Australian and New Zealand Environment and Conservation Council and Agriculture and Resource Management Council of Australia and New Zealand, Canberra, Australia.

Biggs, B.J.F. (2000) Eutrophication of streams and rivers: dissolved nutrient-chlorophyll relationships for benthic algae. *Journal of the North American Benthological Society* 19, 17–31.

Bosch, J.M. and Hewlett, J.D. (1982) A review of catchment experiments to determine the effect of vegetation changes on water yield and evapotranspiration. *Journal of Hydrology* 55, 3–23.

Boul, H.L., Garnham, M.L., Hucker, D., Baird, D. and Aislabie, J. (1994) Influence of agricultural practices on the levels of DDT and its residues in soil. *Environmental Science and Technology* 28, 1397–1402.

Carson, R.L. (1962) *Silent Spring*. Houghton Mifflin, Boston, Massachusetts.

Chambers, P.A., Bothwell, M.L. and Hamilton, H.R. (1989) Roots versus shoots in nutrient uptake by aquatic macrophytes in flowing waters. *Canadian Journal of Fisheries and Aquatic Sciences* 46, 435–439.

Collins, R., McLeod, M., Hedley, M., Donnison, A., Close, M., Hanly, J., Horne, D., Ross, C., Davies-Colley, R., Bagshaw, C. and Matthews, L. (2007) Best management practices to mitigate faecal contamination by livestock of New Zealand waters. *New Zealand Journal of Agricultural Research* 50, pp. 267–278 (in press).

Davie, T.J.A. (2004) Soil water, runoff and streamflow generation. In: Harding, J., Mosley, P., Pearson, C. and Sorrell, B. (eds) *Freshwaters of New Zealand*. New Zealand Hydrological Society and New Zealand Limnological Society. Caxton Press, Christchurch, New Zealand, pp. 4.1–4.10.

Davies-Colley, R.J. (1997) Stream channels are narrower in pasture than in forest. *New Zealand Journal of Marine and Freshwater Research* 31, 599–608.

Davies-Colley, R.J. and Nagels, J.W. (2002) Effects of dairying on water quality of lowland streams in Westland and Waikato. *Proceedings of the New Zealand Grasslands Association* 64, 107–114.

Davies-Colley, R.J. and Smith, D.G. (2001) Turbidity, suspended sediment, and water clarity: a review. *Journal of the American Water Resources Association* 37, 1085–1101.

Davies-Colley, R.J., Nagels, J.W., Smith, R.A., Young, R.G. and Phillips, C.J. (2004) Water quality impact of a dairy cow herd crossing a stream. *New Zealand Journal of Agricultural Research* 38, 569–576.

Davies-Colley, R.J. and Wilcock, R.J. (2004) Water quality and chemistry in running waters. In: Harding, J., Mosley, P., Pearson, C. and Sorrell, B. (eds) *Freshwaters of New Zealand*. New Zealand Hydrological Society and New Zealand Limnological Society. Caxton Press, Christchurch, New Zealand, pp. 11.1–11.17.

Davies-Colley, R.J., Lydiard, E. and Nagels, J.W. (2007) Faecal bacterial dynamics and yields from an intensively dairy-farmed catchment. Oral presentation at the 11th International IWA Specialist Group Conference on Diffuse Pollution. 26–31 August, 2007, Belo Horizonte, Brazil.

Dawson, F.H. and Haslam, S.M. (1983) The management of river vegetation with particular reference to shading effects on marginal vegetation. *Landscape Planning* 10, 147–169.

Donnison, A.M. and Ross, C.M. (1999) Animal and human faecal pollution in New Zealand rivers. *Zealand Journal of Marine and Freshwater Research* 33, 119–128.

Donnison, A., Ross, C. and Thorrold, B. (2004) Impact of land use on the faecal microbial quality of hill-country streams. *Zealand Journal of Marine and Freshwater Research* 38, 845–855.

Duncan, M. and Woods, R. (2004) Flow regimes. In: Harding, J., Mosley, P., Pearson, C. and Sorrell, B. (eds) *Freshwaters of New Zealand*. New Zealand Hydrological Society and New Zealand Limnological Society. Caxton Press, Christchurch, New Zealand, pp. 7.1–7.14.

Elliott, A.H. and Sorrell, B.K. (2002) *Lake Managers' Handbook – Land Water Interactions*. Ministry for the Environment, Wellington, New Zealand.

Elliott, A.H., Alexander, R.B., Schwartz, G.E., Shankar, U., Sukias, J.P.S. and McBride, G.B. (2005) Estimation of nutrient sources and transport for New Zealand using the hybrid mechanistic-statistical model SPARROW. *Journal of Hydrology (NZ)* 44, 1–27.

Erickson, R.J. (1985) An evaluation of mathematical models for the effects of pH and temperature on ammonia toxicity to aquatic organisms. *Water Research* 19, 1047–1058.

Eyster, C. (1964) Micronutrient requirements for green plants, especially algae. In: Jackson, D.F. (ed.) *Algae and Man*. Plenum Press, New York, pp. 86–119.

Findlay, S., Quinn, J.M., Hickey, C.W., Burrell, G. and Downes, M. (2001) Effects of land use and riparian flowpath on delivery of dissolved organic carbon to streams. *Limnology and Oceanography* 46, 345–355.

Gianessi, L.P. and Peskin, H.M. (1984) *An Overview of the RFF Environment Data Inventory: Methods, Sources and Preliminary Results. Vol. 1*. Resources for the Future, Washington, DC.

Gillingham, A.G. and Thorrold, B.S. (2000) A review of New Zealand research measuring phosphorus in runoff from pasture. *Journal of Environmental Quality* 29, 88–96.

Graczyk, T.K., Girouard, A.S., Tamang, L., Napier, S.P. and Schwab, K.J. (2006) Recovery, bioaccumulation, and inactivation of human waterborne pathogens by the Chesapeake Bay nonnative oyster, Crassostrea ariakensis. *Applied and Environmental Microbiology* 72, 3390–3395.

Grau, F.H. (1988) *Campylobacter jejuni* and *Campylobacter Hyointestinalis* in the intestinal tract and on the carcasses of calves and cattle. *Journal of Food Protection* 51, 857–861.

Groffman, P.M., Gold, A.J., Kellogg, D.Q. and Addy, K. (2002) Mechanisms, rates and assessment of N_2O in groundwater, riparian zones and rivers. In: van Ham, J., Baede, A.P.M., Guicherit, R. and Williams-Jacobse, J.G.F.M. (eds) *Proceedings of the Third International Symposium on Non-CO_2 Greenhouse Gases.* Millpress, Rotterdam, The Netherlands, pp. 159–166.

Hamilton, D., Hawes, I. and Davies-Colley, R. (2004) Physical and chemical charcteristics of lake water. In: Harding, J., Mosley, P., Pearson, C. and Sorrell, B. (eds) *Freshwaters of New Zealand.* New Zealand Hydrological Society and New Zealand Limnological Society. Caxton Press, Christchurch, New Zealand, pp. 21.1–21.19.

Hecky, R.E. and Kilham, P. (1988) Nutrient limitation of phytoplankton in freshwater and marine environments. A review of recent evidence on the effects of enrichment. *Limnology and Oceanography* 33, 796–822.

Hewlett, J.D. and Hibbert, A.R. (1967) Factors affecting the response of small watersheds to precipitation in humid areas. In: Sopper, W.E. and Lull, H.W. (eds) *Forest Hydrology*, Pergamon Press, New York, pp. 275–290.

Hickey, C.W. and Vickers, M.L. (1994) Toxicity of ammonia to nine native New Zealand freshwater invertebrate species. *Archives of Environmental Contamination and Toxicology* 26, 292–298.

Hlaváčová, E., Rulík, M., čáp, L. and Mach, V. (2006) Greenhouse gas (CO_2, CH_4, and N_2O) emissions to the atmosphere from a small lowland stream in Czech Republic. *Archiv für Hydrobiologie* 165, 339–353.

Horton, R.E. (1933) The role of infiltration in the hydrological cycle. *Transactions of the American Geophysical Union* 14, 446–460.

Johnes, P.J. (2007) Meeting ecological restoration targets in European waters. A challenge for animal agriculture. In: Swain, D., Charmley, E., Steel, J. and Coffey, S. (eds) *Redesigning Animal Agriculture.* CAB International, Wallingford, UK, pp. 185–203.

Laubel, A., Svendsen, L.M., Kronvang, B. and Larsen, S.E. (1999) Bank erosion in a Danish lowland stream system. *Hydrobiologia* 410, 279–285.

Lintelmann, J., Katayama, A., Kurihara, N., Shore, L. and Wenzel, A. (2003) Endocrine disruptors in the environment. *Pure and Applied Chemistry* 75, 631–681.

Loehr, R.C. (1979) Potential pollutants from agriculture – an assessment of the problems and possible control approaches. *Progress in Water Technology* 11, 169–193.

McDowell, R.W. and Wilcock, R.J. (2004) Particulate phosphorus transport within stream flow of an agricultural catchment. *Journal of Environmental Quality* 33, 2111–2121.

McDowell, R.W. and Wilcock, R.J. (2007) Sources of sediment and phosphorus in stream flow of a highly-productive dairy farmed catchment. *Journal of Environmental Quality* 36, 540–548.

Monaghan, R.M., Wilcock, R.J., Smith, L.C., Tikkisetty, B., Thorrold, B.S. and Costall, D. (2007) Linkages between land management activities and water quality in an intensively farmed catchment in southern New Zealand. *Agriculture, Ecosystems and Environment* 118, 211–222.

Morgan, E.R. and Brunson, M.W. (2002) *Toxicities of Agricultural Pesticides to Selected Aquatic Organisms.* SRAC Publication No. 4600, Southern Regional Aquaculture Center, Stoneville, Mississippi.

Nagels, J.W., Davies-Colley, R.J., Donnison, A.M. and Muirhead, R.W. (2002) Faecal contamination over flood events in a pastoral agricultural stream in New Zealand. *Water Science and Technology* 45, 45–52.

Novotny, V. (1986) Review of hydrologic and water quality models used for simulation of agricultural pollution. In: Jørgensen, S.R. (ed.) *Agricultural Nonpoint Source Pollution: Model Selection and Application.* Elsevier, Amsterdam, pp. 9–35.

Novotny, V. (2003) Water quality. *Diffuse Pollution and Watershed Management.* Wiley, New York.

Odum, E.P. (1971) *Fundamentals of Ecology*. W.B. Saunders, Philadelphia, Pennsylvania.

Parkyn, S. and Wilcock, R.J. (2004) Impacts of agricultural land use. In: Harding, J., Mosley, P., Pearson, C. and Sorrell, B. (eds) *Freshwaters of New Zealand*. New Zealand Hydrological Society and New Zealand Limnological Society. Caxton Press, Christchurch, New Zealand, pp. 34.1–34.16.

Peterson, B.J., *et al.* (2001) Control of nitrogen export from watersheds by headwater streams. *Science* 292, 86–90.

Petrovic, M., Solé, M., López de Alda, M.J. and Damià Barceló, D. (2002) Endocrine disruptors in sewage treatment plants, receiving river water, and sediments, Integration of chemical analysis and biological effects on feral carp. *Environmental Toxicology and Chemistry* 21, 2146–2156.

Quinn, J.M. and Stroud, M.J. (2002) Water quality and sediment and nutrient export from New Zealand hill-land catchments of contrasting land use. *New Zealand Journal of Marine and Freshwater Research* 36, 409–429.

Quinn, J.M., Cooper, A.B., Davies-Colley, R.J., Rutherford, J.C. and Williamson, R.B. (1997) Land use effects on habitat, water quality, periphyton, and benthic invertebrates in Waikato, New Zealand, hill-country streams. *New Zealand Journal of Marine and Freshwater Research* 31, 579–597.

Rutherford, J.C., Davies-Colley, R.J., Quinn, J.M., Stroud, M.J. and Cooper, A.B. (1999) *Stream Shade: Towards A Restoration Strategy*. Department of Conservation, Wellington, New Zealand.

Rutherford, K., Spigel, B., Howard-Williams, C., Elliott, S. and Morgenstern, U. (2006) Improved tools for modelling nutrient runoff from agricultural land and its effects on water quality. In: *Proceedings of the Water 2006 International Conference, 1–4 August 2006, Auckland*. Events Division, Wellington, New Zealand, 15 p.

Ryan, P.A. (1991) The environmental effects of suspended sediment on New Zealand streams: a review. *New Zealand Journal of Marine and Freshwater Research* 25, 175–189.

Schaller, F.W. and Bailey, G.W. (1983) *Agricultural Management and Water Quality*. Iowa State University Press, Ames, Iowa.

Seitzinger, S.P. and Kroeze, C. (1998) Global distribution of nitrous oxide production and N inputs in freshwater and coastal marine ecosystems. *Global Biogeochemical Cycles* 12, 93–113.

Shallenberg, M. (2004) Primary production in open water. In: Harding, J., Mosley, P., Pearson, C. and Sorrell, B. (eds) *Freshwaters of New Zealand*. New Zealand Hydrological Society and New Zealand Limnological Society. Caxton Press, Christchurch, New Zealand, pp. 22.1–22.16.

Smith, R.A., Schwartz, G.E. and Alexander, R.B. (1997) Regional interpretation of water-quality monitoring data. *Water Resources Research* 33, 2781–2798.

Stanley, K.N., Wallace, J.S., Currie, J.E., Diggle, P.J. and Jones, K. (1998) Seasonal variation of thermophilic campylobacters in lambs at slaughter. *Journal of Applied Microbiology* 84, 1111–1116.

Stumm, W. and Morgan, J.J. (1996) *Aquatic Chemistry*, 3rd edn. Wiley-Interscience, New York.

Vanderholm, D.H. (1985) *Agricultural Waste Manual*. NZAE Project Report 32, Lincoln College, Canterbury, New Zealand.

Virtue, W.A. and Clayton, J.W. (1997) Sheep dip chemicals and water pollution. *The Science of the Total Environment* 194/195, 207–217.

Vollenweider, R.A. (1968) *Scientific Fundamentals of the Eutrophication of Lakes and Flowing Waters, With Particular Reference to Nitrogen and Phosphorus as Factors in Eutrophication*. OECD Technical Report, DAS/CSI/68.27, Paris.

Walling, D.E., Russell, M.A., Hodgkinson, R.A. and Zhang, Y. (2002) Establishing sediment budgets for two small lowland agricultural catchments in the UK. *Catena* 47, 323–353.

Waters, T.F. (1995) *Sediment in Streams: Sources, Biological Effects, and Control*. American Fisheries Society Monograph 7. American Fisheries Society, Bethesda, Maryland.

Wilcock, R.J. (1986) Agricultural runoff, a source of water pollution in New Zealand. *New Zealand Agricultural Science* 20, 98–103.

Wilcock, R.J. (2006) *Assessing the Relative Importance of Faecal Pollution Sources in Rural Catchments*. Technical Report TR 2006/41. Environment Waikato, Hamilton, New Zealand.

Wilcock, R.J. and Chapra, S.C. (2005) Diel changes of inorganic chemistry in a macrophyte dominated, softwater stream. *Marine and Freshwater Research* 56, 1165–1174.

Wilcock, R.J. and Nagels, J.W. (2001) Effects of aquatic macrophytes on physico-chemical conditions of three contrasting lowland streams, a consequence of diffuse pollution from agriculture. *Water Science and Technology* 43, 163–168.

Wilcock, R.J. and Sorrell, B.K. (2008) Emissions of greenhouse gases CH_4 and N_2O from low-gradient streams in agriculturally developed catchments. *Water, Air, and Soil Pollution,* 188 (1–4), 155–170.

Wilcock, R.J., Northcott, G.L. and Nagels, J.W. (1994) Mass losses and changes in concentration of chlorpyrifos, *cis*- and *trans*-permethrin applied to the surface of a stream. *Bulletin of Environmental Contamination and Toxicology* 53, 337–343.

Wilcock, R.J., Champion, P.D., Nagels, J.W. and Croker, G.F. (1999) The influence of aquatic macrophytes on the hydraulic and physico-chemical properties of a New Zealand lowland stream. *Hydrobiologia* 416, 203–214.

Wilcock, R.J., Monaghan, R.M., Thorrold, B.S., Meredith, A.S., Betteridge, K. and Duncan, M.J. (2007) Land–water interactions in five contrasting dairying watersheds, issues and solutions. *Land Use and Water Resources Research* 7, 2.1–2.10.

Wischmeier, W.H. and Smith, S.D. (1978) *Predicting Rainfall-Erosion Losses. A Guide to Conservation Planning.* Agricultural Handbook No. 537. US Department of Agriculture, Washington, DC.

4 Socio-economic Issues in Pasture-based Farming

D. Bewsell[1] and C. Dake[2]

[1]AgResearch Limited, Lincoln Research Centre, Christchurch 8140, New Zealand; [2]AgResearch Limited, Grasslands Research Centre, Palmerston North 4442, New Zealand

Introduction

The focus in this chapter is on three countries and one region – Australia, New Zealand and the USA, and the European Union (EU). The EU currently comprises an intergovernmental union of 26 states in Europe. Most of the comparative analysis in this chapter uses the EU before 1 May, 2004 (EU15), which consisted of 15 countries: Austria, Belgium, Denmark, Finland, France, Germany, Greece, Ireland, Italy, Luxembourg, the Netherlands, Portugal, Spain, Sweden and the UK. These three countries and one region were chosen because they provide an excellent overview of the socio-economic trends in developed industrialized countries, and illustrate the pressures faced by pastoral agriculture. The trends in Australia, New Zealand, the USA and the EU15 reveal that there is pressure on farmers to intensify production in response to the growing global demand for pastoral products. However, as outlined in this book, much of this has a significant impact on the environment and socio-economic viability of communities. To address these environmental issues, there is a growing number of extension and policy programmes designed to promote the adoption of practices intended to mitigate the impact of pastoral farming on the environment and social sustainability. In Australia, New Zealand and to some extent in the USA, the response has been focused on voluntary change or to meet market demands. However, in the USA and in Europe, considerable emphasis is placed on the role that agriculture plays to meet environmental and social goals, in addition to providing livelihoods to farmers. As such, considerable amounts of financial support are afforded to producers to provide environmental and social services.

In this chapter, the drivers of demand and trends in demand and production are outlined. Issues around accounting for the positive and negative externalities of pastoral agriculture are discussed as well as mechanisms used to provide financial support for pastoral agriculture. Next, approaches to promoting the adoption of environmental practices are outlined and implications for change summarized. Finally, three case studies designed to illustrate the issues are presented.

Overview of the Socio-economic Landscape of Pastoral Agriculture

Growing global demand for pastoral products

The increasing global demand for livestock products has driven countries and pastoral farmers in Europe, the USA, Australia and New Zealand to increase production to maintain and grow their share of the global livestock product trade. The world consumption of bovine and ovine meat products increased by between 0.7% and 2.1% per year over the last few decades, with consumption in the developing countries increasing by between 3.2% and 3.7% per year (FAO, 2006b). Similar growth rates in milk and dairy product consumption were also recorded in the same period. The consumption of livestock products has been projected to continue to increase over the next decade but at a slightly decreased rate (FAO, 2006b). The growth in demand for dairy products is expected to increase due mainly to growth in developing countries.

The main drivers of demand for livestock products have been the increasing growth in incomes and population and increasing urbanization. The gross domestic product (GDP) per capita, a measure of economic activity and prosperity, is increasing worldwide. Between Australia, New Zealand, the USA and the EU15, GDP is currently between US$26,000 and US$42,000 and growing at 1–4% per year (Table 4.1). As a comparison, the GDP per capita of some developing economies such as Latin America and the Caribbean (US$4,454), South Asia (US$677) and sub-Saharan Africa (US$830), even though lower than the GDP in Australia, New Zealand, the USA and the EU15, was growing at between 4% and 9% per year in 2005 (The World Bank, 2006).

Table 4.1. Gross domestic product and trends in population growth.

	Australia	EU15	New Zealand	USA
Gross domestic product per capita 2005 (US$)[a]	34,480	31,594	26,531	42,007
Gross domestic product growth (annual %)	3	1	2[c]	4
Population at 2005 (million)[b]				
Total population	20	392	4	300
Rural population	1	82	1	58
Urban population	19	310	3	242
Population forecast 2030 (Index 2005 = 100%)				
Total population	119	100	113	123
Rural population	65	74	89	84
Urban population	123	107	117	133

[a]The World Bank, 2006.
[b]FAO, 2006a.
[c]Statistics New Zealand, available at: www.stats.govt.nz/store/2006/05/gross-domestic-product-dec05qtr-mr.htm

The highest population in our sample is Europe with 392 million people; the lowest in New Zealand with 4 million people. By 2030, total population in Australia, New Zealand and the USA is expected to increase by about 20% (compared to 2005 levels), while the population in the EU15 is projected to remain unchanged (Table 4.1). The forecast suggests growing urbanization, with the rural population decreasing to between 65% (Australia) and 89% (New Zealand) of their 2005 populations. The declining rural population also translates into a declining agricultural labour force (Fig. 4.1). For example, in 2010, the agricultural population in the EU15 and the USA is projected to decline to 13% and 25%, respectively, compared to 1950. The combined effects of these demographic changes mean the demand for pastoral products can be expected to be high, and in particular the demand and flow of products from rural areas to the urban area within the EU15 and the USA can be expected to increase.

Countries have to compete with new developing exporters such as Brazil to maintain and grow their share of net exports of livestock products, in particular meat exports. Developing economies are expected to double current exports by 2030 (FAO, 2006b). Developing countries are, however, expected to be net importers of dairy products in the foreseeable future. Australia and New Zealand rely on agriculture for a high proportion of their merchandise exports earnings, which comprise mainly of food product exports (meat and dairy products). These are exported to markets in Europe, North America and Asia.

New Zealand has the highest reliance on agricultural exports (60%) for their mercantile export earnings with 49% derived from food product exports. In Australia, the EU15 and the USA, the share of agricultural exports in total merchandise exports is 23%, 11% and 10%, respectively; about a 30% decrease since 1993 (Table 4.2). In Australia, the decrease in exports has been mainly due to the decrease in the export of raw agricultural products.

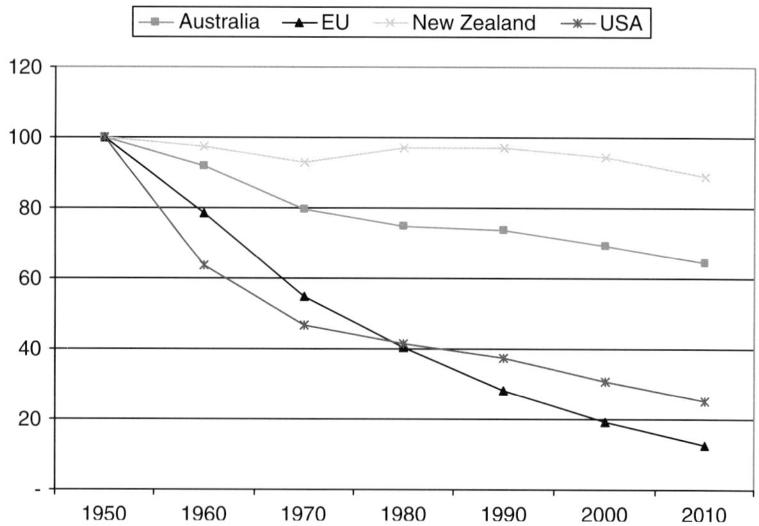

Fig. 4.1. Trend in agricultural population. Index: 1950 = 100%. (Adapted from FAO, 2006a.)

Table 4.2. Tends in agricultural exports. (Adapted from The World Bank, 2006.)

	Australia	EU15	New Zealand	USA
Agricultural exports as a percentage of total merchandise exports	2004	2004	2004	2004
Raw agricultural products exports	4	2	11	2
Food products exports	19	9	49	7
Agricultural exports	23	11	60	10
index 1990 = 100%		Percentage of 1990 level		
Raw agricultural products exports	38	64	63	56
Food products exports	89	80	108	67
Agricultural exports	73	77	96	64

Intensification of pastoral production

Most of the land in agricultural production is occupied by permanent meadows and pastures. The area under pasture has changed little in recent years except in the EU15 where the area has decreased by 8% over the period (FAO, 2006a). In Australia and New Zealand permanent pastures occupy over 80% of the agricultural area (Table 4.3).

Sheep stock numbers have decreased significantly in favour of cattle stock, and in the case of Australia and the USA, sheep numbers are currently 59% and 54% of their 1990 levels, respectively. On a per unit area of pasture basis, the EU15 is the highest producer of meat at about 153 kg/ha, compared to only 7 kg/ha in Australia, 48 kg/ha in the USA and 82 kg/ha in New Zealand. However, production of meat per hectare has declined significantly since 1993. Production in the EU15 has decreased by 89% compared to 1990 (Table 4.3), largely in response to the extensive reform of the Common Agricultural Policy in 1993, which was aimed at decreasing surplus production and dumping of agricultural product in the world market, brought about by a European budget crisis (Garzon, 2005).

The EU15 is also the highest producer of milk on a per unit area of pasture basis at 2223 kg/ha. New Zealand is the next largest producer at 1046 kg/ha. In Australia and New Zealand, dairy milk production increased to 156% and 193% of their 1990 production levels, respectively. The intensity of milk production per area of pasture has also increased dramatically in New Zealand (88%) and Australia (65%) in the past 15 years.

The intensification of pastoral farming in Australia and New Zealand is also reflected in the increasing quantity of fertilizer consumed in these countries (particularly for dairy farming). As of 2002, New Zealand recorded a dramatic increase in fertilizer application rates, growing by more than two and half times the application rates in 1990 (Table 4.3). The application rates of nitrogenous fertilizer in 2005 have increased further to about five times the 1990 levels, most of which is applied to intensive dairy pastures (Parliamentary Commissioner for the Environment, 2004; Austin *et al.*, 2006).

Table 4.3. Trends in land utilization, stock intensification, meat production and fertilizer use. (Adapted from FAO, 2006a.)

	Australia	EU15	New Zealand	USA
Agricultural area (1,000 ha)	445,149	138,324	17,235[a]	414,778
Permanent meadows and pastures (ha)	395,407	54,164	13,863[a]	237,600
Sheep stock (million head)	101	99	40	6
Cattle stock (million head)	28	77	10	96
Cow milk (million t)	10	120	15	80
Beef and sheep meat production (1,000 t)	2,757	8,264	1,140	11,405
Intensity beef and sheep meat production on pastures (kg/ha)	7	153	82	48
Intensity cow milk production on pastures (kg/ha)	26	2,223	1,046	338
Fertilizer consumption, at 2002 (1,000 t plant nutrient equivalent)	2,280	14,846	853	19,298
Change index 1990 = 100%		Percentage of 1990 level		
Agricultural area	96	93	99	97
Permanent meadows and pastures	95	92	103	99
Sheep stock	59	86	69	54
Cattle stock	120	84	120	100
Cow milk	156	95	193	120
Beef and sheep meat production	120	82	113	107
Intensity beef and sheep meat production on pastures	126	89	110	108
Intensity cow milk production on pastures	165	104	188	121
Fertilizer consumption, at 2002	196	76	249	104

[a]At 2003.

Fertilizer application in the EU15 has declined to 76% of their 1990 levels, reflecting the impact of the EU15 policies that reward the adoption of agri-environmental measures (AEM) to decrease environmental pollution (European Commission, 2005).

The area of agricultural land under irrigation has essentially remained unchanged since the late 1990s in the EU15 and the USA at about 13% and 5%, respectively. Less than 2% area of agricultural land is irrigated in New Zealand and Australia. In New Zealand, it is expected that the farmland requiring irrigation will increase as dairying expands into regions previously considered too dry for dairy farming (Parliamentary Commissioner for the Environment, 2004).

Accounting for positive and negative externalities of pastoral agriculture

Pastoral farming is essentially a business that should be financially viable and transferable to future generations. The financial sustainability of the farm business depends on the efficient use of inputs and other resources to supply products that can be traded to provide an acceptable livelihood to farmers. However, pastoral agriculture is 'multifunctional' (European Commission, 2005; Zander *et al.*, 2006), producing not only traded products, such as meat, dairy products and fibre, but also providing non-tradable goods and services, for which the market does not adequately provide compensation. These non-tradable goods may be positive or negative externalities to the production system. The positive externalities include services such as the provision of landscape values, maintenance of the socio-economic viability of rural communities and the provision of enhanced ecosystem services. Negative externalities include leaching of chemicals into sensitive water catchments, emission of green house gasses and soil erosion.

A range of policy approaches is used to minimize negative externalities and compensate for positive environmental impacts of pastoral agriculture. They range from the implementation of market-based instruments in Australia and the USA (King and Kuch, 2003; Young and McColl, 2005) to regulate non-point pollution, to the provision of financial support in the EU and the USA (NRCS, 2004; European Commission, 2005) to compensate for the cost of mitigation strategies and production loss. Market-based instruments are designed to ensure that factors of production such as land, water and other inputs used in pastoral farming system are priced appropriately to reflect their scarcity and the damage they cause to the environment. Market-based instruments also attempt to ensure that the farmer pays for the damage caused to the environment, based on the cost of remedial action, the 'polluter pays' principle, or through the purchase of pollution dischargeable allowances or credits from other farmers.

Calculating and implementing a system to account for the true cost of the environmental damage caused by pastoral farming has great methodological difficulties and legislative challenges (Pretty *et al.*, 2001; Evans, 2006). Systems to date have normally been restricted to computing the direct and the immediate downstream impacts (such as the effect on a stream passing through the farm). Other approaches such as life cycle assessment (LCA) attempt to capture the total impacts on the environment for producing and consuming a product (van der Werf and Petit, 2002; Sonesson, 2005). LCA is used to describe the complex interactions between the product, resources consumed in producing the product and the environment at all stages of the product's life cycle. This includes assessing the environmental impacts of extracting resources used to manufacture capital and intermediate inputs, the provision of services and waste disposal resulting from manufacturing the product.

Accounting for, and attributing, environmental pollution costs to individual pastoral farmers is also particularly challenging since farms are scattered on the rural landscape causing diffuse or non-point sources of pollution, the impacts of which are measured in waterways and lakes used by both the urban and rural communities. Also any mitigation strategies must be implemented by the farm business, which may mean a trade-off between environment and farm income. In this situation, financial and socio-political aspects are likely to take precedence over environmental ones (Filson, 2004; Evans, 2006), requiring that farmers carefully choose enterprises to manage the trade-off between financial and environmental outcomes. The ability of the farm business to successfully manage and accept these trade-offs, and even provide enhanced ecosystem services, depends on the mitigation strategies available to the farmer and the portfolio of available enterprises, the adoption of good farming or best farming practices and the availability of off-farm income-generating possibilities. At the whole-farm level, the farmer would need to match enterprises to land classes to optimize farm returns, minimize environmental emissions, maximize the provision of enhanced ecosystem services and satisfy desired financial targets and risks (Filson, 2004; Dake, *et al.*, 2005).

Providing financial support for producers

Domestic support policies

Assistance given to farmers, known as a producer support estimate (PSE), accounted for 33% and 18% of farm gate returns, respectively, in the EU15 and the USA in 2004. This is estimated to be US$134 billion and US$46 billion for the EU15 and the USA, respectively (OECD, 2005b). PSE for New Zealand and Australia is low at about 3% of farm gate returns.

PSE is:

> an indicator of the annual monetary value of gross transfers from consumers and tax payers to support agricultural producers, measured at farm gate level, arising from policy measures which support agriculture, regardless of their nature, objectives or impacts on farm production or income.
>
> (OECD, 2005a, p. 6)

The level and justification for agricultural support in the EU and the USA have evolved with time and become negotiating points during discussion on trade protection at various multilateral and bilateral trade negotiations (Baylis *et al.*, 2005; Garzon, 2005). In the EU and the USA, farmers may be compensated for additional costs and lost income for implementing environmental enhancing technologies (NRCS, 2004; European Commission, 2005). Currently in the EU, financial support is delivered using the fulfilment of 'agri-environmental measures' (AEM) commitments based on whether the measures are related to: (i) productive land management; or (ii) non-productive land management. They may include a socially based dimension that farmers perform a land stewardship role as a public service and are required in order to maintain rural community cohesion and vitality as the agricultural labour force declines (Garzon, 2005).

The AEM target both mitigating negative and enhancing positive externalities of agriculture. These are:

1. AEM related to productive land management:

Addressing negative externalities

- Input decrease (e.g. fertilizers and plant protection products).
- Extensification of livestock.
- Conversion of arable land to grassland and rotation measures.
- Water decrease measures.

Promote positive externalities

- Organic farming.
- Undersowing and cover crops, strips (e.g. farmed buffer strips) and preventing erosion and fire.
- Actions in areas of special biodiversity/ nature interest.
- Genetic diversity.
- Maintenance of existing sustainable and extensive systems.
- Farmed landscape (e.g. farming systems that enhance the characteristic landscape).

2. AEM related to non-productive land management:

- Set-aside – managed for environmental purposes.
- Upkeep of abandoned farmland and woodland.
- Maintenance of the countryside and landscape features.
- Public access – to agricultural land of environmental interest (Baylis *et al.*, 2005; European Commission, 2005).

The AEM are co-funded by the EU and member states, designated at the local, regional or national levels, and must provide benefits in excess of what is considered to be good farming practice. Funded AEM must meet any statutory or regulatory requirements, in any particular cases that require the adherence to the polluter pays principle.

Support for the environment in the USA is indicated in the environmental quality incentive program (EQIP) of the US Farm Bill, and is currently focused primarily at minimizing negative externalities of farming (NRCS, 2004; Baylis *et al.*, 2005). The national funding priorities of EQIP are:

- Decrease non-point source pollution (nutrients, sediment, pesticides, etc.) consistent with total maximal daily loads.
- Decrease groundwater contamination and conservation of groundwater and surface water resources.
- Decrease emissions.
- Decrease soil erosion.
- Promotion of at-risk species habitat.

International trade implications

Domestic support payments to producers have the potential to change production and prices, and therefore affect the competitiveness of countries in world markets. The regulation of multilateral trade is governed by the General Agreement on Tariffs and Trade (GATT) under the auspices of the World Trade Organization (WTO). In 1994, after an intensive period of negotiations, the WTO negotiated the Uruguay Round Agreement on Agriculture (URAA) to reduce current and future domestic trade distortion policies. In subsequent WTO discussions, countries agreed to include non-trade concerns, such as food security and environmental protection policies, in future trade negotiations (Bohman *et al.*, 1999; Garzon, 2005).

Domestic support policies are classified in 'boxes' based on their degree of trade distortion (WTO, 2007):

- Amber box measures contain trade-distorting policies. The level of distortion is measured by an aggregated measure of support (AMS) across all commodities for a country. Developed countries have committed to reduce their AMS by 20% over a period of 6–20 years from the 1986/88 base period. Developing and least-developed countries have arranged lower exemptions over a longer period.
- Green box measures contain permitted policies without limitation since they are not considered trade-distorting.
- Blue box measures contain policies that are not subject to Amber policies, and cover direct payments to programmes that limit production.

- Exceptions that do not fit any other boxes are called 'de minimis' exemptions where support must be less that 5% of the total value of production.

AMS for the countries and the EU region used here is shown in Table 4.4. The AMS for New Zealand, Australia and the USA is less than 40% of their commitment levels. Support for the EU is below 80%. When a country reaches 100% AMS (i.e. the Amber box is full), then it would not be able to create any new trade-distorting policies.

Creating Change: Promoting the Adoption of Environmental Practices

The issues discussed above have prompted policies and programmes designed to mitigate the environmental issues facing pastoral farmers. Most often this takes the form of encouraging farmers to voluntarily adopt practices developed by researchers to address environmental issues. Fencing waterways to exclude stock is a common practice recommended in the USA (Giuliano, 2006), Australia (Fielding *et al.*, 2005; Natural Resources Science, 2006) and New Zealand (Legg, 2004). Other examples include decreasing the use of phosphorus fertilizer and deferring irrigation of effluent on to pastures for New Zealand farmers (Monaghan *et al.*, 2007b). However, adoption rates for these types of practices vary (Pannell *et al.*, 2006). In many instances the rate of adoption may be considered too low to ensure long-term environmental sustainability. Education and awareness programmes to change landholder attitudes towards the environment and thus their behaviour are often touted as the solution; however, these have varying degrees of success (Vanclay, 1992; Vanclay and Lawrence, 1994; Pannell, 1999).

Factors influencing adoption

Research exploring the influences on adoption of environmental practices in agriculture has typically focused on understanding those characteristics of a farmer or his/her system that indicated that he/she was more or less likely to adopt a

Table 4.4. Actual support (AMS) as a percentage of commitment levels, latest year during 1995–1998 (Amber box). (Adapted from Nelson *et al.*, 2001.)

Commitment levels (%)	Countries
0–19	Canada, Colombia, Costa Rica, Czech Republic, Mexico, Morocco, New Zealand, Poland
20–39	Australia, Brazil, Cyprus, USA, Venezuela
40–59	Hungary
60–79	European Union, Iceland, Japan, Slovak Republic, Switzerland, Thailand
80–100	Argentina, Israel, Korea, Norway, Slovenia, South Africa, Tunisia

particular technology. Depending on the discipline of the researchers these factors range from farm size, education and age, to the environment in which landholders learn about a new technology. For example, Eady and Fisher (2004) cite 'being part of a group devoted to dealing with the problem' and 'learning from each other' as enabling dairy farmers in Queensland, Australia, to change on-farm practices. However, innovations will typically be adopted if they are superior to the practice currently in place and are easy to test (Pannell *et al.*, 2006), if they offer commercial value or will maintain long-term productivity (Guerin and Guerin, 1994), or decrease costs or increase revenue, i.e. have clear economic benefits (Lambert *et al.*, 2006; Macgregor and Warren, 2006). Other major drivers, particularly for adopting some conservation practices, are to meet compliance requirements or for land stewardship reasons (Lambert *et al.*, 2006).

Other researchers have found that landholders' assessment of recommended practices and their own and others' experience of them influence adoption (Cary *et al.*, 2002). Sharing management decisions with their spouse, seeking results from surveys of farmers, talking to other farmers about changes in management on-farm and having the belief that scientific experimentation is appropriate in privately owned catchments also influence adoption (Habron, 2004).

Education is often cited as a factor influencing adoption. Supalla *et al.* (1995), Upadhyay *et al.* (2003) and Fuglie and Kascak (2001) all cite education along with farm size (Fuglie and Kascak, 2001; Upadhyay *et al.*, 2003) and technical and environmental knowledge (Supalla *et al.*, 1995; Upadhyay *et al.*, 2003) as key parameters for adoption of new practices. Personality traits and intelligence (Austin *et al.*, 2001), including locus of control or internal motivation (McNairn and Mitchell, 1992) as well as response to risk (Shrapnel and Davie, 2001), have also been identified as important.

Land characteristics as well as other farm system features, such as having a diversified farm, can also influence adoption of environmentally friendly practices (Kim *et al.*, 2005). Kaine and Lees (1994) found that the stage of farm development dictated the order in which innovations were adopted, reflecting the need to integrate new technologies into an existing farm system. Frank (1995, 1997) also found that North Queensland cattle managers adopt practices in sets, so each practice must 'fit' with the next.

However, some of these studies neglect a basic fact; slow rates of adoption reflect a rational response in circumstances where innovation adoption is undesirable (Frank, 1995). In a recent and comprehensive review of research on the adoption of conservation agriculture, particularly no-till systems among arable farmers, Knowler and Bradshaw (2007) concluded that there is an absence of universally significant factors that influence adoption. This means that it is difficult to develop general policies for voluntarily addressing environmental issues, with Knowler and Bradshaw (2007) suggesting that regionally specific research to guide adoption would be of more use in the future. They question the assumption that a universal explanation of adoption of conservation agriculture is possible.

Barriers to adoption of environmental or conservation practices have also been considered. De Buck *et al.* (2001) found it was hard to distinguish between 'non-adopters' and 'adopters' because of a policy framework in place that meant all farmers had to make some changes to current practice. Other research has revealed differences in perception. Beliefs about whether climate change was a reality influenced the adoption of technology to mitigate this issue (Weber, 1997). Farmers have been found to have no perception of the link between what happens on-farm and what happens in a catchment: for example, the impact of farming on water quality (Dutcher *et al.*, 2004; Macgregor and Warren, 2006). In response to this, Vanclay (2004) questions the way in which environmental degradation is portrayed in the media or in extension material. Often the view is extreme and as many farmers do not see this on their properties they may not believe they have a problem (Vanclay, 1992, 2004).

Farmer attitudes are often cited as needing to change; however, Vanclay (2004, p. 216) asserts that 'farmers' attitudes are not the problem'. Instead, Vanclay (2004) articulates the issue as one of conflicting views on the best way to manage a farm particularly with respect to environmental impacts. Researchers routinely find that there is little difference in terms of environmental attitudes between farmers who have adopted sustainable practices and those who have not (see e.g. Cocklin and Doorman, 1994 and Jones *et al.*, 1995). Barr (1999) and Cary *et al.* (2002) conclude that pro-environmental values have a relatively minor influence on adoption of sustainable practices in farming. In effect, awareness of the environmental issue and potential solutions may not mean that it is economically rational to change current practice (Vanclay, 2004). Cary *et al.* (2002) characterized the available resource management practices in Australia in terms of their environmental sustainability and economic viability, indicating that those practices considered economically sustainable but environmentally unsustainable will have to be replaced with alternative practices, or regulation would be needed to ensure that farmers use these practices. Similarly Pannell *et al.* (2006) concluded that the development of practices that deliver a real advantage to farmers may be the solution to increasing the rate of adoption. They highlight the challenge 'to find or develop innovations that are not only good for the environment, but also economically superior to the practices they are supposed to replace' (Pannell *et al.*, 2006, p. 1421).

Cultural issues

Pastoral agriculture in New Zealand, Australia and to some extent the USA was based on British culture and practices. The way in which indigenous people in these regions lived and used the land was largely ignored. New species were introduced for grazing, although many of these under New Zealand and Australian conditions did not grow or survive as they did in Britain (Barr and Cary, 1992; Star and Brooking, 2006). Gradually other influences from Europe and beyond began to shape agriculture in Australia, New Zealand and the USA and provide successful alternatives. However, in the process indigenous people were 'disempowered by agricultural myths which render them and their interactions with the landscape invisible' (Head, 1999, p. 143).

Europeans arrived in New Zealand to find a farm-based society. Maori had extensive gardens with a range of plants (Banner, 1999; Bassett *et al.*, 2004). This evidence of the potential for civilization, in the eyes of the settlers, heavily influenced the subsequent debate on colonization of New Zealand (Banner, 1999).

> This meant on the one hand, that Maori could be more easily assimilated to the English way of life.... But this same advancement simultaneously meant, on the other hand, that the process of acquiring land would not be as simple as it had been in colonies like Australia, where the perceived absence of agriculture had implied the absence of any basis for recognising aboriginal property rights to land.
>
> (Banner, 1999, p. 818)

However, the assumption that Maori could be directed along the path to civilization masked differences in perceptions of property rights. The English assigned rights on a geographical basis. In contrast, Maori property rights were based on resource use (Banner, 1999). The Maori perspective recognizes that people and the environment are connected (Harada and Glasby, 2000; Ngato, 2004) and that possession of a right to use a resource in one geographic space did not necessarily mean that other rights in that space were also available (Banner, 1999). This set the scene for conflict as European colonizers bought land from Maori and began large-scale land clearing and development, along with the introduction of exotic species (Harada and Glasby, 2000; Ngato, 2004). The solution was to 'civilize' Maori by assigning Maori individual property rights (Banner, 1999).

To that end the Treaty of Waitangi was signed in 1840. The Treaty gave Britain sovereignty over New Zealand, recognized Maori as having ownership of the land but gave the British the right to buy land from Maori if they wished to sell, and granted Maori the same rights and privileges as other British people. However, there were differences between the English and Maori versions of the Treaty which remain sources of conflict. In 1862 the Native Land Court was established, becoming active in 1865 (Banner, 1999; Sissons, 2004). The Native Land Court assigned individual Maori owners to surveyed land (Banner, 1999; Sissons, 2004). This decreased the authority held by traditional leaders and thus traditional rights to land. Over time the area of Maori-owned land decreased significantly, decreasing the influence of the Maori system of rights to land through tribal allegiance.

Conflict over land continues in New Zealand. Recently, with the introduction of the Resource Management Act, Maori interests have been more formally recognized (Ngato, 2004). The Act provides a policy framework for the sustainable management of land, water and air in New Zealand.

In Australia, Aboriginal rights to land have routinely been ignored although Aborigines have been the principal source of labour in the pastoral industry in northern Australia for many years (Head, 1994, 1999). European settlers often failed to acknowledge any Aboriginal rights because they were not agriculturalists and were not considered to be able to take part in any debate on equal terms (Head, 1994; Banner, 1999). However, in 1982, Eddie Mabo, along with others from Murray Island (in the Torres Strait between Papua New Guinea and Australia), brought action against the Queensland government to determine their legal right to the islands (Brennan, 1995). The outcome invalidated the declaration of terra nullius, 'land belonging to no one', effectively recognizing a form of native title (Brennan, 1995; Godden, 1999). In response, the Australian federal government introduced the Native Title Act in 1993, which was designed to outline a process for determining native title claims (Godden, 1999). However, the Mabo decision did not determine the impact of native title on pastoral leases, primarily because there was no pastoral land on the Murray Islands (Brennan, 1995; Godden, 1999).

Subsequently, the Wik people of Cape York Peninsula brought action against the Queensland government to determine the right of access to pastoral land. The judgment, that native title may not be extinguished by pastoral leases and in some cases may coexist (Godden, 1999), resulted in amendments to the Native Title Act. Although there are clear precedents around the world for coexisting rights to land – for example, Godden (1999) cites walkers'access to pathways in England and Wales – pastoralists felt that the Wik decision created uncertainty. The subsequent debate has been acrimonious.

The picture outlined above from Australia and New Zealand illustrates a significant cultural conflict that has occurred over many years. However, more recently the cultural landscape of pastoral farmers, albeit having displaced others in the process of building this landscape in Australia and New Zealand, is increasingly under threat through pressure from different stakeholders. In an environment where farmers are under pressure to intensify, financial and socio-political aspects are likely to take precedence and environmental concerns may become less of a priority (Filson, 2004; Evans, 2006). Pastoral agriculture is facing increasing pressure to demonstrate food safety and quality, as well as respond to environmental concerns as urban populations increase (Hall *et al.*, 2004).

There are many different stakeholders, all with firm ideas of what the farming landscape should look like or encompass. For example, urban dwellers may distinguish between production and 'natural' systems (Egoz *et al.*, 2001). They may find that modern industrial agriculture does not allow for maintenance of the landscape (Soliva, 2007). Therefore, conflict over the role of farming occurs (Moore-Colyer and Scott, 2005). While farmers are concerned that environmental initiatives may threaten the viability of their farm, environmentalists claim that farming is ruining the environment (Egoz and Bowring, 2004).

The contrast between the USA and the EU in this regard illustrates these pressures. American agriculture has been almost entirely developed in the industrial agriculture vein, corporations and subsidies being a large part of the landscape, in order to provide cheap food (Thompson, 2001). Only recently have regulations and laws begun to address some of the environmental issues created. However, in the EU, subsidies provide income for farmers, provided they maintain the landscape (Falconer, 2000; Morris, 2004). This has created different pressures as farmers adjust from being food producers to landscape maintainers: as the value of their profession – farming, producing food – has changed (Deuffic and Candau, 2006).

Implications for change

To date there has been a focus on voluntary adoption of environmental practices using transfer of technology and/or learning approaches; see for example Lawrence *et al.* (2000). There has been mixed success with this approach (Pannell *et al.*, 2006), although a range of factors critical to successful adoption have been identified. The most recent reviews of this (Lambert and Sullivan, 2006; Lambert *et al.*, 2006; Pannell *et al.*, 2006) identify the need to develop appropriate practices, i.e. practices that are economically superior and deliver the on-farm benefits required to make a change.

In the EU, AEM are being used to ensure that environmental practices are adopted, as well as other initiatives in the form of regulations. For example, the Nitrates Directive has facilitated the development of Nitrate Vulnerable Zones where best management practices such as limiting the amount of nitrogen applied have to be adopted (Macgregor and Warren, 2006). However, the success of regulations relies heavily on the integration of the views of farmers, scientists and other stakeholders (Klerkx *et al.*, 2006).

The need to change farming systems to adopt environmental practices has yet to have universal acceptance. Understanding farming systems and contexts in which change needs to occur is essential in order to decrease surprises and increase success (Kaine and Johnson, 2004). Kaine and Higson (2006) conclude that understanding farmers' responses to natural resource policy initiatives is critical because of the potential for unanticipated responses that are not consistent with policy objectives.

The dilemma faced by pastoral agriculture – increasing demand for pastoral products along with increasing pressure to farm in a way that has little or no impact on the environment – is not unique. Similar situations are encountered in other disciplines: for example, managing recreation use of wilderness areas. The dilemma faced by these resources managers is that there is increasing desire to visit wilderness areas, as evidenced by increasing numbers of people entering these areas, but impacts on wilderness areas will begin to occur as soon as people are allowed to visit (McCool and Lime, 2001). For many years resource managers used a carrying capacity model to manage the impact of visitors on wilderness areas (McCool, 1996). However, the carrying capacity model failed to account for the negative impacts on wilderness areas caused by the presence and activity of visitors (McCool, 1996; McCool and Lime, 2001). McCool and Lime (2001) conclude that allowing visitors means there will be a need for some trade-offs.

The limits of acceptable change (LAC) system was developed by Stankey *et al.* (1985) to try and address these trade-offs. The LAC approach allowed resource managers to answer the question: 'what resource and social conditions are appropriate (or acceptable), and how do we attain these conditions?' (McCool, 1996, p. 2). The LAC process has now been used extensively within the USA for managing designated wilderness areas (McCool and Cole, 1998), as well as in other parts of the world such as the Great Barrier Reef World Heritage Area (Shafer and Inglis, 2000). LAC has become a useful tool in situations where two goals are in conflict and trade-offs need to be managed (Cole and McCool, 1998).

The LAC approach does not provide simple answers. However, it does allow for discussion between stakeholders (McCool and Cole, 1998). Those associated with pastoral farming could start a similar discussion on the trade-offs involved. To paraphrase Cole and McCool (1998): 'What environmental conditions are appropriate for, and as a result of, pastoral farming and associated communities and how do we achieve these?' The ensuing discussion could help resolve the dilemma faced by pastoral agriculture and indicate the types of changes required.

Conclusion

This chapter has highlighted that pastoral agriculture in Australia, New Zealand, the USA and the EU faces significant challenges to minimizing detrimental environmental and social impacts of farming while attempting to meet the growing global demand for pastoral products. These challenges range from accounting for positive and negative externalities of pastoral agricultural production to designing cost-effective environmental mitigation technologies that farmers are able to adopt. In situations where farmers are unable to meet acceptable environmental emissions through the use of what is considered to be good or best farming practice, society is faced with the problem of deciding and funding the desirable values that it believes farmers provide. The USA and the EU have decided that a mixture of regulation and significant financial support to producers reflects the high importance they place on the environmental and social services they receive from agriculture. New Zealand and Australia, because of their year-round pasture grazing system, believe it best to farm with the minimum of subsidies.

Whether the EU and the USA are able to sustain their significant financial support to agriculture is unknown. Also, whether New Zealand and Australia are able to expect farmers to continue to farm under increasing environmental scrutiny without compensating them for the environmental and social services they provide is yet to be determined. From our analysis, the policy positions adopted by the various countries can be enhanced by research and the availability of the appropriate technologies that farmers can adopt to meet acceptable environmental and social outcomes.

Case Studies

Case study #1: Eco-efficiency of intensification of milk production in New Zealand – life cycle assessment (LCA)

Claudine Basset-Mens, Stewart Ledgard and Mark Boyes (AgResearch Limited, Ruakura Research Centre, East Street, Private 3123, Hamilton, New Zealand)

Eco-efficiency of milk production measures the ability of a dairy farm system to deliver milk while minimizing its impacts on the environment. This eco-efficiency is presented as the environmental impact per kilogram of milk or per hectare of land use.

LCA is a methodology to assess the eco-efficiency of a product. LCA is used to estimate the potential environmental impact of a product through all stages of production by evaluating resources consumed and emissions lost to the environment. In this case study, we report the environmental impacts in terms of global warming potential, eutrophication potential, acidification potential, energy use and land use. A more detailed analysis can be found in Basset-Mens *et al.* (2008).

The eco-efficiencies of a year-round pasture-based dairy system at three levels of intensification in the Waikato region of New Zealand we re compared with the average New Zealand dairy farm and published data from Europe. The three contrasting intensification systems were: (i) low-input system (LI); (ii) nitrogen fertilizer (NF) system with no brought-in feed; and (iii) nitrogen fertilizer and maize supplemented with brought-in feed system (NFMS).

The processes evaluated in the LCA included: (i) processes up to the farm – production and delivery of crop and pasture farm inputs, all capital items, production of feed supplement and off-farm pasture production for replacement cows; and (ii) milk production on the farm, including on-farm pasture production, herd management and milk extraction.

Data on emissions of inputs used in each of the processes were obtained from a range of sources and models. In particular, models such as the OVERSEER nutrient budget model (Wheeler *et al.*, 2003) and IPCC-NZ methodology (de Klein et *al.*, 2001) were used to create inventory data on nitrogen, phosphorus and greenhouse gas emissions. These were combined with published inventory data sources for industrial processes and fertilizer manufacture.

Due to favourable climate conditions and long-term perennial ryegrass/white clover pasture, the New Zealand dairy farm systems studied in this case study were demonstrated to be very eco-efficient per kilogram of milk compared to European systems. However, per hectare of land use their eco-efficiency was similar for global warming potential. For eutrophication, acidification and energy use, the average New Zealand system still presented some advantages on the per hectare of land-use measure, but those advantages may be severely jeopardized by further intensification as demonstrated by the results for NF and NFMS systems. LI presented a very high eco-efficiency compared to all intensified systems studied, in terms both of milk production and land use. This finding is valuable from an LCA point of view since it is rare to find a systematic advantage to a given system whatever the functional unit, especially when comparing different degrees of intensification (Basset-Mens and van der Werf, 2005). This can be explained by the fact that LI production relies almost exclusively on non-limiting natural resources in the New Zealand context (clover N2 fixation, sun, rainfall, well-structured soils). Both NF and NFMS systems had similar eco-efficiencies for all parameters except for energy use, which was inefficient for the NFMS system. All studied New Zealand systems indicated potential areas for improvement where new technologies available for dairy farms might play a role. Finally, the comparison with European studies deserves a more comprehensive study with harmonized methodology and assumptions across countries.

Case study #2: Tackling declining water quality in New Zealand – dairy best practice catchments

Adoption of environmental best practice on dairy farms is becoming increasingly important in New Zealand as the quality of our fresh water declines (Wilcock *et al.*, 1999). In order to address this issue a large research project investigating the economic and environmental performances of dairy farming in contrasting catchments has been initiated. This project is interdisciplinary and has a focus on developing cost-effective on-farm practices that mitigate any damaging environmental impacts of dairy farming. Five catchments within New Zealand which have a high proportion of their area occupied by dairy farms are part of this research programme (Wilcock *et al.*, 2007). Field measurements, farm management surveys and farm systems modelling have identified particular land management practices that appear to be key sources of contaminants in catchment waterways (Monaghan *et al.*, 2007a).

The project has involved social research to determine those factors that influence the dairy farmers to adopt some of the on-farm practices developed through research. Initial work focused on the following practices: excluding stock from waterways, decreasing phosphorus use, improving soil macroporosity and effluent management. The approach taken is outlined in Bewsell and Kaine (2006). The results obtained are consistent with the conclusions from Pannell *et al.* (2006).

Excluding stock from waterways

A number of factors were identified that influenced farmers' decisions on fencing streams and other waterways. These factors were centred on management of stock, such as stream fencing, because the stream is a boundary to ensure stock stay on the property. Fencing was also undertaken when redeveloping property.

Effluent management, fertilizer use, managing wet soils

Farmers choose an effluent management system based on their herd requirements and location. However, when managing the fertilizer requirement for the property, farmers' perceptions of the difference in pasture yield on areas where dairy shed effluent was irrigated became important. There appeared to be a consistent association between farmers' perceptions of whether effluent made a difference to grass growth in a paddock and their management of fertilizer in every catchment.

Farmers did not tend to see phosphorus as a separate issue deserving special treatment from other fertilizers. Phosphorus was part of the fertilizer mix going on to the farm. Some farmers had responded to advice recommending a decrease in phosphorus application, particularly where this saved money. Working with the fertilizer companies and farm advisors may have more effect than working directly with farmers.

Managing wet soils was an issue for all farmers in all catchments. However, many are faced only with pugging problems in winter and have rules of thumb that work for their property, depending on the timing and severity of waterlogging.

Adoption of the environmental best practices

Social research highlighted the importance of understanding context when considering adoption of environmental best practice and the need to provide practical solutions to environmental problems that also address specific on-farm needs. Farmers' decisions about the environmental practices are primarily based on a systematic evaluation of their production context and the management options that are available. This suggests that the choices farmers make in regards to adoption of these practices are not strongly influenced by their attitudes to sustainability and the environment.

The results from the social research were used to inform the development of an extension programme. Farm plans were identified as useful tools for farmers tackling environmental initiatives on their properties. Monaghan *et al.* (2007a) outline the approach taken to develop farm plans for each catchment. The plans helped define some of the obvious issues to be addressed and allowed farmers to develop a schedule for the tasks identified. For example, in the one of the North Island catchments, a comprehensive riparian fencing and planting planning initiative had been supported by the local regional council, and so farm plans became part of this effort. In another North Island catchment, farm plans were developed with the local regional council, dairy company field staff and research providers. In the South Island catchments, a simpler approach was taken where farmers were encouraged to develop a plan to fence streams and adopt one other relevant best management practice (Monaghan *et al.*, 2007a).

An important result of the project and associated activities carried out within the catchments has been the recognition that 'one size doesn't fit all' vis-à-vis the adoption of appropriate best management practices to minimize contaminant losses to water (Monaghan *et al.*, 2007a). Although monitoring information is yet to show discernible change in stream water quality in most of the catchments, the trends for one of the North Island catchments indicate an improvement over the last 10 years (Wilcock *et al.*, 2007). Ongoing surveys are demonstrating that on-farm practices are gradually changing. For example, phosphorus inputs are gradually declining across the catchments (Monaghan *et al.*, 2007a).

Case study #3: Addressing salinity issues in the dairy industry in Australia – the Macalister Irrigation District

Geoff Kaine (Department of Primary Industries Victoria Australia) and Denise Bewsell (AgResearch Limited, Lincoln New Zealand)

The Macalister Irrigation District (MID) in Gippsland, Victoria, covers 53,000 ha of land in central Gippsland, south-east of Melbourne. Approximately 33,500 ha is irrigated, 90% of which is under pasture (Southern Rural Water, 2007). There are approximately 500 dairy farmers in the area, most of who rely on flood irrigation. The first signs of salinity in the area appeared in 1952, not long after irrigation began (Southern Rural Water, 2007). Since then, drainage and bores have been used to decrease groundwater and alleviate salinity problems.

Salinity problems are prevalent in Australia. Soils in Australia are naturally saline (Pannell, 2001). Many native Australian plants have adapted to these conditions. European settlers cleared large tracts of land for agriculture and began irrigation development in some areas, setting the scene for the emergence of salinity problems (Pannell, 2006). There are two forms of salinity found in Australia: the first is irrigation-induced, and the second is dryland salinity. Both have an impact on the pastoral industry. Irrigation-induced salinity occurs because of over-irrigation or the irrigation of unsuitable soils. This causes the water table to rise, bringing salt into the topsoil and affecting plant growth. Dryland salinity is associated with land clearing; decreasing the number of deep-rooted perennial plants and replacing them with shallow-rooted annual plants that change the hydrology leading to a rising water table.

The MID has irrigation-induced salinity. In 2001, a study was undertaken to investigate the use of flood irrigation in the area (Kaine and Bewsell, 2001, 2002). The research was part of an attempt to improve irrigation management and water-use efficiency on dairy farms in order to address the district's salinity problem, and the emerging problems of decreasing water allocations and a decline in water quality in the region, due to nutrient losses from these systems. The research involved extensive interviewing and a mail survey.

In general, farmers interviewed faced two alternatives when it came to increasing the efficiency of irrigation water use. One was to use less water either by laser grading (to long, wide irrigation bays) or by installing spray irrigation. The other alternative was to use the same amount of water but increase milk production per hectare, perhaps by using feed supplements to increase stocking rates. The choices made depended on a number of factors. The first was the relative importance of flood and spray irrigation on the farm. Where a large proportion of the farm was spray-irrigated, there was little opportunity to decrease water use per hectare. The second factor was the proportion of the dairy farm that could be laser-graded. If large, then there was potential to significantly increase water-use efficiency. However, if small, increased water-use efficiency could be achieved by installing spray irrigation. The most attractive depended on the soil type and the amount of labour available.

Where a large proportion of the farm had been laser-graded, primarily to save time, there was some potential to increase efficiency by installing spray irrigation. Farmers in this situation were unlikely to adopt any form of spray irrigation that was labour-intensive (such as lateral move sprays). Centre pivot or linear move irrigation systems were more attractive. However, the gains in water-use efficiency would need to be quite substantial to justify investing in these systems.

The information gathered from interviews and data collected from 30% of the dairy farmers in the MID via a mail survey was used to classify dairy farmers in the MID into four segments. The first segment represented 50% of survey respondents. These farmers relied on flood irrigation and had undertaken an extensive programme of laser grading on their property. This was motivated by a need to decrease the amount of time spent irrigating (decreasing the number of bays and improving the layout) and a desire to save water by maximizing flow rates down the bay.

The second segment represented 26% of respondents. They also relied on flood irrigation; however, they had only laser-graded a small proportion of their land. The laser grading that had been done was to save time irrigating, improve bay layout and grade land that had not been properly graded. Farmers in the third segment represented 17% of respondents. These farmers had a mix of flood and spray irrigation, as they had a high proportion of land that was unsuitable for laser grading. Farmers in the fourth segment represented 7% of respondents. These farmers had light permeable soils and undulating land that was unsuitable for laser grading and relied on spray irrigation instead.

The results suggested that widespread adoption of spray irrigation in the MID was unlikely. For most farmers, laser grading was and would continue to be the most effective means of decreasing water use. The particular irrigation technology used by a farmer depended on the physical and structural characteristics of their farms. Farmers also believed they had a high degree of competency in managing irrigation on their farms. They believed they could become proficient in managing a new irrigation system within a season; two at the most.

The research illustrated the importance of understanding the context in which farmers were making decisions, in order to understand their response to extension and policies meant to mitigate salinity in the MID. This prevented surprises when responses were not consistent with policy objectives (Kaine and Higson, 2006).

References

Austin, D., Cao, K. and Rys, G. (2006) Modelling nitrogen fertiliser demand in New Zealand. In: *Proceedings of the New Zealand Agriculture and Resource Economics Society Conference.* NZARES, Nelson, New Zealand.

Austin, E.J., Deary, I.J. and Willock, J. (2001) Personality and intelligence as predictors of economic behaviour in Scottish farmers. *European Journal of Personality* 15, S123–S137.

Banner, S. (1999) Two properties, one land: law and space in nineteenth-century New Zealand. *Law and Social Inquiry* 24, 807–852

Barr, N. (1999) Social aspects of rural natural resource management. In: *Proceedings of the Outlook 99 Conference.* ABARE, Canberra, pp. 133–140.

Barr, N. and Cary, J. (1992) *Greening a Brown Land: The Australian Search for Sustainable Land Use.* Macmillan, South Melbourne, Australia.

Basset-Mens, C. and van der Werf, H.M.G. (2005) Scenario-based environmental assessment of farming systems: the case of pig production in France. *Agriculture, Ecosystems and Environment* 105, 127–144.

Basset-Mens, C., Ledgard, S. and Boyes, M. (2008) Eco-efficiency of increasing intensification scenarios of milk production in New Zealand. *Journal of Ecological Economics*, doi: 10.1016/j.ecolecon.2007.11.017. Available at: http://www.sciencedirect.com/science?_ob=ArticleURL&_udi=B6VDY-4RJJRJD-3&_user=883069&_rdoc=1&_fmt=&_orig=search&_sort=d&view=c&_acct=C000047280&_version=1&_urlversion=0&_userid=883069&md5=2cccae6c47c95cf5074af3f24f1124ed (in press).

Bassett, K.N., Gordon, H.W., Nobes, D.C. and Jacomb, C. (2004) Gardening at the edge: documenting the limits of tropical Polynesian kumara horticulture in southern New Zealand. *Geoarcheaology* 19, 185–218.

Baylis, K., Rausser, G. and Simon, L. (2005) The agri-environmental nexus in the US and EU. Institute of Governmental Studies, University of California. Available at: http://igov.berkeley.edu/presentations/series1/No19_Simon.pdf

Bewsell, D. and Kaine, G. (2006) Adoption of environmental best practice amongst dairy farmers. In: *Practice Change for Sustainable Communities: Exploring Footprints, Pathways and Possibilities: APEN 2006 International Conference.* APEN, La Trobe University, Beechworth, Victoria, Australia.

Bohman, M., Cooper, J., Mullarkey, D., Normile, M.A., Skully, D., Vogel, S. and Young, E. (1999) The use and abuse of multifunctionality. Available at: www.ers.usda.gov/briefing/WTO/PDF/multifunc1119.pdf

Brennan, G. (1995) Aboriginal land claims: an Australian perspective. In: *The 7th International Appellate Judges Conference, Ottawa.*

Cary, J.W., Webb, T.J. and Barr, N.F. (2002) Understanding landholders' capacity to change to sustainable practices. Insights about practice adoption and social capacity for change. Bureau of Rural Sciences, Canberra.

Cocklin, C. and Doorman, P. (1994) Ecosystem protection and management in New Zealand: a private land perspective. *Applied Geography* 14, 264–281.

Cole, D.N. and McCool, S.F. (1998) Limits of acceptable change and natural resources planning: when is LAC useful, when is it not? In: McCool, S.F. and Cole, D.N. (eds) *Limits of Acceptable Change and Related Planning Processes: Progress and Future Directions.* Department of Agriculture, Forest Service, Rocky Mountain Research Station, Missoula, Montana, pp. 69–71.

Dake, C.K.G., Mackay, A.D. and Manderson, A.K. (2005) Optimal trade-offs between financial and environmental risks in pastoral farming. In: Zerger, A. and Argent, R.M. (eds) *MODSIM 2005 International Congress on Modelling and Simulation.* Modelling and Simulation Society of Australia and New Zealand, Melbourne, Australia, pp. 2421–2427.

de Buck, A.J., van Rijn, I. and Wossink, G.A.A. (2001) Farmers' reasons for changing or not changing to more sustainable practices: an exploratory study of arable farming in the Netherlands. *Journal of Agricultural Education and Extension* 7, 153–166.

de Klein, C.A.M., Sherlock, R.R., Cameron, K.C. and van der Weerden, T.J. (2001)

Nitrous oxide emissions from agricultural soils in New Zealand – A review of current knowledge ad directions for future research. *Journal of the Royal Society of New Zealand* 31, 543–574.

Deuffic, P. and Candau, J. (2006) Farming and landscape management: how French farmers are coping with the ecologization of their activities. *Journal of Agricultural and Environmental Ethics* 19, 563–585.

Dutcher, D.D., Finley, J.C., Luloff, A.E. and Johnson, J. (2004) Landowner perceptions of protecting and establishing riparian forests: a qualitative analysis. *Society and Natural Resources* 17, 319–332.

Eady, J. and Fisher, B. (2004) On-farm adoption model resulting in improved environmental actions and outcomes. *Australian Journal of Dairy Technology* 59, 142–144.

Egoz, S. and Bowring, J. (2004) Beyond the romantic and naïve: the search for a complex ecological aesthetic design language for landscape architecture in New Zealand. *Landscape Research* 29, 57–73.

Egoz, S., Bowring, J. and Perkins, H.C. (2001) Tastes in tension: form, function, and meaning in New Zealand's farmed landscapes. *Landscape and Urban Planning* 57, 177–196.

European Commission (2005) Agri-environment measures. Overview on general principles, types of measures, and application. Available at: http://ec.europa.eu/agriculture/publi/reports/agrienv/rep_en.pdf

Evans, R. (2006) Sustainable practices to limit soil erosion: a review and discussion. *CAB Reviews: Perspectives in Agriculture Veterinary Science Nutrition and Natural Resources* 1, no. 30.

Falconer, K. (2000) Farm-level constraints on agri-environmental scheme participation: a transactional perspective. *Journal of Rural Studies* 16, 379–394.

FAO (2006a) Statistical databases. Available at: http://faostat.fao.org/default.aspx

FAO (2006b) World agriculture: towards 2030/2050. Prospects for food, nutrition, agriculture and major commodity groups. Available at: http://www.fao.org/es/esd/AT2050web.pdf

Fielding, K.S., Terry, D.J., Masser, B.M., Bordia, P. and Hogg, M.A. (2005) Explaining landholders' decisions about riparian zone management: the role of behavioural, normative, and control beliefs. *Journal of Environmental Management* 77, 12–21.

Filson, G.C. (2004) *Intensive Agriculture and Sustainability: A Farming Systems Analysis. Sustainability and the Environment*, UBC Press, Vancouver, Canada.

Frank, B.R. (1995) Constraints limiting adoption in the North Queensland beef industry. I: a socio-economic means of maintaining a balanced lifestyle. *Agricultural Systems* 47, 291–321.

Frank, B.R. (1997) Adoption of innovations in the North Queensland beef industry. III: implications for extension management. *Agricultural Systems* 55, 347–358.

Fuglie, K.O. and Kascak, C.A. (2001) Adoption and diffusion of natural-resource-conserving agricultural technology. *Review of Agricultural Economics* 23, 386–403.

Garzon, I. (2005) Multifunctionality of agriculture in the European Union: is there substance behind the discourse's smoke? Available at: http://igov.berkeley.edu/presentations/series1/No20_Garzon.pdf

Giuliano, W.M. (2006) Should I fence the streams, ponds, and wetlands in my pastures. *Rangelands* 28, 29–31.

Godden, D. (1999) Attenuating indigenous property rights: land policy after the Wik decision. *Australian Journal of Agricultural and Resource Economics* 43, 1–33.

Guerin, L.J. and Guerin, T.F. (1994) Constraints to the adoption of innovations in agricultural research and environmental management. *Australian Journal of Experimental Agriculture* 34, 549–571.

Habron, G.B. (2004) Adoption of conservation practices by agricultural landowners in three Oregon watersheds. *Journal of Soil and Water Conservation* 59, 109–115.

Hall, C., McVittie, A. and Moran, D. (2004) What does the public want from agriculture and the countryside? A review of evidence and methods. *Journal of Rural Studies* 20, 211–225.

Harada, K. and Glasby, G.P. (2000) Human impact on the environment in Japan and New Zealand: a comparison. *The Science of the Total Environment* 263, 79–90.

Head, L. (1994) Aborigines and pastoralism in north-western Australia: historical and contemporary perspectives on multiple use of the rangelands. *Rangeland Journal* 16, 167–183.

Head, L. (1999) The northern myth revisited? Aborigines, environment and agriculture in the Ord River Irrigation Scheme, stages one and two. *Australian Geographer* 30, 141–158.

Jones, D., Cocklin, C. and Cutting, M. (1995) Institutional and landowner perspectives on wetland management in New Zealand. *Journal of Environmental Management* 45, 143–161.

Kaine, G. and Bewsell, D. (2001/2002) Managing irrigation and fertiliser on dairy farms. School of Marketing and Management, University of New England, Armidale, Australia.

Kaine, G. and Higson, M. (2006) Understanding variety in landholders' responses to resource policy. *Australiasian Agribusiness Review* 14, Paper 6.

Kaine, G. and Johnson, F. (2004) Water policy in New Zealand and Australia: we're not playing for tiddlywinks. *Proceedings of the New Zealand Grasslands Association* 66, 25–29.

Kaine, G.W. and Lees, J.W. (1994) Patterns in innovation. An analysis of practices in beef cattle breeding. The Rural Development Centre, Armidale, Australia.

Kim, S., Gillespie, J.M. and Paudel, K.P. (2005) The effect of socioeconomic factors on the adoption of best management practices in beef cattle production. *Journal of Soil and Water Conservation* 60, 111–120.

King, D.M. and Kuch, P.J. (2003) Will nutrient credit trading ever work? An assessment of supply and demand problems and institutional obstacles. Available at: http://www.envtn.org/docs/ELR_trading_article.PDF#search = %22king%20kuch%20will%20nutrient%20trading%22

Klerkx, L., de Grip, K. and Leeuwis, C. (2006) Hands off but strings attached: the contradictions of policy-induced demand-driven agricultural extension. *Agriculture and Human Values* 23, 189–204.

Knowler, D. and Bradshaw, B. (2007) Farmers' adoption of conservation agriculture: a review and synthesis of recent research. *Food Policy* 32, 25–48.

Lambert, D. and Sullivan, P. (2006) Use of conservation-compatible practices varies by farm type. *Amber Waves* 4, 7.

Lambert, D., Sullivan, P., Claassen, R. and Foreman, L. (2006) Conservation-compatible practices and programs: who participates? Economic Research Report No. (ERR14). Available at: http://www.ers.usda.gov/publications/err14/

Lawrence, D.N., Cawley, S.T. and Hayman, P.T. (2000) Developing answers and learning in extension for dryland nitrogen management. *Australian Journal of Experimental Agriculture* 40, 527–539.

Legg, A. (2004) Clean streams. A guide to managing waterways on Waikato farms. Environment Waikato, Hamilton, New Zealand.

Macgregor, C.J. and Warren, C.R. (2006) Adopting sustainable farm management practices within a nitrate vulnerable zone in Scotland: the view from the farm. *Agriculture, Ecosystems and Environment* 113, 108–119.

McCool, S.F. (1996) Limits of acceptable change: a framework for managing national protected areas: experiences from the United States. In: *Workshop on Impact Management in Marine Parks.* Maritime Institute of Malaysia, Kuala Lumpur.

McCool, S.F. and Cole, D.N. (1998) Experiencing limits of acceptable change: some thoughts after a decade of implementation. In: McCool, S.F. and Cole, D.N. (eds) *Limits of Acceptable Change and Related Planning Processes: Progress and Future Directions.* Department of Agriculture, Forest Service, Rocky Mountain Research Station, Missoula, Montana, pp. 72–78.

McCool, S.F. and Lime, D.W. (2001) Tourism carrying capacity: tempting fantasy or useful reality. *Journal of Sustainable Tourism* 9, 372–388.

McNairn, H.E. and Mitchell, B. (1992) Locus of control and farmer orientation: effects on conservation adoption. *Journal of Agricultural and Environmental Ethics* 5, 87–101.

Monaghan, R.M., Wilcock, R.J., Bramley, M., Houlbrooke, D.J., Bewsell, D., Smith, L.C., McGowan, A.W., Quinn, J.M., Robinson, B. and Betteridge, K. (2007a) Farm management plans and mitigation options relevant to farms in four catchments used for intensive dairy farming. In: *The 20th Annual Fertiliser and Lime Research Centre Workshop,* Palmerston North, New Zealand.

Monaghan, R.M., Wilcock, R.J., Smith, L.C., Tikkisetty, B., Thorrold, B.S. and Costall, D.

(2007b) Linkages between land management activities and water quality in an intensively farmed catchment in southern New Zealand. *Agriculture, Ecosystems and Environment* 118, 211–222.

Moore-Colyer, R. and Scott, A. (2005) What kind of landscape do we want? Past, present and future perspectives. *Landscape Research* 30, 501–523.

Morris, C. (2004) Networks of agri-environmental policy implementation: a case study of England's countryside stewardship scheme. *Land Use Policy* 21, 177–191.

Natural Resources Science (2006) Managing stock in and around waterways. Fact sheets, vol. QNRM02097. Available at: www.nrw.qld.gov.au/factsheets/pdf/river/r33.pdf

Nelson, F., Young, E., Liapis, P. and Schnepf, R. (2001) AoA issues series: domestic support policies. Economic Research Service, USDA. Available at: www.ers.usda.gov/briefing/wto/domestic.htm

Ngato, N. (2004) Management of water and soil from a Maori perspective. In: Hurley, E. (ed.) *Te Ohu Whenua Hui A Tau, Maori Succeeding in Agribusiness Conference.* Massey University, Palmerston North, New Zealand.

NRCS (2004) Farm Bill 2002. Environmental quality incentives program. Key points. Available at: www.nrcs.usda.gov/programs/farmbill/2002/pdf/EQIPKyPt.pdf

OECD (2005a) Producer and consumer support estimates. OECD database 1986–2005. User's guide. Available at: www.oecd.org/dataoecd/60/57/37034570.pdf

OECD (2005b) PSE/CSE database. Available at: www.oecd.org/document/55/0,2340,en_2825_494504_36956855_1_1_1_1,00.html

Pannell, D.J. (1999) Economics, extension and the adoption of land conservation innovations in agriculture. *International Journal of Social Economics* 26, 999–1014.

Pannell, D.J. (2001) Dryland salinity: economic, scientific, social and policy dimensions. *Australian Journal of Agricultural and Resource Economics* 45, 517–546.

Pannell, D.J. (2006) Dryland salinity: new knowledge with big implications. Connections: farm, food and resource issues. Available at: www.agrifood.info/connections/2006/Pannell.html

Pannell, D.J., Marshall, G.R., Barr, N., Curtis, A., Vanclay, F. and Wilkinson, R. (2006) Understanding and promoting adoption of conservation practices by rural landholders. *Australian Journal of Experimental Agriculture* 46, 1407–1424.

Parliamentary Commissioner for the Environment (2004) Growing for good: intensive farming, sustainability and New Zealand's environment. Parliamentary Commissioner for the Environment, Wellington, New Zealand.

Pretty, J., Brett, C., Gee, D., Hine, R., Mason, C., Morison, J., Rayment, M., Van der Bijl, G. and Dobbs, T. (2001) Policy and practice. Policy challenges and priorities for internalizing the externalities of modern agriculture. *Journal of Environmental Planning and Management* 44, 263–283.

Shafer, C.S. and Inglis, G.J. (2000) Influence of social, biophysical, and managerial conditions on tourism experiences within the Great Barrier Reef World Heritage Area. *Environmental Management* 26, 73–87.

Shrapnel, M. and Davie, J. (2001) The influence of personality in determining farmer responsiveness to risk. *Journal of Agricultural Education and Extension* 7, 167–178.

Sissons, J. (2004) Maori tribalism and post-settler nationhood in New Zealand. *Oceania* 75, 19–31.

Soliva, R. (2007) Landscape stories: using ideal type narratives as a heuristic device in rural studies. *Journal of Rural Studies* 23, 62–74.

Sonesson, U. (2005) Environmental assessment of future dairy farming systems – quantifications of two scenarios from the FOOD 21 synthesis work. SIK Rapport, 2005, no. 741, Swedish Institute for Food and Biotechnology (SIK), Göteborg, Sweden.

Southern Rural Water (2007) Macalister irrigation district. Available at: www.srw.com.au/irrigation/maca.html

Stankey, G.H., Cole, D.N., Lucas, R.C., Petersen, M.E. and Frissell, S.S. (1985) The limits of acceptable change (LCA) system for wilderness planning. General Technical Report INT-176, United States Department of Agriculture, Forest Service, Ogden, Utah.

Star, P. and Brooking, T. (2006) Fescue to the rescue: chewing fescue, paspalum, and the application of non-British experience to pastoral practice in New Zealand, 1880–1920. *Agricultural History* 80, 312–335.

Supalla, R.J., Selley, R.A., Bredeweg, S. and Watts, D. (1995) Adoption of nitrogen and water management practices to improve water quality. *Journal of Soil and Water Conservation* 50, 77–82.

The World Bank (2006) 2006 World development indicators online. Available at: http://publications.worldbank.org/ecommerce/catalog/product?item_id = 631625

Thompson, P.B. (2001) The reshaping of conventional farming: a North American perspective. *Journal of Agricultural and Environmental Ethics* 14, 217–229.

Upadhyay, B.M., Young, D.L., Wang, H.H. and Wandschneider, P. (2003) How do farmers who adopt multiple conservation practices differ from their neighbours. *American Journal of Alternative Agriculture* 18, 27–36.

Vanclay, F. (1992) The social context of farmers' adoption of environmentally sound farming practices. In: Lawrence, G., Vanclay, F.M. and Furze, B. (eds) *Agriculture, Environment and Society*. Macmillan, South Melbourne, Australia, pp. 94–121.

Vanclay, F. (2004) Social principles for agricultural extension to assist in the promotion of natural resource management. *Australian Journal of Experimental Agriculture* 44, 213–222.

Vanclay, F. and Lawrence, G. (1994) Farmer rationality and the adoption of environmentally sound practices: a critique of the assumptions of traditional agricultural extension. *Journal of Agricultural Education and Extension* 1, 59–89.

van der Werf, H.M.G. and Petit, J. (2002) Evaluation of the environmental impact of agriculture at the farm level: a comparison and analysis of 12 indicator-based methods. *Agriculture, Ecosystems and Environment* 93, 131–145.

Weber, E.U. (1997) Perception and expectation of climate change. Precondition for economic and technological adaptation. In: Bazerman, M., Messick, D. and Tenbrunsel, A. (eds) *Environment, Ethics and Behaviour.* Josey-Bass, San Francisco, California, pp. 314–341.

Wheeler, D.M., Ledgard, S.F., De Klein, C.A.M., Monaghan, R.M., Carey, P.L., Mcdowell, R.W. and Johns, K.L. (2003) OVERSEER nutrients budgets – moving towards on-farm resource accounting. *Proceeding of the New Zealand Grasslands Association* 65, 191–194.

Wilcock, R.J., Nagels, J.W., Rodda, H.J.E., O'Connor, M.B., Thorrold, B.S. and Barnett, J.W. (1999) Water quality of a lowland stream in a New Zealand dairy farming catchment. *New Zealand Journal of Marine and Freshwater Research* 33, 683–696.

Wilcock, R.J., Monaghan, R.M., Thorrold, B.S., Meredith, A.S., Betteridge, K. and Duncan, M.J. (2007) Land–water interactions in five contrasting dairying catchments: issues and solutions. *Land Use and Water Resources Research* 7, 2.1–2.10.

WTO (2007) World Trade Organization, rue de Lausanne 154, CH-1211 Geneva 21, Switzerland. Available at: www.wto.org/english/tratop_e/agric_e/ag_intro03_domestic_e.htm

Young, M.D. and McColl, J.C. (2005) Defining tradable water entitlements and allocations: a robust system. *Canadian Water Resources Journal* 30, 65–72.

Zander, P., Groot, J.C.J., Knierim, A. and Rossing, W.A.H. (2006) Multifunctionality of agriculture: tools and methods for impact assessment and valuation. *Agriculture, Ecosystems and Environment* 120, 1–92.

5 Impact of Livestock Grazing on Extensively Managed Grazing Lands

W.H. Schacht[1] and P.E. Reece[2]

[1]*University of Nebraska-Lincoln, Department of Agronomy and Horticulture, Nebraska, USA;* [2]*Prairie & Montane Enterprises, LLC, Gering, Nebraska, USA*

Introduction

This chapter focuses on the environmental effects of livestock grazing on extensively managed grazing lands. Land resources in these regions used by livestock producers are predominantly rangeland with relatively small amounts of seeded, grassland pasture. Rangelands are uncultivated lands with native and often species-rich plant communities dominated by grasses, grass-like plants, forbs or shrubs. Subcategories of rangeland include natural grasslands, savannahs, shrublands, most deserts, tundra, alpine communities and wet meadows. Rangeland is the world's most common land type and accounts for about 50% of the world's land area (Holechek *et al.*, 2004). A relatively large percentage of farmland was rangeland before cultivation. Consequently, remaining rangeland often has significant physical limitations for farming (e.g. low and erratic precipitation and rough topography). Extensively managed livestock production is the most sustainable and common form of agriculture on rangelands; however, there is a multitude of other possible uses including wildlife habitat, recreation, ground water recharge and other environmental services. Today, rangelands have relatively low agricultural value but have significant overall economic and social value because of the geographic magnitude of these natural catchments, their vast atmospheric interface, their inherent biodiversity and their immense open spaces – which are increasingly coveted by growing human populations.

Grassland pasture is usually established on marginal farmland. Initially these pastures are composed of high-yielding forage species capable of responding to high levels of agronomic inputs (e.g. fertilization, irrigation and weed control). Historically, overgrazing and decreased use of agronomic inputs have degraded much of these pastures. The resulting dominance of less-productive plant species has compromised ecosystem function and made many grassland pastures less responsive to agronomic inputs and proper management. Consequently, degraded pastureland usually is managed extensively. Additionally, millions of hectares of semi-arid shrubland throughout the world, particularly in the Great Basin of western USA, were converted to seeded pasture (e.g. crested wheatgrass (*Agropyron cristatum*)) in the 20th century. These grassland pastures have always been extensively managed. Reference to pastureland in this chapter will be specific to extensively managed grassland pasture.

Because livestock production is the most common use of forage resources on rangelands and grassland pastures of the world, understanding the influence of livestock enterprises on grazing land is critical for the development of management strategies needed to optimize ecosystem functions. Livestock production on extensively managed grazing lands depends on nutrients within the ecosystem, local precipitation and hydrologic conditions, and the ability of plants to use soil water and produce leaf area needed for photosynthesis. Management is the principal input in these systems and the primary grazing management tools are timing, intensity and frequency of grazing. Grazing management strategies have been developed to maintain or improve the structure and function of plant communities known to provide sustainable forage resources for livestock. For instance, implementation of management strategies involving deferred-rotation grazing and limiting use of above-ground net primary production to 30% by grazing livestock seems to be sustainable and compatible with maintenance of productive plant communities on semi-arid grasslands of the Great Plains, USA (Reece *et al.*, 1996, 2007). The impacts of exceeding sustainable levels of use, i.e. overgrazing, on grazing land ecosystems is a topic in the remainder of this chapter. Overgrazing can be defined as a level of defoliation beyond which plants are unable to recover sufficiently before a subsequent period of grazing. Grazing management strategies affect many ecosystem components besides livestock and forage production. Grazing also influences plant community heterogeneity, soil chemical and physical properties and the distribution and cycling of nutrients within the plant–soil–water continuum.

Plant Community Production and Composition

Livestock's effect on plant production is a function of environmental variables, selective herbivory, and timing and intensity of grazing. Grazing or defoliation can increase net primary production (NPP) in many grassland environments (McNaughton, 1985; Williamson *et al.*, 1989; Holland *et al.*, 1992). The grazing optimization hypothesis of McNaughton (1979, 1983) states that NPP increases with increasing grazing intensity up to some moderate level of grazing because of plant compensatory growth mechanisms. This hypothesis has been the basis for much discussion since the late 1970s. Reports of compensatory growth that support the grazing optimization hypothesis are common, but there is a substantial literature debunking the hypothesis (Taylor, 1989; Belsky *et al.*, 1993). NPP of grasslands commonly is reported to be maximum at light to moderate levels of use, whereas productivity is relatively low when grasslands are not grazed or hayed for several years or when they are overgrazed during multiple drought years. Little or no use in sub-humid or more moist environments results in significant accumulation of dead plant material that: (i) creates habitat for disease pathogens that decrease the vigour of the plants; (ii) sequesters nutrients in the dead plant material, making them unavailable for growth; (iii) decreases photosynthetic efficiency of leaves that must grow through the thick layer of litter or that are shaded by a relatively dense canopy; and (iv) intercepts a significant amount of precipitation and decreases the amount of water that reaches the soil surface and infiltrates into the soil (McNaughton, 1985; Williamson *et al.*, 1989; Holland *et al.*, 1992).

Application of the grazing optimization hypothesis is problematic because compensatory growth is not seen under conditions common to arid and semi-arid environments. Additionally, livestock often remove more than 50% of the herbage from preferred species and preferred locations. Compensatory growth is most commonly reported in mesic to sub-humid environments where periods of active plant growth extend through much of the growing season, and good soil moisture conditions are common throughout the growing season (see review by Lauenroth *et al.*, 1994). In semi-arid and arid environments, plants rapidly grow for relatively short periods of time, as few as 30–45 days (Reece *et al.*, 2007), and respond to defoliation with growth only during those periods. Moreover, precipitation is relatively low and erratic; therefore, good soil moisture after defoliation is not dependable. Because compensatory growth is not likely to occur during dry periods (Williamson *et al.*, 1989), developing grazing systems and selecting stocking rates that assume compensatory growth can be risky in semi-arid and arid environments (Heitschmidt *et al.*, 1982; Taylor, 1989). Given the inherent variation in grazing distribution on extensively managed grazing lands, it is likely that vegetative growth can be 'optimized' only on a small percentage of area available, even when environmental conditions are favourable for plant growth. Furthermore, above-ground production in semi-arid and arid environments is variable within and between years and relatively low on average. Increasing managerial input in an effort to take advantage of small and uncertain incremental increases in herbage production is not justifiable in these ecosystems. The effects of overestimating the grazing intensity that can be tolerated by a plant community can be devastating to the long-term productivity and stability of a site. Semi-arid or arid grazing lands are particularly susceptible to intense and/or frequent defoliations, resulting in deterioration of plant communities and hydrologic condition (Noy-Meir, 1976).

Plant community composition is responsive to timing and intensity of grazing (see review by Lauenroth *et al.*, 1994). Grazing lands dominated by grazing-resistant plant species show only minimum shifts in species composition when grazing pressure is light to moderate. Grazing-induced changes in botanical composition on grazing lands are more likely to occur when grazing pressure is high for a number of years during which drought stress is common. Precipitation regime was the overwhelming factor affecting changes in species composition and productivity on shortgrass prairie in the central Great Plains (Klipple and Costello, 1960).

Biodiversity

Natural grazing lands are biologically diverse because of their inherent species richness and the spatial and temporal variability in topo-edaphic features, environmental conditions, species interactions and disturbances (Patten and Ellis, 1995;

Fuhlendorf and Smeins, 1999). For rangelands, biodiversity is reported to be a reflection of healthy ecosystem function and is positively related to the primary production, stability and resilience of a system, as well as nutrient retention and nutrient-use efficiency (Tillman *et al.*, 1996; Loreau, 2000). Different plant species exploit soil water and nutrients at different times and soil depths. It follows that nutrient uptake and nutrient-use efficiency of vegetation increases with increasing species diversity. Mobile soil nutrients such as nitrate are rapidly taken up by the densely distributed root systems of diverse plant communities, thus decreasing the likelihood of losing nutrients to leaching or volatilization (Reich *et al.*, 2001; Scherer-Lorenzen *et al.*, 2003). Grazing lands with a diversity of native plant species also tend to be less susceptible to weed invasion (van Ruijven *et al.*, 2003).

Grazing vegetation on extensively managed grazing lands has the potential to optimize biodiversity and is critical to the sustainability and maintenance of these ecosystems (Bakker, 1994). Herbivores do not uniformly graze a management unit but selectively graze, resulting in a mosaic of use intensity across the landscape. Intensity and frequency of grazing are high in favoured areas (e.g. near water and flat to gently rolling topography) while vegetation on other areas will be only lightly used or not grazed at all. Some ecologists suggest that species-rich grazing lands maintain their ecosystem function and production more effectively if they are periodically disturbed or stressed through defoliation, disease infection and non-invasive weed infestation (Knops *et al.*, 1999; Pywell *et al.*, 2004). In sub-humid ecosystems, landscape units with this patchy framework of use and plant species distribution have resilience and are sustainable as long as spatial-use patterns shift and/or heavily used sites are allowed to recover from time to time (Fuhlendorf and Engle, 2001). However, there is no clear evidence that diverse systems lead to a lower variability in temporal and spatial distribution of production than that found in less-diverse plant communities on the same ecological site (Tracy *et al.*, 2004). Furthermore, there are examples where long-term (70 years) protection from large animal grazing does not have any detrimental effect on plant communities or soil quality (Willms *et al.*, 2002).

When site characteristics are relatively homogeneous, stocking rate is the primary driver of patchiness of grazing and use at the local and landscape scale. Patchiness of use and the resulting mosaic of different plant communities are most common at light to moderate stocking rates. At a local scale, light grazing increases sward structural heterogeneity and patchiness of plant communities because of selective grazing, trampling, nutrient cycling and propagule dispersal (Correll *et al.*, 2003; Isselstein *et al.*, 2003). These patches are relatively stable. For instance, in tall grass prairie, patches of tall grass become dominant in areas of light grazing and good soil nutrient cycling and hydrologic conditions, whereas short grasses are dominant in patches of intense grazing, poorer nutrient cycling and poorer hydrologic conditions. At the landscape level, large area differences become apparent based on favoured plant communities, water and shade location, topography, fence orientation and habitat types.

Increasing stocking rate increases grazing pressure and evenness of use spatially and temporally. The homogeneity of use at the local and landscape levels, characteristic of heavily stocked, continuously grazed systems, results in low biodiversity because of uniformity of grazing (Fuhlendorf *et al.*, 2006). Resulting plant communities are those that are most resistant to heavy grazing pressure, thus decreasing diversity. Non-grazed landscapes in sub-humid to humid ecosystems also tend to have low diversity because of lack of disturbance and similar micro-environmental conditions across the entire landscape.

Grazing system can also affect patchiness of use and biodiversity on rangelands (Fuhlendorf and Engle, 2001; Kempema *et al.*, 2006). Short-duration grazing systems tend to favour monocultures and simple mixtures because a basic objective of these systems is to optimize harvest efficiency by improving grazing distribution at the local and landscape level. The timely movement of high concentrations of animals (i.e. high stocking density (animal units/ha)) through multiple grazing units results in high grazing pressure. This method gives the manager effective control in uniformly applying the intensity and frequency of grazing that favours the key forage species. In this way, patchiness of use is greatly decreased and plant species not tolerant to the grazing management are uncommon.

Increasing spatial and temporal heterogeneity of disturbance in grasslands increases variability in vegetation structure that results in greater variability at other trophic levels, e.g. grassland birds (Vickery *et al.*, 2001, Fuhlendorf *et al.*, 2006). Using livestock to develop and maintain biodiversity on grassland/rangeland systems is a new target in many regions of the world (Isseltein, 2005). Ecologists suggest that the relationship between grazing and diversity should be better exploited to the benefit of many degraded grasslands or rangelands. Defining desirable diversity and identifying critical functional groups of species become increasingly important as biodiversity gains momentum as a primary goal of ecosystem restoration programmes. Additional prerequisites for restoring rangeland include the ability to measurably decrease stocking rates and alternate season of defoliation.

Hydrology and Soils

Grazing land soils vary greatly in physical and chemical properties. Grassland soils in sub-humid and semi-arid regions tend to be the deepest, most fertile soils of the world; whereas soils in arid regions covered with a mixture of plant types tend to be shallow and have low fertility. Soil formation occurs at variable rates depending on such factors as climatic conditions, topography and parent material. Rate of soil formation in naturally occurring and well-managed grazing lands is greater than the erosion rate. Accelerated erosion rates, of course, adversely affect many soil properties that promote plant growth. Grazing land soils are susceptible to soil erosion, both wind and water, during periods of disturbance caused by drought, fire and heavy defoliation (Schlesigner *et al.*, 1996).

As stated earlier, the most common goal on grazing lands is to maximize livestock production or profitability on a sustained basis. Sustainability is considered to be the fundamental goal of grazing land management, and sustainable management of these lands largely depends on soil management and conservation. Erosion associated with long-term overgrazing generally occurs at accelerated rates. It removes mineral particles, organic matter and nutrients from the soil, thus decreasing depth of topsoil, fertility and water-holding capacity (Dormaar and Willms, 1998). Soil loss on grazing lands is a major concern throughout the world. The National Land and Water Resources Audit in Australia (NLWRA, 2001) showed that erosion on native pasture, principally semi-arid woodlands and grazing lands in northern regions, accounted for 76% of the continent's total soil erosion. The rate of soil loss is several times greater than the average rate of soil formation across these grazing lands. In the USA, as many as 100M ha of rangeland is considered highly erodible with an annual sediment loss from grazing lands approaching 2.4 pg (USDA-NRCS, 1992). Arid and semi-arid lands in China account for about 37% of the nation's total land. Around 60% of this land has been degraded, largely due to wind erosion (Zhu, 1991; Dong *et al.*, 2000). The effects of this degree of erosion with time are cumulative and resulting in deterioration of grazing lands. On the other hand, reports of minimum sediment and soil nutrient losses in overland flow (surface runoff) from grazing lands are common (Emmerich and Heitschmidt, 2002; Haan *et al.*, 2006). Properly managed grazing lands generally have little soil loss because of: (i) good protective plant cover that buffers raindrop impact, filters sediment from surface runoff and enhances water infiltration; and (ii) good soil structure that enhances infiltration and water-holding capacity.

Water is the critical factor in soil development and soil erosion on grazing lands. Maximizing water infiltration and limiting overland flow are key for enhancing plant production and soil development and minimizing soil loss (Thurow *et al.*, 1988). Movement of precipitation into, through, and over the landscape is controlled by hydrologic condition which is a function of vegetation, soil, topography and climate. Standing herbage and plant litter on the soil surface decrease the physical impact of raindrops on surface soil structure and water infiltration, and retard surface flow of water when heavy rains occur. Water infiltration is decreased when surface soil structure is destroyed by the force of raindrop impact and the filling of surface pore spaces with sediment. With less water entering the soil and increased soil/nutrient loss through overland flow, plant growth potential is decreased, both above and below ground. Roots are important as the primary source of organic matter (i.e. dead roots and exudates) for soil structure formation and maintenance. Soil organic matter not only leads to improved soil structure and infiltration, but also increases water-holding capacity, cation exchange capacity and nutrient retention (Paustian *et al.*, 1997).

Livestock grazing can be detrimental to hydrologic conditions by removing plant cover and potentially decreasing the vigour of the grazed plants. As stocking rate increases, water infiltration generally decreases while runoff and sediment loss increase (Alderfer and Robinson, 1947; Warren *et al.*, 1986b). Heavily defoliated plants have decreased root production and leak less exudates; therefore, heavily grazed plant communities have less ground cover and poorer soil structure resulting in less infiltration and less plant production. Downward cyclic interactions of hydrologic conditions and plant vigour can be insidious (Fig. 5.1). Decreased water infiltration (effective precipitation) creates a less-productive plant community that is characteristic of a lower-precipitation zone. Soil formation slows and the susceptibility of the system to soil loss increases. This process increases the potential for desertification.

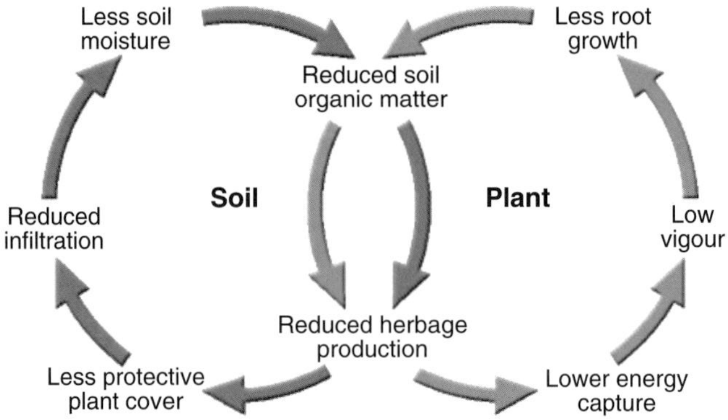

Fig. 5.1. Cyclic interaction of hydrologic conditions and plant growth on grazing lands.

Grazing, often coupled with drought, has resulted in the invasion or increase of shrubs and/or trees on rangeland (Buffington and Herbel, 1965; Frederickson *et al.*, 1998). Soil and nutrient (e.g. N and P) loss from areas with recent shrub invasion is greater than from associated grasslands (Schlesinger *et al.*, 1999) when stocking rates are not decreased to account for less forage production. Throughout the world (Scholes and Archer, 1997; Dye and Jarmain, 2004), losses of grass cover are associated with lower rates of soil water infiltration and increased overland flow. The overall amount of surface runoff is greater in shrublands than in grasslands because bare soils compose as much as 60% of the surface area of shrublands (Schlesinger *et al.*, 1999). The physical impact of grazing also can decrease the cover and species richness of cryptogams in desert shrubland communities. These N-fixing, non-vascular plants, e.g. algae, lichens and mosses, form soil surface crusts between widely spaced vascular plants and enhance plant-available N, surface soil stability and water infiltration, and decrease sediment in runoff (Brotherson and Rushforth, 1983). Heavily grazed areas in southern Utah, USA, supported only about 22% as much cover and 25% as many species of lichen and mosses per unit area as non-grazed exclosures (Anderson *et al.*, 1982). The loss of soil and soil nutrients through erosion on shrublands exacerbates desertification in many arid areas.

Cattle treading, or trampling, not only decreases soil surface cover, but can also destroy soil surface structure and increase soil bulk density and surface roughness (Betteridge *et al.*, 1999; Russell *et al.*, 2001), especially of moist soils. The effects of decreased cover and increased bulk density decrease water infiltration, increase surface water flow and sediment loss (Thurow *et al.*, 1986, 1988; Elliott *et al.*, 2002) and increase P and N losses by two- to threefold (McDowell *et al.*, 2003; Haan *et al.*, 2006). Increased surface roughness generally decreases water runoff and sediment loss by slowing water movement and acting as a trap for detached soil particles (Warren *et al.*, 1986a; Russell *et al.*, 2001). Soils covered with vigorous plant communities in semi-arid and sub-humid grasslands generally are not affected by livestock grazing because soil surface conditions are favourable to infiltration and soil structure recovers fairly rapidly from compaction effects. Livestock density also is relatively light in extensively managed systems, except in high-use areas around watering and supplementation sites and shade. Potential for damaging soil structure or compacting soil generally is greater on wet compared to dry soils and greater on fine-textured clayey or silty soils compared to coarse-textured sandy soils (Warren *et al.*, 1986a). Grazing land soils can show rapid recovery from the detrimental effects of grazing animals within the growing season (Warren *et al.*, 1986b; Betteridge *et al.*, 1999) because of the disruptive action of roots and soil fauna, and the wetting and drying or shrinking and swelling of soils.

Riparian areas

Riparian areas are usually small compared to the remainder of the catchment, but represent critical components of rangeland ecosystems. They are characterized by high biotic diversity, water storage capacity and NPP, and act as biogeochemical buffers between uplands and streams (Belsky *et al.*, 1999; Blank *et al.*, 2006). Sound catchment management involves upland and riparian sites. The perimeter of riparian areas along waterways can serve as vegetation filter strips that trap sediment derived from upland sources before it reaches a waterway (Pearce *et al.*, 1998; McEldowney *et al.*, 2002). Loss of sediments and nutrients from poorly managed grazing land not only represents lost productivity, but also constitutes the second leading cause of stream impairment by non-point pollutants in many grazing land regions. Excessive sediment loads in runoff water can cause radical changes in streambed morphology, loss of aquatic habitat, nutrient loading and decrease in storage capacity of reservoirs (Novotny and Olem, 1994; McEldowney *et al.*, 2002). Management for high soil organic matter and vigorous plant communities with high root-length densities is critical for sustaining the structure and function of riparian systems (Wheeler *et al.*, 2002; Blank *et al.*, 2006).

Grazing by livestock has damaged as much as 80% of the streams and riparian ecosystems in arid regions of the western USA (Belsky *et al.*, 1999). Cattle favour riparian areas and congregate there because of excellent forage and water availability, shade, topography and general lack of physical constraints to grazing as compared to the drier and often rougher characteristics of upland areas (Pinchak *et al.*, 1991). When poor grazing management, e.g. high stocking rates or inappropriate timing of grazing, causes species composition shifts from high-producing plants with extensive root systems to species with less root biomass and length, streambank instability and channel lateral expansion often occur – detrimentally affecting aquatic habitat and lowering flood plain water tables. Plant communities in good condition have vigorous rooting activity and ensure that readily available soil nutrients are sequestered in plant tissue during the growing season, lessening the possibility of stream water degradation (Wheeler *et al.*, 2002; Blank *et al.*, 2006).

Carbon Sequestration

Rangelands are considered important carbon (C) sinks on a global scale (Svejcar *et al.*, 1997; Follett and Schuman, 2005). Grasslands alone contain some of the largest stocks of soil organic C (SOC; about 20 kg C/m^2), rivaling soil C stocks in arctic tundra and composing nearly 15% of the terrestrial (non-wetland) global C stocks (Schlesinger, 1997; Johnson and Matchett, 2001). Estimates of C sequestration in grassland soils on a worldwide basis range from 0.2 to 0.6 Mg C/ha/year (see review by Jones and Donnelly, 2004). Follet *et al.* (2001) reported that intensively managed grassland in the USA has the potential to sequester 10.5–4.3 million t C/year, which represents 25–80% of the CO_2-C emissions from all US agriculture (42.9 million t C/year; Lal *et al.*, 1998). Estimates of C sequestration on extensively managed grazing land are available (Follett and Schuman, 2005). However, predicting C sequestration on a regional or worldwide basis is complicated by several variables, including types, species and numbers of grazing animals; management practices; spatial and temporal climate variation; complexity of plant communities; and presence and proportions of N-fixing plants. Because of relatively low moisture and nutrient availability, extensively managed grazing lands have low C sequestration potential per hectare. However, they cover vast areas of the globe and so have a great C sequestration potential. Follett *et al.* (2001) estimated that improved management practices (e.g. implementation of proper stocking rates and grazing systems) on rangeland in the USA could result in additional sequestration of 5.4–16.0 Tg C/year. Ways of increasing C sequestration include improved grazing management, introduction of legumes and control of undesirable species (Follett and Schuman, 2005). The rate at which C is sequestered is usually greatest in the first several decades following implementation of improved management (Schlesinger, 1990).

Overgrazing generally is the cause of degradation and can result in a loss of more than 20% of the SOC found in properly managed grazing lands (Kimble *et al.*, 2001). Proper grazing can have a neutral or positive effect on C storage. Defoliation and treading facilitate litter decomposition. Conversely, excluding grazing can lead to excessive accumulation of plant litter, thus increasing the amount of C immobilization and potential volatilization. Proper grazing is also key in maintaining diverse, productive plant communities (above and below ground) (see section on Biodiversity) that are better able to use available soil nutrients and water for more days during the growing season. Highly productive plant communities have high shoot turnover, high root biomass and hydrologic conditions that optimize water infiltration (see section on Hydrology and Soils); therefore, net primary productivity remains high and C 'pumped' into the soil is optimized in the long term. The principal SOC inputs to grazing land soils are roots through death and decomposition, exudation from living roots (and soil microbes), mucilage production and sloughing from living roots (Reeder *et al.*, 2001).

Moderate to heavy stocking rates (high grazing pressure) applied during the growing season generally have a negative impact on NPP resulting in decreased C sequestration. Intensity and timing of grazing also can cause shifts in species composition. In perennial grass systems, annual herbaceous plants can become dominant either in long-term rested areas or in repeatedly overgrazed areas. Areas dominated by annuals have low root/shoot ratios and lack the dense fibrous rooting systems conducive to soil organic matter formation and accumulation (Reeder and Schuman, 2002). In the central Great Plains of North America where there are mixtures of C_4 and C_3 grasses, intense spring grazing annually can result in

C_4-grass dominance. C_4 grasses can be more productive with greater root biomass and depth than C_3 grasses (Mousel *et al.*, 2007; Fig. 5.2), thus sequestering more C (Frank *et al.*, 1995; Reeder and Schuman, 2002). In other situations, a change from grassland to woody plant dominance (because of intensity and timing of grazing) can result in an increase of C sequestration (Hibbard *et al.*, 2001), although a transition from low-quality shrublands to grasslands or grass/shrublands is associated with net increases in the SOC pool (Sobecki *et al.*, 2001).

Adding legumes to grazing lands may enhance their C sequestration potential. Interseeding *Medicago sativa* ssp. *falcata* in the northern mixed prairie of South Dakota, USA, increased SOC (Mortenson *et al.*, 2004). Mortenson *et al.* (2004) suggested that N fixation by the legume likely led to significant increases in total soil N, increased forage production and increased soil organic matter. However, the potential for accumulating more SOC with legumes appears to be limited because organic matter from grass/legume mixtures has relatively low C/N ratios and is rapidly decomposed (Schnabel *et al.*, 2001).

Grazing intensity and frequency appear to be the principal management tools to affect SOC concentration. They can be incorporated into grazing systems designed to maintain diverse plant communities capable of producing expansive and dense root systems and characterized by high NPP and a greater degree of C sequestration in the soil. Encouraging practices that enhance C sequestration on

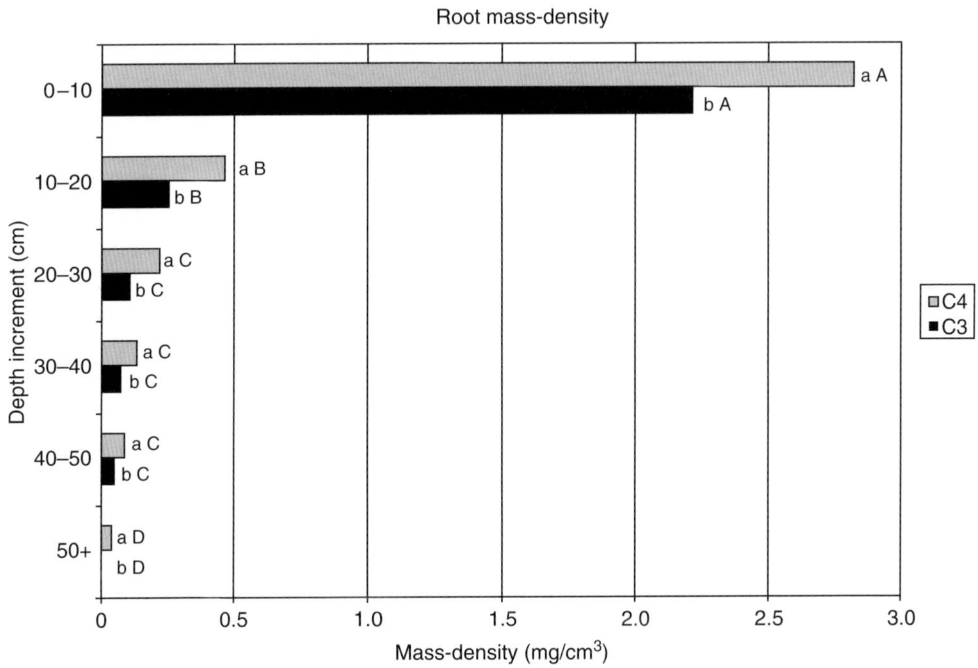

Fig. 5.2. Total root mass-density (mg/cm^3) of vegetation in cool-season (C_3) and warm-season (C_4) subirrigated meadows in the Sandhills of Nebraska, USA. Different lowercase letters between meadow types are different at $P < 0.1$. Different uppercase letters within meadow types are different at $P < 0.1$.

grazing lands has several advantages. Sequestering practices can be implemented quickly with existing technology and with minimum impact on the economic system. These practices also result in improved soil and water quality, efficient water use, less soil erosion, and sustained, high-level productivity of plant communities for grazing livestock production (Follett and Schuman, 2005).

Air Quality

Methane (CH_4) and nitrous oxide (N_2O) are the principal greenhouse gases (GHG) derived from agriculture, with agriculture accounting for >55% and >75% of the world's anthropogenic CH_4 and N_2O emissions, respectively (IPCC, 2001). The primary source of CH_4 from grazing lands is fermentation of feed in the rumen of ruminants. By-products of rumen fermentation are volatile fatty acids (VFA), CH_4 and CO_2. The VFAs are absorbed across the rumen wall, transported to body tissue by the circulatory system and metabolized as an energy source. The CH_4 and CO_2 are not used productively by the animal but released through eructation. CH_4 emissions from ruminants on grazing lands may represent as much as 20% of all agricultural CH_4 emissions (Clark *et al.*, 2005).

On extensively managed grazing lands, the principal source of N_2O is excreta of grazing livestock. N content of diets of grazing livestock generally is much greater than livestock requirements, resulting in relatively large quantities of N being deposited as excreta on grazing lands. Nitrification and denitrification then convert N in the excreta to N_2O. Fertilization can be a significant source of N_2O on more intensively managed grazing land, but extensively managed grazing land is rarely fertilized. N_2O production from grazing lands as a whole is estimated to be between 16% and 33% of total estimated agricultural N_2O emissions (Clark *et al.*, 2005).

As stated earlier, livestock density and agronomic inputs (e.g. fertilization) on extensively managed grazing lands are relatively low; therefore, CH_4 and N_2O emissions per unit area from these grazing lands is much lower than from intensively managed grazing lands. Although data specific to grazing land type are not available, a large part of the total grazing land GHG emissions is from intensively managed grazing lands. Grazing lands in general may represent a significant source of CH_4 and N_2O, but levels of emissions are minor relative to total CO_2 equivalent emissions. The CH_4 and N_2O emissions from agriculture as a whole are generally less than 10% of total CO_2 equivalent emissions in much of the world (UNFCCC, 2004). Extensively managed grazing lands likely have a very small role in GHG production and associated effects on atmospheric conditions.

Nutrient Cycling

Grazers can have a major impact on nutrient cycling by affecting inputs, outputs and transformations of nutrients. Herbage consumption routes a portion of above-ground biomass through the animal rather than into the litter component, affecting the microclimate for both the plant community and soil microbes.

Frequent defoliation, especially at moderate to high intensities, can also impact the below-ground nutrient dynamics by decreasing root initiation and extension and increasing root mortality (Johnson and Matchett, 2001; Mousel *et al.*, 2005). These grazing-induced changes in nutrient availability, forms and cycles feed back to the grazing land system and contribute to shifts in species composition. Long-term changes in botanical composition and diversity affect litter quality, mass and seasonal dynamics of decomposition such that a positive feedback loop develops (Archer and Smeins, 1991). Grazing-induced nutrient losses from grazing land systems can be significant. The relatively small amounts of nutrients that are lost when livestock are sold or moved to other grazing units (e.g. 5–10% of feed N in beef cattle) are generally offset with natural inputs within these ecosystems. In contrast, nutrient losses can be great under conditions where erosion and runoff are accelerated because of overgrazing.

N is the soil nutrient most commonly limiting plant growth on grazing lands. High levels of plant production are dependent on rapid cycling of N, including transformation of organic forms to mineral forms that can be taken up again by plants (although N also can be taken up in some organic forms (Schimel and Bennett, 2004)). Light to moderate levels of grazing intensity are thought to increase N mineralization (Frank and Zhang, 1997; McNaughton *et al.*, 1997; Biondini *et al.*, 1998) and N availability to plants (Bauer *et al.*, 1987; Holland and Detling, 1990). Research indicates that large herbivores increase N cycling in rangelands by: (i) redistributing N in forms more available to plants and soil microbes (urine and dung) than in the litter or standing dead of non-grazed plant communities; (ii) incorporating surface organic matter (litter) into the soil through trampling; and (iii) lowering the C/N ratios of plant litter, roots and soil organic matter.

N in urine and dung is in more labile forms that are readily available for uptake by plants, instead of the recalcitrant forms in plant litter that require time for decomposition and mineralization to occur. Urinary N is largely soluble and in the form of urea (60–90%), making it subject to rapid cycling or loss. Most of the urea in urine is generally hydrolysed to ammonium within a few days (Stillwell and Woodmansee, 1981), and much of the ammonium is nitrified to nitrate within a few weeks (Whitehead, 1995). Nitrification can be slowed by such things as high soil pH and ammonium concentration, resulting in increased ammonia volatilization. Warm, dry soils of semi-arid grasslands favour ammonia volatilization from urine spots during much of the growing season. Although reports are extremely variable, about 15% of urinary N is volatilized as ammonia (Whitehead, 1995) and much of the ammonia emitted is deposited within 2 m of the urine spot (Ross and Jarvis, 2001). Although nitrates are susceptible to leaching (Stillwell and Woodmansee, 1981), conditions generally are not conducive to leaching in semi-arid grasslands except during periods of high soil moisture and/or slow plant growth, such as early spring in temperate areas when rainfall is relatively high and plants are initiating growth. There can be significant losses of N in non-grazed grassland through volatilization of ammonia from plants, denitrification in the cooler and wetter soil conditions that may occur on non-grazed sites (Bauer *et al.*, 1987), decomposition within the dead-shoot component of the canopy and photochemical decomposition of the litter layer (Coupland and Van Dyne, 1979). N loss through ammonia volatilization from urine patches on grazed grassland is an order of magnitude lower than the loss due to ammonia volatilization from non-consumed or senescent vegetation on non-grazed grassland (Schimel *et al.*, 1986).

Nutrient cycling via grazing animals can be important in enhancing or maintaining soil fertility (Floate, 1981) and helps keep a pool of readily mineralizable organic nutrients in the upper soil profile where they are more accessible to plants and microbes (Botkin *et al.*, 1981). The turnover rate of nutrients in the above-ground shoots is greater for grazed than non-grazed grassland. Animal traffic enhances physical breakdown, soil incorporation and rate of decomposition of litter. In a non-grazed situation, C and N are immobilized in above-ground standing dead and litter material. Removal of grazing also decreases soil microbial turnover rates and net soil N mineralization (Holland and Detling, 1990). Schuman *et al.* (1999) reported that 15–25 kg/ha more N is immobilized in dead plant material in non-grazed grassland than grazed grassland of the northern mixed-grass prairie of the USA. This slow rate of turnover decreases accrual of C and N in the soil (McNaughton *et al.*, 1988; Seagle *et al.*, 1992; Schuman *et al.*, 1999). Less N being made available each year for primary production means less production with time and less C and N available for future production, all of which compromise ecosystem function (McNaughton *et al.*, 1988; Seagle *et al.*, 1992).

Grazed plants tend to have higher N concentration, in both the shoots and roots, because C is limiting. Shoot tissue growth following defoliation is better quality than that of the original tissue. Moderate to heavy grazing generally results in decreased C allocation to roots, less root growth and a lower C/N ratio. The increased tissue quality feeds back to affect N cycling. Higher-quality shoot and root tissue (with lower C/N) leads to lower microbial immobilization of N and more rapid N mineralization and greater N availability (Johnson and Matchett, 2001). Microbial activity also can be stimulated by the exudation of high-quality organic compounds from roots of defoliated plants (Ingham *et al.*, 1985). However, there is not the consensus that root production declines in response to grazing in all cases (Milchunas and Laurenoth, 1993).

Influence of grazing on plant species composition is another consideration of importance because plants with high C/N ratios and/or that are unpalatable tend to immobilize nutrients. The reported increase in N quality of roots of grazed plant communities in some cases may be a result of shifts in botanical composition in response to grazing. A shift from C_3 to C_4 grassland, resulting from changes in timing and intensity of grazing, affects below-ground plant biomass and C/N ratios. The C_4 grasses typically have greater root biomass (Mousel *et al.*, 2007) and lesser N concentration in their root tissue (Johnson and Matchett, 2001). Schuman *et al.* (1999) reported that heavy stocking rates on cool-season mixed prairie in the northern Great Plains, USA, favoured an increase in the C_4 grass, blue grama (*Bouteloua gracilis*). Blue grama had increased production rates and greater C allocation to the below-ground portions of the system, resulting in greater C/N ratio in roots, more soil organic matter and greater potential immobilization of N. As stated earlier, grazing also can decrease the cover and species richness of cryptogams in desert shrubland communities. These N-fixing, non-vascular plants significantly influence N cycling in semi-desert and desert shrublands and grasslands (Anderson *et al.*, 1982).

P loss from grasslands is primarily through overland flow and closely tied to sediment movement (Leinweber *et al.*, 2002). Unit increase in runoff P concentration with a unit soil P increase is greater for disturbed (e.g. overgrazed) grasslands with much exposed surface soil than for properly managed grasslands with a limited amount of bare soil. Many other factors, such as P application (rate, method, timing and form of P added as fertilizer or manure) and runoff and erosion potential, influence the concentration of P in surface runoff (Sharpley *et al.*, 2001). P loss from grasslands is generally low unless management practices (e.g. overgrazing) or soil and topographic conditions (e.g. steep slopes) are conducive for accelerated erosion (Schlesinger *et al.*, 1999).

Overall, the regulation of N and C dynamics in soils is affected by a wide range of environmental and management factors, including herbivory (Holland and Detling, 1990; Ritchie *et al.*, 1998). These interacting factors may be a reason for a lack of total agreement in the literature on the relationship between grazing and soil N. Some literature indicates that grazing does not suppress immobilization of N on grazing lands (Verchot *et al.*, 2002), but actually decelerates N cycling (Ritchie *et al.*, 1998). Ritchie *et al.* (1998) and others (e.g. Wedin, 1994) argue that herbivores feed selectively on plant species with nutrient-rich tissue and increase the dominance of plant species with tissue that is nutrient-poor or defended by secondary compounds. Litter from these latter species decomposes slowly and decreases nutrient turnover and availability. A modelling exercise and analysis conducted by Pineiro *et al.* (2006) indicates that SOC and soil N storage of the Rio de la Plata grasslands of southern South America decreased by about 20% during the last four centuries of livestock grazing (at an estimated stocking rate of 178–302 kg/ha/year). Continued research in this area is certainly warranted.

Socio-economic Considerations

The worldwide demand for livestock products and meat production from grazing lands is increasing with human populations and as incomes in less-affluent countries increase. Ruminant livestock are expected to account for 27% of the increase in global meat consumption between 2003 and 2020, up from 23% over the previous two decades (Delgado, 2005). This increased ruminant livestock production demand will coincide with increasing societal pressure for a variety of other, often competing, uses of grazing lands (e.g. wildlife habitat, recreation and open space) and for larger-scale environmental resource management objectives (e.g. C sequestration, maintaining air and water quality and repository of biodiversity). Since the late 1960s, research, education and incentives for food production have improved rangeland and pastureland production systems and doubled worldwide beef and veal production (FAOSTAT data, 2004). Meat production from the grazing lands of most developed countries is not expected to increase in response to increasing demands. Increases in beef and veal production are more likely to occur in developing countries, especially in the subtropics and tropics, with improved management (FAOSTAT data, 2004). The likelihood of overgrazing arid and semi-arid grazing lands in countries may increase as their own demand for meat increases, transportation infrastructure improves and access to markets increases.

Only recently have we focused on the geographic magnitude of services provided by grazing land ecosystems. On extensively managed grazing lands, where productivity is limited by precipitation, plant vigour and composition of plant communities, managers must select stocking rates and grazing systems that yield satisfactory production and income, but do not have adverse environmental impacts (Fig. 5.3). Integrative thinking is critical for optimum management of grazing lands because timing and intensity of grazing impact effectiveness of precipitation, NPP and sustainability of livestock enterprises. Management of rangelands and pasturelands will become increasingly complex as the societal value of benefits or uses other than livestock production increases in many developed countries.

Government, community or absentee ownership of grazing lands can preclude increases in livestock production or the opportunity to improve range condition. The amount of NPP allocated for livestock on public lands (government owned) may be measurably decreased if herbage is also needed for wildlife habitat or catchment management objectives. Methods of allocating use of 'grazing commons' in tribal or communal regions may preclude potentially beneficial nomadic or deferred-rotation grazing systems. During the past 50 years, a growing percentage of privately owned grazing land in North America has been inherited or purchased by people who live

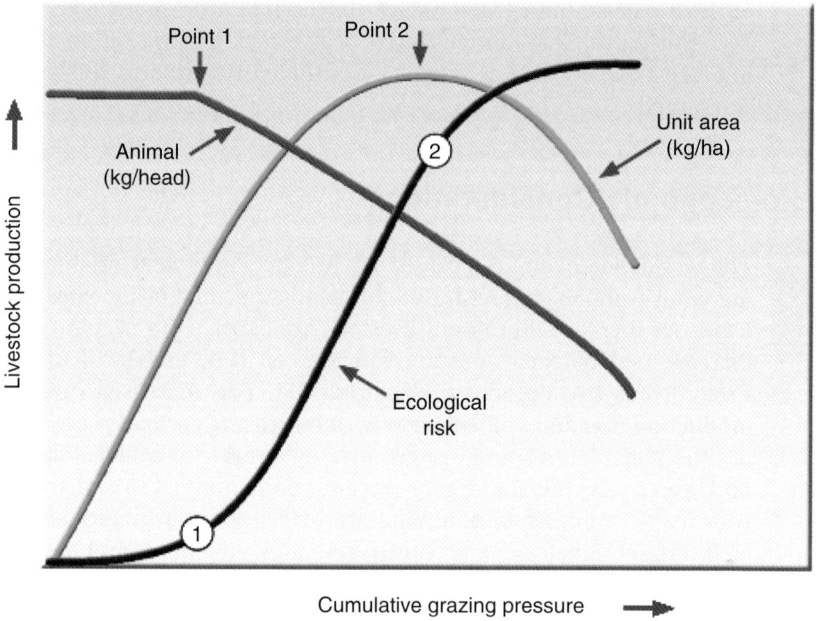

Fig. 5.3. Effects of cumulative grazing pressure on average animal performance, total livestock production per hectare and ecological risk at the end of the 'summer' grazing season. Maximum production per unit land area (Point 2) is always associated with relatively low average animal performance, which begins to decline at the critical cumulative grazing pressure (Point 1).

great distances from the land. Additionally, retiring ranchers often move to town and become interim absentee owners. Absentee ownership is near 40% in many counties in the semi-arid region of the Great Plains, USA. In contrast to their relatives, who developed a strong land management ethic over many years, subsequent generation absentee owners are rarely involved in agriculture. These geographically distant owners often appreciate the beauty and vastness of the grazing lands, but they have little knowledge of the infrastructure and management needed for efficient livestock production. Collectively, absentee-owned grazing land is at risk to overgrazing without trustworthy and knowledgeable management. Good grazing land stewardship requires a working knowledge of the critical plant–animal–environment interactions needed to minimize the cumulative effects of the grazing and drought stresses. Efficient livestock producers are economically astute, but consider the ecological integrity of the grazing lands to be their highest priority.

Likelihood of accomplishing societal-environmental objectives will depend on the ability of governments to mandate changes or provide enough economic incentive for grazing land owners to comply with desired management practices. In fact, many societies now expect agriculture to be caretakers of the environment. Although grazing land owners generally profit from good management, communities and society as a whole benefit from sound land management practices. Direct community payments to land owners to support good environmental management are practised in many developed countries. An ever-increasing number of agri-environment schemes have been developed and financially supported by governments in European countries since the mid-1980s. Similarly, grassland conservation and habitat enhancement programmes in the USA have been integral parts of farm bills passed by the Congress since the mid-1980s. The resulting government agencies and programmes offer cash payments and other incentives to land owners for various conservation practices, including restoring plant and wildlife habitat. Agri-environment and conservation programmes recognize that environmental objectives can be achieved much more effectively with perennial vegetation than annual crops. Additionally, annual cropping systems are characterized by relatively high levels of agronomic inputs and nutrient leakage, frequent and significant disturbance of the soil surface and net losses of SOC. In contrast, plant cover and species diversity are maintained and water, nutrient and energy cycles are sustained on well-managed grazing lands, while achieving suitable levels of livestock production.

Grazing is increasingly viewed as a valuable tool for achieving vegetation management objectives and environmental services in non-agricultural areas. As reviewed in this chapter, grazing can be used to enhance biodiversity, nutrient cycling, hydrologic conditions, C sequestration and wildlife habitat on rangelands and other land types. In the San Francisco Bay area of the USA, domestic livestock are used by government agencies to graze rangelands and other open spaces surrounding cities to optimize biodiversity, control invasion of unwanted plant species, and decrease fuel loads and the risk of devastating wildfire (Sulak *et al.*, 2007). Grazing is also used in Australia's grassland reserves to maintain native grassland communities and to control invasion of unwanted plant species (Lunt, 2003). Additionally, periodic grazing is permitted on the 15 million ha of perennial grasslands established as part of the Conservation Reserve Program in the USA because grazing is seen as a means of maintaining grassland health.

Conclusions

Most grazing lands have relatively low NPP potential; however, they have significant economic and societal value because of a number of factors including their inherent biophysical diversity, expansive catchments and immense open space. Properly managed grazing lands are significant contributors to global C sequestration, air and water quality, aquifer recharge, nutrient cycling, wildlife habitat and biodiversity as well as food and fibre production. Because of the diversity of benefits and possible uses, integrative and holistic approaches are needed to efficiently manage the multiple resources of grazing lands. For these approaches to be sustainable, they must be based on models that clearly identify the effect of grazing and other uses on environmental quality and ecosystem services. Public demands for benefits and non-consumptive uses of grazing lands now trump livestock production on many public lands worldwide where livestock grazing is viewed as a tool and not the number one priority. However, increasing demands for food production in developing countries will continue to put grazing lands at risk of overgrazing and soil erosion. Communities and governments will likely become more involved in guiding use of grazing land resources regardless of land ownership.

References

Alderfer, R.B. and Robinson, R.R. (1947) Runoff from pastures in relation to grazing intensity and soil compaction. *Journal of the American Society of Agronomy* 39, 948–958.

Anderson, D.C., Harper, K.T. and Holmgren, R.C. (1982) Factors influencing development of cryptogamic crusts in Utah deserts. *Journal of Range Manage* 35, 180–185.

Archer, S. and Smeins, F.E. (1991) Ecosystem-level processes. In: Heitschmidt, R.K. and Stuth J.W. (eds) *Grazing Management: An Ecological Perspective*. Timber Press, Portland, Oregon, pp. 109–139.

Bakker, J.P. (1994) Nature management in Dutch grasslands. In: Haggar, R.J. and Peel, S. (eds) *Grassland Management and Nature Conservation*. BGS Occasional Symposium No. 28, Reading, UK, pp. 115–124.

Bauer, A., Cole, C.V. and Black, A.L. (1987) Soil property comparisons in virgin grasslands between grazed and nongrazed management systems. *Soil Science Society of America Journal* 51, 176–182.

Belsky, A.J., Carson, W.P., Jense, C.L. and Fox, G.A. (1993) Overcompensation by plants: herbivore optimization or red herring? *Evolutionary Ecology* 7, 109–121.

Belsky, A.J., Matzke, A. and Uselman, S. (1999) Survey of livestock influences on stream and riparian ecosystems in the western United States. *Journal of Soil and Water Conservation* 54, 419–431.

Betteridge, K., Mackay, A.D., Barker, D.J., Shepherd, T.G., Budding, P.J., Devantier, B.P. and Costall, D.A. (1999) Effect of cattle and sheep treading on surface configuration of a sedimentary hill soil. *Australian Journal of Soil Resources* 37, 743–760.

Biondini, M.E., Patton, B.D. and Nyren, P.E. (1998) Grazing intensity and ecosystem processes in a northern mixed-grass prairie, USA. *Ecological Applications* 8, 469–479.

Blank, R.R., Svejcar, T. and Riegel, G. (2006) Soil attributes in a Sierra Nevada riparian meadow as influenced by grazing. *Rangeland Ecology and Management* 59, 321–329.

Botkin, M.J.M., Melillo, J.M. and Wu, L.S.Y. (1981) How ecosystem processes are linked to large mammal population dynamics. In: Fowler, C.W. and Smith, T.D. (eds) *Dynamics of Large Mammal Populations*. Wiley, New York, pp. 373–387.

Brotherson, J.D. and Rushforth, S.R. (1983) Influence of cryptogamic crusts on moisture relationships of soils in Navajo National Monument, AZ. *Great Basin Naturalist* 43, 73–78.

Buffington, L.C. and Herbel, C.H. (1965) Vegetation changes on a semidesert grassland range from 1858 to 1963. *Ecological Monographs* 35, 139–164.

Clark, H., Pinares-Patino, C. and de Klein, C. (2005) Methane and nitrous oxide emissions from grazed grasslands. In: McGilloway, D.A. (ed.) *Grassland: A Global Resource*. Wageningen Academic Publishers, Wageningen, the Netherlands, pp. 279–293.

Correll, O., Isselstein, J. and Pavlu, V. (2003) Studying spatial and temporal dynamics of sward structure at low stocking densities: the use of an extended rising-plate-meter method. *Grass and Forage Science* 58, 450–454.

Coupland, R.T. and VanDyne, G.M. (1979) Systems synthesis. In: Coupland, R.T. (ed.) *Grassland Ecosystems of the World: Analysis of Grasslands and Their Uses*. International Biological Programme 18, Cambridge University Press, Cambridge, pp. 97–106.

Delgado, C.L. (2005) Rising demand for meat and milk in developing countries: implications for grasslands-based livestock production. In: McGilloway, D.A. (ed.) *Grassland: A Global Resource*. Wageningen Academic Publishers, Wageningen, the Netherlands, pp. 29–39.

Dong, Z., Wang, X. and Liu, L. (2000) Wind erosion in arid and semiarid China: an overview. *Journal of Soil and Water Conservation* 55, 439–444.

Dormaar, J.F. and Willms, W.D. (1998) Effect of forty-four years of grazing on fescue grassland soils. *Journal of Range Management* 51, 122–126.

Dye, P. and Jarmain, C. (2004) Water use by black wattle (*Acacia mearnsii*) implications for the link between removal of invading trees and catchment streamflow response. *South African Journal of Science* 100, 40–44.

Elliott, A.H., Tian, Y.Q., Rutherford, J.C. and Carlson, W.T. (2002) Effect of cattle treading on interrill erosion from hill pasture: modelling concepts and analysis of rainfall simulator data. *Australian Journal of Soil Research* 40, 963–976.

Emmerich, W.E. and Heitschmidt, R.K. (2002) Drought and grazing: II. Effects on runoff and water quality. *Journal of Range Management* 55, 229–234.

FAOSTAT data. (2004) Available at: http://apps.fao.org/faostat/collections?version=ext&hasbulk=0

Floate, M.S.J. (1981) Effects of grazing by large herbivores on nitrogen cycling in agricultural ecosystems. In: Clark, F.E. and Rosswall, T. (eds) *Terrestrial Nitrogen Cycles*. Ecological Bulletin No. 33, Swedish Natural Science Research Council, Stockholm, Sweden, pp. 585–601.

Follett, R.F. and Schuman, G.E. (2005) Grazing land contributions to carbon sequestration. In: McGilloway, D.A. (ed.) *Grassland: A Global Resource*. Wageningen Academic Publishers, Wageningen, the Netherlands, pp. 265–277.

Follett, R.F., Kimble, J.M. and Lal, R. (2001) The potential of U.S. grazing lands to sequester carbon. In: Follett, R.F., Kimble, J.M. and Lal, R. (eds) *The Potential of U.S. Grazing Lands to Sequester Carbon and Mitigate the Greenhouse Effect*. Lewis Publishers, Boca Raton, Florida, pp. 401–430.

Frank, A.B., Tanaka, D.L., Hofmann, L. and Follet, R.F. (1995) Soil carbon and nitrogen of northern Great Plains grasslands as influenced by long-term grazing. *Journal of Range Management* 48, 470–474.

Frank, D.A. and Zhang, Y. (1997) Ammonia volatilization from a seasonally and spatially variable grazed grassland: Yellowstone National Park. *Biogeochemistry* 36, 189–203.

Frederickson, E., Havstad, K.M., Estell, R. and Hyder, P. (1998) Perspectives on desertification: south-western United States. *Journal of Arid Environments* 39, 191–207.

Fuhlendorf, S.D. and Engle, D.M. (2001) Restoring heterogeneity on rangelands: ecosystem management based on evolutionary grazing patterns. *Bioscience* 51, 625–632.

Fuhlendorf, S.D. and Smeins, F.E. (1999) Scaling effects of grazing in a semi-arid savanna. *Journal of Vegetation Science* 10, 731–738.

Fuhlendorf, S.D., Harrell, W.C., Engle, D.M., Hamilton, R.G., Davis, C.A. and Leslie, Jr. D.M. (2006) Should heterogeneity be the basis for conservation? Grassland bird response to fire and grazing. *Ecological Applications* 16, 1706–1716.

Haan, M.M., Russell, J.R., Powers, W.J., Kovar, J.L. and Benning, J.L. (2006) Grazing management effects on sediment and phosphorus in surface runoff. *Rangeland Ecology and Management* 59, 607–615.

Heitschmidt, R.K., Price, D.L., Gordon, R.A. and Frasure, J.R. (1982) Short duration grazing at the Texas experimental ranch: effects on aboveground net primary production and seasonal growth dynamics. *Journal of Range Management* 35, 367–372.

Hibbard, K.A., Archer, S., Schimel, D.S. and Valentine, D.W. (2001) Biogeochemical changes accompanying woody plant encroachment in a subtropical savannah. *Ecology* 82, 1999–2011.

Holechek, J.L., Pieper, R.D. and Herbel, C.H. (2004) *Range Management: Principles and Practices*. Pearson/Prentice-Hall, Upper Saddle River, New Jersey.

Holland, E.A. and Detling, J.K. (1990) Plant response to herbivory and belowground nitrogen cycling. *Ecology* 71, 1040–1049.

Holland, E.A., Parton, W.J., Detling, J.K. and Coppock, L. (1992) Physiological responses of plant populations to herbivory and their consequences for ecosystem nutrient flow. *American Naturalist* 140, 685–706.

Ingham, E.R., Klein, D.A. and Trlica, M.J. (1985) Responses of microbial components of the rhizosphere to plant management strategies in a semiarid rangeland. *Plant and Soil* 85, 65–76.

Intergovernmantal Panel on Climate Change (IPCC). (2001) *IPCC Third Assessment Report: Climate Change 2001: The Scientific Basis*. Contribution of Working Group I to the Third Assessment Report of the Intergovernmental Panel on Climate Change (IPCC). In: Houghton, J.T., Ding, Y., Griggs, D.J., Noguer, M., van der Linden, P.J. and Xiaosu, D. (eds) Cambridge University Press, Cambridge.

Isseltein, J. (2005) Enhancing grassland biodiversity and its consequences for grassland management and utilization. In: McGilloway, D.A. (ed.) *Grassland: A Global Resource*. Wageningen Academic Publishers, Wageningen, the Netherlands, pp. 305–320.

Isseltein, J., Correll, O., Strodthoff, J., Zhao, F. and Hofmann, M. (2003) Variability of sward structure and plant species composition of pastures at low stocking rates. *Grassland Science in Europe* 8, 606–609.

Johnson, L.C. and Matchett, J.R. (2001) Fire and grazing regulate belowground processes in tallgrass prairie. *Ecology* 82, 3377–3389.

Jones, M.B. and Donnelly, A. (2004) Carbon sequestration in temperate grassland ecosystems and the influence of management, climate and elevated CO_2. *New Phytologist* 164, 423–439.

Kempema, S.L.F., Powell, L.A. and Schacht, W.H. (2006) *Influence of Grazing Systems on Grassland Bird Density and Productivity on Private Rangeland in the Nebraska Sandhills*. 67th Midwest Fish and Wildlife Conference, Omaha, Nebaraska (3–6 December, 2006). (Abstract).

Kimble, J.M., Follett, R.F. and Lal, R. (2001) Introduction: the characteristics and extent of U.S. grazing lands. In: Follett, R.F., Kimble, J.M. and Lal, R. (eds) *The Potential of U.S. Grazing Lands to Sequester Carbon and Mitigate the Greenhouse Effect*. Lewis Publishers, Boca Raton, Florida, pp. 3–19.

Klipple, G.E. and Costello, D.F. (1960) Vegetation and cattle responses to different intensities of grazing on short-grass ranges on the Central Great Plains. *USDA Technical Bulletin 1216*, Washington, DC.

Knops, J.M.H., Tillman, D., Haddad, N.M., Naeem, S., Mitchell, C.E., Haarstad, J., Ritchie, M.E., Howe, K.M., Reich, P.B., Siemann, E. and Groth, J. (1999) Effects of plant species richness on invasion dynamics, disease outbreaks, insect abundances and diversity. *Ecology Letters* 2, 286–293.

Lal, R., Kimble, J.M., Follett, R.F. and Cole, C.V. (1998). *The Potential of U.S. Cropland to Sequester Carbon and Mitigate the Greenhouse Effect*. Ann Arbor Press. Chelsea, Michigan, pp. 128.

Lauenroth, W.K., Milchunas, D.G., Dodd, J.S., Hart, R.H., Heitschmidt, R.K. and Rittenhouse, L.R. (1994) Effects of grazing on ecosystems of the Great Plains. In: Vavra, M., Laycock, W.A. and Pieper, R.D. (eds) *Ecological Implications of Livestock Herbivory in the West*. Society for Range Management, Denver, Colorado, pp. 69–100.

Leinweber, P., Turner, B.L. and Meissner, R. (2002) Phosphorus. In: Haygarth, P.M. and Jarvis, S.C. (eds) *Agriculture, Hydrology and Water Quality*. CAB International, Wallingford, UK, pp. 29–55.

Loreau, M. (2000) Biodiversity and ecosystem functioning: recent theoretical advances. *Oikos* 91, 3–17.

Lunt, I.D. (2003) A protocol for integrated management, monitoring, and enhancement of degraded *Themeda triandra* grasslands based on plantings of indicator species. *Restoration Ecology* 11, 223–230.

McDowell, R.W., Drewry, J.J., Paton, R.J., Carey, P.L., Monaghan, R.M. and Condron, L.M. (2003) Influence of soil treading on sediment and phosphorus losses in overland flow. *Australian Journal of Soil Resources* 41, 949–961.

McEldowney, R.R., Flenniken, M., Frasier, G.W., Trlica, M.J. and Leininger, W.C. (2002) Sediment movement and filtration in a riparian meadow following cattle use. *Journal of Range Management* 55, 367–373.

McNaughton, S.J. (1979) Grazing as an optimization process: grass-ungulate relationships in the Serengeti. *American Naturalist* 11, 691–703.

McNaughton, S.J. (1983) Compensatory plant growth as a response to herbivory. *Oikos* 40, 329–336.

McNaughton, S.J. (1985) Ecology of a grazing ecosystem: the Serengeti. *Ecological Monographs* 55, 259–294.

McNaughton, S.J., Ruess, R.W. and Seagle, S.W. (1988) Large mammals and process dynamics in African ecosystems. *Bioscience* 38, 794–800.

McNaughton, S.J., Banyikwa, F. and McNaughton, M. (1997) Promotion of the cycling of diet-enhancing nutrients by African grazers. *Science* 278, 1798–1800.

Milchunas, D.G. and Laurenoth, W.K. (1993) Quantitative effects of grazing and soils over a global range of environments. *Ecological Monographs* 63, 327–366.

Mortenson, M.C., Schuman, G.E. and Ingram, L.J. (2004) Carbon sequestration in rangelands interseeded with yellow-flowering alfalfa (*Medicago sativa* ssp. *falcata*). *Environmental Management* 33 (Supplement), S475–S481.

Mousel, E.M., Schacht, W.H., Zanner, C.W. and Moser, L.E. (2005) Effects of summer grazing strategies on organic reserves and root characteristics of big bluestem. *Crop Science* 45, 2008–2014.

Mousel, E.M., Schacht, W.H., Zanner, C.W. and Wedin, D.A. (2007) Comparison of botanical composition, soil carbon content, and root distribution of subirrigated meadows in the Nebraska Sandhills. *Great Plains Research* 17, 47–60.

National Land and Water Resources Audit (NLWRA). (2001) *Australian Agriculture Assessment 2001*. National Land and Water Resources Audit, Department of Environment and Water Resources. Canberra, Australia.

Novotny, V. and Olem, H. (1994) *Water Quality: Prevention, Identification, and Management of Diffuse Pollution*. Van Nostrand Reinhold, New York.

Noy-Meir, I. (1976) Rotational grazing in a continuously growing pasture: a simple model. *Agricultural Systems* 1, 87–112.

Patten, R.S. and Ellis, J.E. (1995) Patterns of species and community distributions related to environmental gradients in an arid tropic ecosystem. *Vegetation* 117, 69–79.

Paustian, K., Andren, O., Janzen, H.H., Lal, R., Smith, P., Tian, G., Tiessen, H., Van Noordwijk, M. and Woomer, P.L. (1997) Agricultural soils as a sink to mitigate CO_2 emissions. *Soil Use and Management* 13, 230–244.

Pearce, R.A., Frasier, G.W., Trlica, M.J., Leininger, W.C., Stednick, J.D. and Smith, J.L. (1998) Sediment filtration in a montane riparian zone under simulated rainfall and overland flow. *Journal of Range Management* 51, 309–314.

Pinchak, W.E., Smith, A., Hart, R.H. and Waggoner, Jr. J.W. (1991) Beef cattle distribution patterns on foothill range. *Journal of Range Management* 44, 267–275.

Pineiro, G., Paruelo, J.M. and Oesterheld, M. (2006) Potential long-term impacts of livestock introduction on carbon and nitrogen cycling in grasslands of southern South America. *Global Change Biology* 12, 1267–1284.

Pywell, R.F., Bullock, J.M., Walker, K.J., Coulson, S.J., Gregory, S.J. and Stevenson, M.J. (2004) Facilitating grassland diversification using the hemiparasitic plant *Rhinanthus minor*. *Journal of Applied Ecology* 41, 880–887.

Reece, P.E., Brummer, J.E., Engel, R.K., Northup, B.K. and Nichols, J.T. (1996) Total organic reserves in prairie sandreed and sand bluestem after four years of grazing treatments. *Journal of Range Management* 49, 112–116.

Reece, P.E., Schacht, W.H. and Volesky, J.D. (2007) *Building Blocks for Skillful Grazing Management on Semi-Arid Rangelands*. Extension Circular IANR, University of Nebraska, Lincoln, Nebraska.

Reeder, J.D. and Schuman, G.E. (2002) Influence of livestock grazing on C sequestration in semi-arid mixed-grass and short-grass rangelands. *Environmental Pollution* 116, 457–463.

Reeder, J.D., Franks, C.D. and Milchunas, D.G. (2001) Root biomass and microbial processes. In: Follett, R.F., Kimble, J.M. and Lal, R. (eds) *The Potential of U.S. Grazing Lands to Sequester Carbon and Mitigate the Greenhouse Effect*. Lewis Publishers, Boca Raton, Florida, pp. 139–166.

Reich, P.B., Knops, J., Tilman, D., Craine, J., Ellsworth, D., Thoelker, M., Lee, T., Wedin, D., Naeem, S., Bahauddin, D., Hendry, G., Jose, S., Wrage, K., Goth, J. and Bengston, W. (2001) Plant diversity enhances ecosystem responses to elevated CO_2 and nitrogen deposition. *Nature* 410, 809–810.

Ritchie, M.E., Tillman, D. and Knops, J.M.H. (1998) Herbivore effects on plant and nitrogen dynamics in oak savanna. *Ecology* 79, 165–177.

Ross, C.A. and Jarvis, S.C. (2001) Measurement of emission and deposition patterns of ammonia from urine in grass swards. *Atmospheric Environment* 35, 867–875.

Russell, J.R., Betteridge, K., Costall, D.A. and Mackay, A.D. (2001) Cattle treading effects on sediment loss and water infiltration. *Journal of Range Management* 54, 184–190.

Scherer-Lorenzen, M., Palmborg, C., Prinz, A. and Schulze, E. (2003) The role of plant diversity and composition for nitrate leaching in grasslands. *Ecology* 84, 1539–1552.

Schimel, J.P. and Bennett, J. (2004) Nitrogen mineralization: challenges of a changing paradigm. *Ecology* 85, 591–602.

Schimel, D.S., Parton, W.J., Adamsen, F.J., Woodmansee, R.G., Senft, R.L. and Stillwell, M.A. (1986) The role of cattle in the volatile loss of nitrogen from a shortgrass steppe. *Biogeochemistry* 2, 39–52.

Schlesinger, W.H. (1990) Evidence from chronosequence studies for low carbon-storage potential of soil. *Nature* 348, 267–276.

Schlesinger, W.H. (1997) *Biogeochemistry: An Analysis of Global Change, Second Ed.* Academic Press, New York.

Schlesigner, W.H., Raikes, J.A., Hartley, A.E. and Cross, A.F. (1996) On the spatial pattern of soil nutrients in desert ecosystems. *Ecology* 77, 364–374.

Schlesinger, W.H., Abrahams, A.D., Parsons, A.J. and Wainwright, J. (1999) Nutrient losses in runoff from grassland and shrubland habitats in southern New Mexico: I. rainfall simulation experiments. *Biogeochemistry* 45, 21–34.

Schnabel, R.R., Franzluebbers, A.J., Stout, W.L., Sanderson, M.A. and Stuedeman, J.A. (2001) The effects of pasture management practices. In: Follett, R.F., Kimble, J.M. and Lal, R. (eds) *The Potential of U.S. Grazing Lands to Sequester Carbon and Mitigate the Greenhouse Effect*. Lewis Publishers, Boca Raton, Florida, pp. 291–322.

Scholes, R.J. and Archer, S.R. (1997) Tree–grass interactions in savannas. *Annual Review of Ecological Systems* 28, 517–544.

Schuman, G.E., Reeder, J.D., Manley, J.T., Hart, R.H. and Manley, W.A. (1999) Impact of grazing management on the carbon and nitrogen balance of a mixed-grass rangeland. *Ecological Applications* 9, 65–71.

Sharpley, A.N., McDowell, R.W., Weld, J.L. and Kleinman, P.J.A. (2001) Assessing site vulnerability to phosphorus loss in an agricultural watershed. *Journal of Environmental Quality* 30, 2026–2036.

Seagle, S.W., McNaughton, S.J. and Ruess, R.W. (1992) Simulated effects of grazing on soil nitrogen and mineralization in contrasting Serengeti grasslands. *Ecology* 73, 1105–1123.

Sobecki, T.M., Moffitt, D.L., Stone, J., Franks, C.D. and Mendenhall, A.G. (2001) A broad-scale perspective on the extent, distribution, and characteristics of U.S. Grazing Lands. In: Follett, R.F., Kimble, J.M. and Lal, R. (eds) *The Potential of U.S. Grazing Lands to Sequester Carbon and Mitigate the Greenhouse Effect*. Lewis Publishers, Boca Raton, Florida, pp. 21–63.

Stillwell, M.A. and Woodmansee, R.G. (1981) Chemical transformations of urea-nitrogen and movement of nitrogen in a shortgrass prairie soil. *Soil Science Society of America Journal* 45, 893–898.

Sulak, A., Huntsinger, L. and Barry, S. (2007). *Public Grazing Program Goals, Policies, and Consequences in an Urbanizing Landscape*. Abstract. 60th Annual Meeting Society for Range Management, Reno, Nevada.

Svejcar, T., Mayeux, J. and Angell, R. (1997) The rangeland carbon dioxide flux project. *Rangelands* 19, 16–18.

Taylor, C.A. (1989) Short duration grazing; experiences from the Edwards Plateau region in Texas. *Journal of Soil and Water Conservation* 44, 297–302.

Thurow, T.L., Blackburn, W.H. and Taylor, C.A. (1986) Hydrologic characteristics of vegetation types as affected by livestock grazing systems, Edwards Plateau, Texas. *Journal of Range Management* 39, 505–509.

Thurow, T.L., Blackburn, W.H. and Taylor, C.A. (1988) Infiltration and interill erosion responses to selected livestock grazing strategies, Edwards Plateau, Texas. *Journal of Range Management* 41, 296–302.

Tillman, D., Wedin, D. and Knops, J. (1996) Productivity and sustainability influenced by biodiversity in grassland ecosystems. *Nature* 379, 718–720.

Tracy, B.F., Renne, I.J., Gerrish, J. and Sanderson, M.A. (2004) Effects of plant diversity on the invasion of weed species in experimental pasture communities. *Basic and Applied Ecology* 5, 543–550.

UNFCCC. (2004) Greeenhouse gas inventory data. Available at: http://unfccc.int/ghg_emissions_data/items/3800.php

USDA-NRCS (1992) National resource inventory, data base. *USDA*, Washington, DC.

van Ruijven, J., de Deyn, G.B. and Berendse, F. (2003) Diversity reduces invasibility in experimental plant communities: the role of plant species. *Ecology Letters* 65, 84–94.

Verchot, L.V., Groffman, P.M. and Frank, D.A. (2002) Landscape versus ungulate control of gross mineralization and gross nitrification in semi-arid grassland of Yellowstone National Park. *Soil Biology and Biochemistry* 34, 1691–1699.

Vickery, J.A., Tallowin, J.R.B., Febers, R.E., Asteraki, E.J., Atkinson, P.W., Fuller, R.J. and Brown, V.K. (2001) The management of lowland neutral grasslands in Britain: effects of agricultural practices on birds and their food resources. *Journal of Applied Ecology* 38, 647–664.

Warren, S.D., Nevill, M.B., Blackburn, W.H. and Garza, N.E. (1986a) Soil response to trampling under intensive rotation grazing. *Soil Science Society of America Journal* 50, 1336–1341.

Warren, S.D., Nevill, M.B., Blackburn, W.H. and Garza, N.E. (1986b) The influence of livestock trampling under intensive rotation grazing on soil hydrologic characteristics. *Journal of Range Management* 39, 491–495.

Wedin, D.A. (1994) Species, nitrogen and grassland dynamics: the constraints of stuff. In: Jones, C. and Lawton, J.H. (eds) *Linking Species and Ecosystems*. Chapman Hall, New York, pp. 253–262.

Wheeler, M.A., Trlica, M.J., Frasier, G.W. and Reeder, J.D. (2002) Seasonal grazing affects soil quality physical properties of a montane riparian community. *Journal of Range Management* 55, 49–56.

Whitehead, D.C. (1995) *Grassland Nitrogen*. CAB International, Wallingford, UK, pp. 82–107.

Williamson, S.C., Detling, J.K., Dodd, J.L. and Dyer, M.I. (1989) Experimental evaluation of the grazing optimization hypothesis. *Journal of Range Management* 42, 149–152.

Willms, W.D., Dormaar, J.F., Adams, B.W. and Douwes, H.E. (2002) Response of the mixed prairie to protection from grazing. *Journal of Range Management* 55, 210–216.

Zhu, Z. (1991) Fragile ecological zones and land desertification in China. *Journal of Desert Research* 11, 11–22.

6 Environmental Effects of Sheep Farming

J.T. Scott[1] and F.A. Robertson[2]

[1]Landcare Research, Manaaki Whenua, Hamilton 3240, New Zealand;
[2]Future Farming Systems Research, Department of Primary Industries, Hamilton, Victoria 3300, Australia

Introduction

Sheep constitute the world's second largest livestock sector after cattle. In 2004, more than 1000 million sheep grazed arid to humid and arctic to tropical landscapes (FAO, 2007). Some claim this large animal industry has caused and still causes significant environmental harm. Wright (2004) and Diamond (2005) suggest that past societies have collapsed due in part to their adverse effects on, or overuse of, the environment. With increasing intensity of pastoral farming, this is an opportune time to discuss the environmental effects of sheep farming. Although we will discuss international phenomena, the emphasis will be on Australia and New Zealand, since both countries have significant sheep farming industries and systems spanning a wide range of climatic zones, landscape classes, soil types and degrees of farming intensity. These represent many of the conditions found elsewhere in the world and consequently many of the environmental issues discussed with specific reference to these two countries are replicated elsewhere. This chapter gives a detailed presentation of grazing-based sheep farming, emphasizing the central role animal and farm management practices have on the broader environment. Detailed discussion of impacts on soil, water, above-ground and atmospheric environments will be presented. Mitigation measures that may reduce the negative environmental impacts of a sheep farming enterprise or practice will be reviewed, as will implications for public administrations involved with formulating relevant policy. While considering the results of research to date, we will also suggest a broad area of future research that may further our understanding of underlying processes.

Sheep Farming Systems

Sheep are farmed in almost all climatic and geographic areas of the world (FAO, 2007). However, the range in sheep numbers from one area to the next is considerable. Table 6.1 shows total sheep numbers for regions of the world and the 12 countries with the greatest number of sheep.

Sheep farming systems can be classified by climatic type, geographical region, level of sophistication, intensity and socio-economics. Wright (2005) presented the environmentally based classification of livestock grazing systems of de Haan *et al.* (1997), namely: arid, semi-arid, sub-humid tropical, temperate, mixed farming systems. Wright (2005) considered the form of environmental impact as largely a function of the environmental conditions in which the grazing system is located. He recognized that acceptable levels of animal performance are dependent on high-quality forage and high levels of pasture utilization. However, these may cause deleterious impacts on soil, water, atmosphere and wildlife habitats. Wright (2005) further noted that a pastoralist can alter the species of livestock, timing or seasonality of grazing and grazing pressure. Furthermore, within these management choices farmers apply a range of inputs to achieve their target production levels.

Australia and New Zealand provide good examples of the range of possible sheep farming systems. Since the late 1950s, Australia has mostly had the largest sheep population in the world, and today it contains the second largest (after China; Cottle, 1991; Australian Wool Innovation, 2006). The Australian sheep industry can be divided along climatic lines into the pastoral zone, the wheat-sheep zone and the high-rainfall zone. The pastoral zone, which represents more than 50% of the area grazed by sheep but contains only 12% of the flock, is mainly in the arid/semi-arid inland, with annual rainfall less than 300 mm (Cottle, 1991; Australian Wool Innovation, 2007). The industry is based on extensive grazing of unfertilized native grassland and shrubland, much of which has been invaded by introduced pasture species. Carrying capacity varies between 3 and 25 ha per sheep, depending on availability of water as much as on the quantity and quality of feed. Variation in rainfall between and within seasons can be extreme (Cottle, 1991; Squires, 1991).

Table 6.1. Sheep numbers in regions of the world and the 12 most populous countries in 2004. (From FAO, 2007.)

Countries	Numbers	Region	Numbers
China	157,330,000	Africa	246,284,000
Australia	94,500,000	North and Central America	17,618,000
India	62,500,000	South America	70,237,000
Iran	54,000,000	Central Asia	52,679,000
Sudan	47,000,000	Far East Asia	267,775,000
New Zealand	40,000,000	Near East Asia	125,358,000
UK	35,500,000	Europe	123,166,000
South Africa	29,100,000	Oceania	134,574,000
Turkey	25,000,000	Total	1,037,691,000
Pakistan	24,700,000		
Spain	24,000,000		
Nigeria	23,000,000		

The wheat-sheep zone, which contains 55% of the country's flock, is characterized by the rotation of cropping and improved pastures, each phase lasting up to 5 years. Annual rainfall averages 300–600 mm. The pastures are annual or perennial grasses with a dominant legume component, and have a carrying capacity of around 0.6–1.5 sheep/ha (Cottle, 1991; Pratley and Godyn, 1991).

The high-rainfall zone (annual rainfall >600 mm) is the most productive, with an average carrying capacity of more than 3 sheep/ha. It contains 33% of Australia's sheep flock. Pastures are generally sown perennial grass-legume mixtures with a dominant grass component. Many pastures retain a minor presence of native species. As in the wheat-sheep zone, pasture production in the high-rainfall zone is highly seasonal (Cottle, 1991; Pratley and Godyn, 1991; Australian Wool Innovation, 2006).

In all zones, the most common grazing management is set stocking (where grazing is continuous and the number of animals per unit area changes very little throughout the year), but various forms of rotational grazing (where grazing is intermittent and animal density is varied according to feed availability and target animal performance) are also practised in the wheat-sheep and high-rainfall zones. In the pastoral zone, there is usually no supplemental feeding, and chemical inputs are restricted to ectoparasiticides to treat lice and blowfly strike. In the wheat-sheep and high-rainfall zones, feed may be supplemented by forage crops, conserved fodder or grain at times of low pasture production (Pratley and Godyn, 1991). In addition to ectoparasiticides, chemical inputs in the wheat-sheep and high-rainfall zones include anthelmintics for control of internal parasites; herbicides for pasture establishment and general weed control (mainly glyphosate, paraquat, triazines); and insecticides for control of pasture pests. Soil amendments include lime and fertilizer (mainly superphosphate (10–30 kg P/ha/year), and possibly occasional applications of micronutrients, potassium (<100 kg K/ha) or nitrogen (20–100 kg N/ha; Quinn *et al.*, 2005; McCaskill and Quigley, 2006). Fire may occur occasionally in any of the sheep farming zones, either through wildfire or controlled burning to aid pasture establishment and growth.

Intensive sheep production systems, with high inputs of fertilizer, irrigation, high pasture production and grazing pressure, or housed animals, represent a very small proportion of the Australian sheep industry (Dawe, 1991).

New Zealand has the highest density of sheep per unit area in the world, closely followed by England (FAO, 2007). Brief descriptions of sheep farming in New Zealand have been presented by Mathews *et al.* (1999), Hodgson *et al.* (2005) and Agritech (2007). Agritech outlined a simple description of three broad sheep farming systems in New Zealand: extensive rangeland grazing, often sub-alpine, producing fine and mid-micron wool; Romney and Romney-derived sheep grazing low-productivity hill country pastures producing coarse wool and lamb, where lambs are usually sold to higher-producing farms to be finished for slaughter; and intensive grazing of lowland pastures with high productivity producing prime lambs, primarily for export. Stocking rates (in stock units (su), which are defined as a 50 kg ewe rearing one lamb, their feed intake being 550 kg dry matter (DM)/year) range from a mean of 0.7 su/ha in the high country to 7.5 su/ha in the hill country to 14 su/ha on flat and rolling land. Very intensive irrigated flat land can exceed 20 su/ha. Hill and high country can exceed 25° slopes while rolling land tends to be less than 15°.

Approximately 16,000 sheep and beef farms contain 45 million sheep, including 30 million ewes that produce 36 million lambs each spring. About 24 million lambs are finished and exported each year, making New Zealand the world's largest lamb meat exporter. The remaining stock are used to replace ewes and rams sent to slaughter and lost by natural attrition (Mathews *et al.*, 1999; Agritech, 2007).

Of the farm types noted above, extensive rangeland systems are practised elsewhere around the world, notably in Australia, South Africa, China and South America, and intensive fat lamb systems in Europe and North America. However, hill country farming, with low fertilizer inputs and easy-care sheep, mostly unshepherded, is most common to New Zealand and the UK. Low-cost production is based on large numbers of sheep per labour unit; self-sufficiency by foraging for food from pastures; unsupervised lambing, with hardy, active lambs; low rates of use of animal remedies, selection for resistance to disease and infertility; and the use of sheep dogs, yards, motor bikes, dipping facilities and shearing sheds (Agritech, 2007).

MacLoed and Moller (2006) noted the increased intensity of New Zealand agriculture including sheep farming since the late 1960s. This has occurred despite the decline in the number of sheep and sheep-farmed land since 1980. The productivity of New Zealand hill country sheep farming, being the dominant farming type, has increased quickly in the past 20 years because of, but not limited to, the use of improved pasture species, more fertilizer, weed control, closer paddock subdivision, breeding for disease resistance and more fertile and resilient sheep breeds (Agritech, 2007).

Environmental Effects of Sheep Farming

Soil

Sheep farming can negatively impact the soil environment by damaging soil structure through trampling, facilitate increased erosion, decrease soil biodiversity and organic C and cause excessive soil N, P, insecticide and pharmaceutical concentrations (Greenwood and McKenzie, 2001; Drewry, 2006). Furthermore, the impact of disturbance by farmed sheep may have a greater impact on old soils compared with younger soils and landscapes (Walker *et al.*, 2001). Conversely, the grazing of sheep can enhance the soil environment by increasing soil biodiversity, soil C and soil nutrient levels (Bardgett, 2005).

The concept of soil quality has been considered by some as a tool to evaluate the state of 'soil health'. The Soil Science Society of America (Karlen *et al.*, 1997) defined soil quality as, 'the capacity of a specific kind of soil to function within natural or managed ecosystem boundaries, to sustain plant and animal productivity, maintain or enhance water and air quality, and support human health and habitation'. However, it is acknowledged that a soil property may have two mutually exclusive functions simultaneously (Letey *et al.*, 2003). Sparling and Schipper (2002, 2004) and Sparling *et al.* (2004) carried out an extensive soil quality measurement programme over 222 sites in New Zealand that included 44 sheep farming sites. Unlike soil quality assessment in some other countries that focused on production (Karlen *et al.*, 2001), Sparling and Schipper's (2002, 2004) and Sparling *et al.*'s (2004) objective was to collect data for reporting on the state of the environment at a regional scale. They were cognizant of the need for relevance to specific land use. After extensive evaluation they limited themselves to seven soil properties that explained 87–88% of the variance in their data (Lilburne *et al.*, 2004). The seven soil properties were total C, total N, mineralizable N, pH, Olsen P, bulk density and macroporosity. They did not choose some additional useful properties because of difficulty of measurement or because of high correlations with the core seven soil properties. For example, soil microbial biomass and activity is well correlated with anaerobically mineralizable N. We discuss the effect of sheep farming on roughly similar and related soil properties.

Soil physical properties

The relevance of research in this area is not only to understand the effects on production (Edmond, 1964; Gillingham and During, 1973; Proffitt *et al.*, 1993) but also the wider environment. Although sheep exert lower static pressure (83 kPa) than cattle (160–192 kPa), sheep hooves can exert up to 200 kPa of pressure (Willatt and Pullar, 1983), which can be greater than some tractors (Soane, 1970). Such pressures under sheep suggest the potential for degradation of soil physical quality that could adversely affect soil function and increase water surface runoff and erosion. This has indeed been found to occur (Lambert *et al.*, 1985; Greenwood and McNamara, 1992; Eldridge, 1998). Sheep treading can cause soil compaction and soil homogenization (poaching, puddling and pugging; Drewry, 2006), although to a lesser extent than cattle (Drewry *et al.*, 2000). Treading damage decreases soil permeability through reduced pore space and disrupted soil pore networks, and increased bulk density (Drewry, 2006).

The effect of an applied load by sheep to soil will not only depend on the state of the soil (soil water content) but also on the inherent physical nature of the soil being grazed upon. Climo and Richardson (1984) found macroporosity (soil pores >30 μm) before compaction was well correlated with susceptibility to treading damage. There was less treading damage in soil with greater structural stability and degree of drainage (lower soil water content). Surveying 97 sheep farms across five different soil types in southern New Zealand, Drewry *et al.* (2000) found those sheep farms on soils with greater inherent macroporosity, air permeability, hydraulic conductivity and low bulk density maintained this advantage. Nevertheless, within the constraints or structural advantages of different soils, sheep grazing can reduce soil macroporosity, and high stocking rate-intensive grazing has a greater effect than low stocking rates (Greenwood and McNamara, 1992). Furthermore, they found air permeability and hydraulic conductivity to be good indicators of the degree of soil compaction.

Geology, rainfall and topography are the main factors controlling sediment output from New Zealand catchments (Hicks and Griffiths, 1992). However, the replacement of forest by grazed pastures and sheep treading contributes to sediment loss to waterways. For example, Fig. 6.1 shows steep hill country in the North Island of New Zealand that has been partially converted from native forest to rough sheep grazing. Areas of soil erosion can be seen. The compaction and tread-

Fig. 6.1. Steep hill country, in the North Island of New Zealand, that has been partially converted from native forest to rough sheep grazing. Areas of soil erosion can be seen.

ing damage caused by sheep has been shown to increase soil erodibility because of decreased infiltration rates and soil surface damage (Eldridge, 1998; Elliot and Carlson, 2004), changes to surface roughness (Hairsine *et al.*, 1992; Betteridge *et al.*, 1999), and loss of ground cover (Lang and McCaffrey, 1984; Lambert *et al.*, 1985; Elliot and Carlson, 2004). These can be mitigated to varying degrees by management within the constraints of soil characteristics. Using rainfall simulators on hill country (12–19°) grazed by sheep, Elliot and Carlson (2004) found that sediment and particulate nutrient concentrations in overland flow increased by a factor of 13–16 after intensive winter grazing and rainfall applied at high intensity; in summer by a factor of 3. Sediment and particulate nutrients in overland flow were highly correlated with percentage bare ground. In winter, infiltration rate was reduced to a greater extent than summer because of greater smearing of the soil surface and blocking of macropores. The same trends were observed under cattle grazing, except cattle created more bare ground and soil damage than sheep, taking >2 months to recover compared with 6 weeks under sheep grazing (Elliot *et al.*, 2002). In lower rainfall areas the greatest treading effect can be in summer and autumn with a higher risk of bare ground (McColl and Gibson, 1979).

Carbon

Current atmospheric CO_2 concentrations and emissions to the atmosphere contribute to climate change (Forster *et al.*, 2007), and changes in land use affect a soil's contribution to this C balance (Guo and Gifford, 2002). Many consider it desirable to increase rather than decrease soil C concentrations (Sparling *et al.*, 2006). Increased soil C concentrations enhance nutrient retention and soil structure and general soil resilience. Consequently, any land use that affects soil C dynamics requires careful evaluation. Soil C stocks have shown increases and decreases in response to different levels of grazing within the wide range of habitats that support sheep grazing (Tate *et al.*, 1995; Stewart and Metherell, 1999a,b; Schuman *et al.*, 2002; Zhou *et al.*, 2007). The effects of sheep grazing on soil C concentrations are complex, reflecting interactions among pasture growth, pasture utilization by the animal, management and environmental factors. Underlying the impact of managed grazing are the implications of the inherent soil properties (Walker *et al.*, 2001).

Sheep are grazed on grassland converted from forest or scrub and on natural grasslands and rangelands, some of which have been modified by cultivation, inputs of fertilizers and introduced forage species. These land-use changes affect soil C concentrations. In a lowland summer dry area of New Zealand, conversion of unimproved dryland grassland to higher-producing border-dyke irrigated pasture by cultivation, land forming, fertilization, liming and sowing of ryegrass and clover showed an initial decline in soil C concentrations to 27 g/kg in the surface 75 mm, followed by a subsequent rise over 11 years to 37 g/kg (Nguyen and Goh, 1990; Metherell, 2003). Little change was observed in the subsequent 30 years. Fifty years after development there was no difference in soil C concentrations and bulk density in 0, 188 or 376 kg superphosphate/ha/year treatments (Metherell, 2003). However, an adjacent experiment with irrigated and unirrigated treatments has shown soil C to be 36 g/kg in the 0–10 cm surface soil with irrigation and 42 g/kg without irrigation, despite greater herbage production in the irrigated treatments. These C responses reflect treatment effects on root production, litter quality, earthworm activity and mineralization rates (Fraser and Piercy, 1996; Stewart and Metherell, 1999a). Soil C concentrations were maintained in a steep (27°) and semi-arid high country tussock grassland site, where grazing management and stocking rate were compared after over-sowing and top dressing with fertilizer and improved pasture species. Interestingly, there was some increase in soil C concentrations at low stocking rates of 1.9 sheep/ha (Metherell, 2003). In other semi-arid areas, grazing was found to increase soil C in an alpine meadow in Wyoming (Povirk, 1999). Henderson (2000) reported that on some sites in a southern Canadian prairie soil, organic C content in the surface 0–10 cm was greater under grazing (herbivore not specified) than ungrazed, but no significant difference to 105 cm depth. This was in contrast to Zhou *et al.* (2007) who, in semi-arid (385 mm) northern China with severe degradation by sheep overgrazing, reported uncontrolled grazing with approximately 75% utilization resulted in total soil organic C to 100 cm being 60% less than when stock had been excluded for 3–4 years. Henderson (2000) reported that in semi-arid sites (mean precipitation 328–390 mm) topsoil C concentration was greater under grazing than ungrazed enclosures, whereas at wetter sites (476 mm) topsoil C concentration was greater without grazing than with grazing. These results are somewhat consistent with those found on a humid site in New Zealand converted from forest to pasture with annual rainfall of approximately 1200 mm. There was a decline in soil C concentrations in the surface 75 mm from 54 to 48 g/kg (Lambert *et al.*, 2000). Again, in spite of large increases in pasture production with increased soil fertility, soil C concentrations either did not increase compared with lower fertility sites or decreased. It was thought this was a consequence of conversion from forest to pasture and an increased proportion of net primary production (NPP) being consumed by animals and lost to the atmosphere (Lambert *et al.*, 2000; Clark *et al.*, 2001).

Grazing can increase NPP, dung and urine deposition, which enhances nutrient cycling, stimulates root respiration, root exudates and C allocation below ground (Wardle and Bardgett, 2004), all of which may enhance soil C storage (Schuman *et al.*, 2002). This interpretation is consistent with less-intensive grazing management in sub-humid environments. In a summer dry region of the Canterbury plains, New Zealand, Hoglund (1985) found a linear increase in soil C and N with increasing residual DM left after grazing of a newly sown ryegrass white clover pasture. Stewart and Metherell (1999a,b) and Metherell (2003) showed that unfertilized and unirrigated pasture ecosystems produced less above-ground biomass, greater root biomass, below-ground net primary production (BNPP), root C allocation and longer root turnover times than did fully irrigated and fertilized pasture ecosystems. This process is consistent with lower soil C concentrations in intensively managed, high-fertility, humid environment or irrigated pastures (Lambert *et al.*, 2000; Bardgett *et al.*, 2001; Metherell, 2003) and high forage utilization or overgrazed sub-humid rangelands (Zhou *et al.*, 2007). New Zealand's pastoral agriculture has intensified production significantly over the past 20 years (MacLoed and Moller, 2006). Not surprisingly, Schipper *et al.* (2007), after re-sampling sites on dairy and sheep/beef farms 17–30 years later, found significant declines in soil C concentration under both land uses. Sankaran and Augustine (2004) suggest grazer stimulation of production is highest at intermediate grazing intensities and this production stimulation may offset consumptive losses and produce a net increase in C inputs to soil.

Nitrogen and Phosphorus

Sheep are farmed in a range of landscapes, from those that may receive no fertilizer, e.g. extensively grazed rangelands, to landscapes that receive significant quantities of fertilizer, e.g. where soil moisture is generally not limiting and animal performance and production per unit area are high. At both extremes, sheep farming affects soil nutrient dynamics that in turn can affect the wider environment.

NITROGEN Sheep farming based on sown pastures with inputs of N from fertilizer or legumes will almost invariably increase soil N concentrations compared with those existing under native vegetation (Russell, 1960). In P-limited environments, application of superphosphate to otherwise unfertilized pastures can dramatically increase soil N through stimulation of legume growth, N fixation and increased deposition of N in dung and urine (McCaskill and Cayley, 2000). Increased soil N is not likely, however, where sown pastures are neither fertilized nor contain a legume component (Bligh, 1990). In extensive grazing systems based on unfertilized native vegetation, increases in soil N may also occur through the invasion of exotic legumes (Pratley and Godyn, 1991).

The usual consequence of increased soil N is an increased concentration of N in plant material and in the urine of the grazing sheep. The concentration of N in the dung is relatively insensitive to diet (Barrow and Lambourne, 1962).

The proportion of ingested N that is voided in urine may vary from around 40% where feed N content is low to around 80% where feed N content is high (Sears, 1950; Barrow and Lambourne, 1962; Whitehead, 2000).

About 70% of the N in urine is urea; most of the remainder being other forms of organic N (Doak, 1952). The urea is rapidly hydrolysed, resulting in extremely high concentrations of ammonium (>200 mg N/kg) in the surface soil of the urine patch within several days of urination (Sherlock and Goh, 1984; Haynes and Williams, 1993). While some of this N can be utilized by plants and microorganisms or stabilized in soil clays or organic matter (Haynes and Williams, 1993; Sakadevan *et al.*, 1993; Thomsen *et al.*, 2003), the N surplus becomes vulnerable to loss through volatilization, denitrification, leaching or surface runoff (Ball and Keeney, 1983; Carran *et al.*, 1982; Silva *et al.*, 1999; Decau *et al.*, 2004). These losses are discussed elsewhere in this chapter. Sheep dung contains around 20–55% of its N in soluble, mainly organic, forms (Haynes and Williams, 1993; Whitehead, 2000). The bulk of the dung is insoluble material that is relatively recalcitrant and more resistant to decomposition than the plant material from which it is derived (Floate, 1970a; Thomsen *et al.*, 2003). Release of N from decomposing dung can, nevertheless, lead to high concentrations of soil inorganic N (>100 mg N/kg) at the dung patch that may also contribute to N loss from the system (Haynes and Williams, 1993).

Data reviewed by Haynes and Williams (1993) suggest an individual sheep may return between 5 and 14 kg N/year as urine and between 0.3 and 9 kg N/year as dung. This return of excreta contributes substantially to N cycling and maintenance of N availability in pastoral systems (Lambert *et al.*, 1982; Haynes and Williams, 1993). However, because it also promotes loss of N and heterogeneous distribution of N leading to inefficient utilization (Ledgard, 2001), its benefit may be much less than the quantities returned would suggest.

PHOSPHORUS Soil P is an essential macronutrient for plant and animal growth and function. It is also a pollutant. Soil P is found in inorganic (Pi) and organic (Po) forms, with Po constituting from 5% to 90% of total soil P. Sheep farming in poorly producing landscapes, such as dry rangelands, often depends on the productive capacity of unfertilized soil. In landscapes with promising attributes for potential increases in forage production (e.g. climate, topography, infrastructure, etc.), P fertilizers and an N source (clovers and/or N fertilizers) become standard inputs. The addition of inorganic P in mineral fertilizers fuels the highly productive sheep farming areas of the world. An example of a P cycle on steep hill country grazed by sheep and beef is shown in Fig. 6.2 (Gillingham *et al.*, 1984).

Different forms of soil P have differing levels of solubility and mobility in water and plant availability (Stewart and Tiessen, 1987). Furthermore, physical location and protection of soil P within the structural soil matrix have been shown to differ (He *et al.*, 1995; Six *et al.*, 2001; Scott *et al.*, 2005; McDowell *et al.*, 2006), which has implications for P loss to waterways. Characterizing these forms under different land uses and management regimes can give us an insight into biogeochemical cycles, plant nutrition and P loss to waterways (Frossard *et al.*, 2000; Leytem *et al.*, 2002; Turner *et al.*, 2003a; Condron *et al.*, 2006; McDowell and Stewart, 2006).

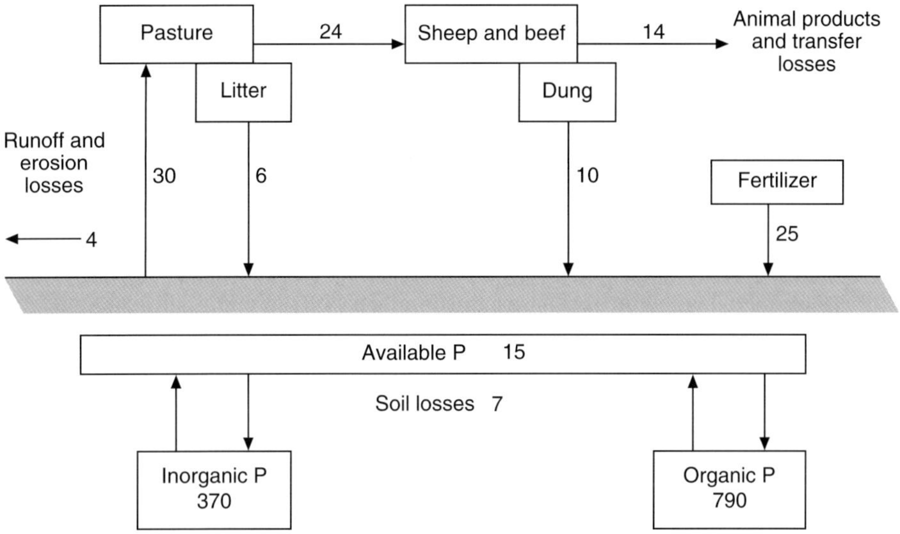

Fig. 6.2. The P cycle for steep hill country pasture, grazed by sheep and beef in the North Island of New Zealand. The numbers represent kg P/ha/year. (Adapted from Gillingham *et al.*, 1984.)

During the development phase of a sheep-grazed pasture, the addition of superphosphate resulted in an increase in various forms of inorganic and organic soil P with time (Condron and Goh, 1989). Condron and Goh (1989) found that with lower rates of superphosphate, lower or no increases in various forms of chemically extracted forms of P occurred. However, even if no fertilizer was applied, Condron and Goh (1989) found organic P increased with time. This was attributed to biological transformation of soil Pi via plant, animal and microbial residues. Given a particular amount of Pi input, with time, soil Po accumulation will decline with a concomitant increase in Pi accumulation. This can be attributed to a steady state being reached between organic inputs and decomposition (Anderson, 1980) and possible limits of organic C accumulation (Six *et al.*, 2002).

Soil biota

Cattle and sheep grazing in British upland grasslands has been shown to increase litter quality that in turn stimulates soil microbial activity, soil mineralization processes, N cycling and hence plant production and therefore leads to increased herbivore-carrying capacity (Floate, 1970a,b; Bardgett *et al.*, 1997; Bardgett, 2005). In contrast, increased fertility on irrigated and intensively sheep-grazed pasture at Winchmore in New Zealand showed a decline in microbial biomass, even though soil C concentrations were similar between 0 and 376 kg/ha/year superphosphate application rates (Metherell, 2003) This was a consequence of a decline in C/N ratio, and enhanced decomposition and nutrient cycling through improved litter quality. Furthermore, Metherell *et al.* (2002) reported that in spite of increased herbage production with irrigation compared with dryland, soil C concentration and the soil C/N ratio declined under irrigation, whereas soil microbial biomass C and N increased. Long-term irrigation increased earthworm (Fraser and Piercy, 1996) and microbial activity. There was no difference in three measures of labile C (K_2SO4, hot water soluble and cold water soluble) or C input. Consequently, Metherell *et al.* (2002) considered that soil moisture retention is more closely related to biological activity than to total soil organic matter or even their measures of labile C.

Bardgett *et al.* (1993; 1997) showed that grazing by sheep on temperate grasslands in the UK can increase microbial biomass and activity, and also the abundance of microbial consumers, namely nematodes and microarthropods. However, as reported earlier, intensive sheep grazing systems well supplied with moisture tend in the long term to result in lower soil C concentrations and can, but not always, result in decreased soil microbial biomass concentrations. This is consistent with the findings of Bardgett *et al.* (2001), who, when considering the structure of microbial communities under long-term grazing by sheep, showed that microbial biomass of soil is maximal at low to intermediate levels of grazing and the phenotypic evenness (a component of diversity) of the microbial community declines as intensity of grazing increases. This was indicated by an ester-linked phospholipid fatty acid (PLFA) evenness index: a low index indicates a dominance of certain microbial groups. They reported that bacterial-based energy channels of decomposition dominate communities of heavily grazed sites, while in systems that are less intensively grazed, or completely unmanaged, fungi have a proportionally greater role.

The possible negative effects of sheep grazing alluded to above have been reported elsewhere. In Australia as stocking rates increased (10–30 sheep/ha) Collembola numbers in the surface soil declined significantly (King and Hutchinson, 1976; King *et al.*, 1976). Likewise, under perennial ryegrass (*Lolium perenne*) grassland, decreases in Collembola numbers were associated with increased sheep stocking rate (Walsingham, 1976). These were attributed to declines in soil pore space and surface litter with increased sheep grazing intensity.

There seems to be a high degree of species redundancy at any particular trophic level within soil food webs but some species are more redundant than others (Bardgett, 2005). Consequently research on intensive production systems which involve the effects of high levels of soil fertility, non-limiting soil moisture levels and high levels of pasture utilization by sheep on soil biology is required.

Pharmaceuticals, pesticides, heavy metals and other contaminants

Sheep farming involves a range of inputs, the quantity of which increases with production. These inputs include fertilizers (which contain heavy metals), pesticides and veterinary pharmaceuticals.

HEAVY METALS AND FLUORINE Phosphate fertilizers are often applied to sheep pastures to increase or maintain production. Loganathan *et al.* (2003) reviewed fertilizer contaminants in New Zealand's grazed pasture. As fertilizers in New Zealand are produced from the same raw materials as other parts of the world, this review has general applicability. The starting material, phosphate rock, may contain toxic elements such as: arsenic (As), cadmium (Cd), chromium (Cr), fluorine (F), strontium (Sr), thorium (Th), uranium (U) and zinc (Zn). All these metals and F have the potential to accumulate in soil (Sauerbeck, 1992). Cadmium is the element of most concern as it can be absorbed into animal and human bodies where it accumulates in the kidneys and liver and can be toxic in body tissues (Black, 1988). Decreasing the ingestion of Cd by sheep may be achieved by using low-Cd fertilizers, and cultivating paddocks to mix surface soil and soil without much Cd from lower in the profile. Avoiding hard grazing decreases soil ingestion on high-Cd sites.

Even though F is an essential element for animals and humans it can be toxic. Phosphate rock can contain 3–4% F, superphosphates 1–2.4%, and diammoinium phosphate 1.2–3.0% (Loganathan *et al.*, 2003). Fluorosis in sheep occurs and therefore it is recommended that stock be withheld from P-fertilized pastures until 25 mm of rain has fallen and minimize soil ingestion.

Continued application of rock phosphate-based fertilizers, however, will cause a continual increase in soil contaminant levels on sheep farms, which is likely to result in metal sensitive organisms being replaced by tolerant organisms within each functional group. Nevertheless, lower soil microbial diversity (Lakzian *et al.*, 2002) may reduce the resilience of a soil community to an additional stress or disturbance (Degens *et al.*, 2001; Almås *et al.*, 2004).

PESTICIDES Pesticides include herbicides, insecticides and fungicides. Many pesticides used in sheep farming are common to other farming systems and therefore a detailed review will not be attempted. Debates about the environmental effects of pesticide use in sheep farming are similar to debates about other farming systems. While there is a dearth of literature specifically focusing on pesticides, sheep farming and the environment, several relevant reviews have been published recently (e.g. Sarmah *et al.*, 2004).

Pesticides used in sheep farming can have clear environmental effects. For example, the insecticide DDT was applied extensively between 1947 and 1970 to pastures in New Zealand to control grass grub (*Costelytra zealandica*) and porina (*Wiseana spp*) until its use was banned (Boul, 1994). Residues still persist and restrict land-use options. Soil contamination at sheep dip sites is common both in New Zealand and elsewhere (Hooda *et al.*, 2000). The herbicide 245-T, used to control woody weeds on sheep farms in New Zealand, is another example of a banned pesticide with continuing environmental implications, primarily for human health. Advice on safe use of pesticides is usually provided in developed countries. There are many web sites that attempt to provide advice on the use and dangers of pesticide use in sheep farming. For example, the Veterinary Medicines Directorate (2007) in the UK states: 'Less than one teaspoon of cypermethrin sheep dip can wipe out insect life for hundreds of metres and may ruin fishing.' Not surprisingly, the British Department for the Environment, Food and Rural Affairs issued a report considering financial instruments that could be used to decrease the environmental impact of pesticide use (defra, 2004).

VETERINARY PHARMACEUTICALS Veterinary pharmaceuticals include antibiotics (anti-microbials) and parasiticides such as anthelmintics (Tolls, 2001; McKellar, 2006). McKellar (2006) recently published a short review of the medicines used in the sheep industry. Similar to pesticides, veterinary antibiotics (VAs) used on sheep farms are common to other livestock farming systems. Sarmah *et al.* (2006) reviewed extensively a global perspective on the use, sales, exposure pathways, occurrence, fate and effects of VAs in the environment. Their review is a source of, and cites, many relevant topics. Although consideration of antibiotics in the environment is important, the sheep industry is a very low user relative to other animal industries. Conversely, anthelmintics are widely used. McKellar (1997) reported on the ecotoxicology of anthelmintic compounds and their residues. The greatest ecotoxicological risks were associated with the avermectins and milbemycins, which adversely affected a number of insect species. Yeates *et al.* (2007) found fewer earthworms in soil where dung from sheep was treated with an ivermectin bolus compared with albendazole or the fungus *Duddingtonia flagrans* (Duddington) and a control.

Water

Water flows

Sheep production systems are generally small users of water. Pastures are rarely irrigated, and water use is often restricted to stock watering and occasional spraying and dipping. Use of surface and groundwater supplies is therefore unlikely to be of environmental concern. An exception may be the use of surface water during drought conditions. Sheep farming can, however, affect all the major pathways of soil water movement – overland flow, interflow and drainage – with potential consequences for soil erosion and the quantity and quality of underground and surface water.

Grazing by sheep can damage surface soil structure and decrease ground cover, both of which tend to decrease soil infiltration and promote overland flow, i.e. infiltration-excess overland flow (Willatt and Pullar, 1983; Proffitt *et al.*, 1993; Elliot and Carlson, 2004). These effects are particularly marked at high stocking densities and in high-traffic areas of the paddock such as sheep tracks and camping areas, and on upper slopes (Dougherty *et al.*, 2004). In fragile environments such as semi-arid rangelands, even relatively low stocking rates (e.g. 0.5 sheep/ha) can have these adverse effects (Greene *et al.*, 1994). Whether infiltration-excess overland flow will be initiated depends on rainfall intensity as well as on soil and pasture conditions, but on bare soil with a surface seal it may occur rapidly at light rainfall intensity (Greene *et al.*, 1994). A difficulty in many extensive sheep farming systems is setting stocking rates that will be non-destructive during periods of low forage availability (Pratley and Godyn, 1991; Squires, 1991).

Overland flow from sheep-grazed land often occurs as saturation excess during the wettest parts of the year (McColl and Gibson, 1979; Costin, 1980; Murphy *et al.*, 2004). Under these conditions, surface soil characteristics and ground cover become comparatively unimportant for overland flow generation (Murphy *et al.*, 2004). Saturation-excess overland flow often occurs on small down-slope parts of the paddock that are the first to become saturated (Ward, 1984). It may include interflow and lateral subsurface flow from further up the slope (Ward, 1984).

That small areas of the landscape contribute disproportionately to overland flow (through either infiltration or saturation excess, or both) has been observed in several sheep farming systems (Smith, 1987; McCaskill *et al.*, 2003; Melland, 2003). Interflow may be more or less significant than overland flow in different situations (Cox *et al.*, 2005), and will be similarly spatially variable. Both are potentially important conduits of contaminants to surface waterways (Stevens *et al.*, 1999).

Deep drainage may convey contaminants to groundwater and surface water. Drainage may occur rapidly through bypass flow, in which case the concentrations in the drainage water reflect solute conditions at the soil surface. Alternatively, drainage may occur slowly through matrix flow, reflecting solute conditions in the bulk soil (Chittleborough *et al.*, 1992).

The partitioning of water flows between drainage, overland flow and interflow depends on environment and management factors, and in sheep-grazed systems has been shown to vary enormously between similar sites and between years or even between rainfall events (White *et al.*, 2000; Ridley *et al.*, 2003). Climate is of primary importance and interacts with other factors. Soil properties such as texture, porosity, infiltration and hydraulic conductivity also have a major influence (Stevens *et al.*, 1999). Management that maximizes pasture growth and water use (e.g. plant species selection or application of soil amendments) will tend to reduce surface and subsurface water movement (White *et al.*, 2000). This technique is used in places like Australia to decrease recharge to groundwater in an attempt to combat dryland salinity (White *et al.*, 2003). Vigorous permanent pastures may, on the other hand, promote infiltration and drainage because of the accumulation of organic matter in the surface soil (Connolly *et al.*, 1998) and the development of large soil pores through the subsoil (McCallum *et al.*, 2004). Installation of subsurface drains can dramatically reduce overland flow (Sharpley and Syers, 1976).

Water quality

Sheep farming is often considered environmentally benign in comparison with forms of agriculture such as dairying (Wilkins, 2002), and the relative lack of environmental research in sheep-grazed systems reflects this position. There is, nevertheless, evidence that sheep farming can contribute to contamination of waterways.

Issues of possible concern for water quality in sheep farming areas include the concentrations of sediment, nutrients (P and N), C, dissolved oxygen, pesticides and undesirable microorganisms. P and N are primary factors controlling algal and plant growth in freshwater aquatic systems, and excessive concentrations (eutrophication) can promote algal blooms and plant infestations (Hatch *et al.*, 2002; Leinweber *et al.*, 2002). Suspended sediment may be a pollutant in its own right due to its physical effects in watercourses or may be the vector for transfer to water of chemical or microbiological contaminants (Harrod and Theurer, 2002).

Degraded water quality has been reported in areas where the predominant land use is sheep farming, e.g. P concentrations in lakes in New Zealand (Ministry for the Environment, 1997), P and N concentrations in streams and rivers in southern Australia (Victorian Catchment Management Council, 2002), faecal coliforms in the UK (Vinten *et al.*, 2004), sediment in Australia (Beeton *et al.*, 2006) and pesticides in the UK (Virtue and Clayton, 1997).

Loss of P from farmland is rarely enough to be significant in terms of productivity, even if it may have detrimental consequences for receiving waters (McDowell and Catto, 2005). The major pathways of P loss in many environments are overland flow and interflow, sometimes collectively referred to as runoff. Export of P in overland flow and interflow from sheep-grazed land in New Zealand and Australia has been reported in the range of 0.01–1.6 kg P/ha/year, with runoff P concentrations of 0.01–2.0 mg P/l (Bargh, 1978; McColl and Gibson, 1979; Costin, 1980; Smith, 1987; Nelson *et al.*, 1996; McCaskill *et al.*, 2003; Melland, 2003; Ridley *et al.*, 2003; Parfitt *et al.*, 2007). It is now recognized that P may also be lost through deep drainage (McDowell and Monaghan, 2002; Toor *et al.*, 2005; Condron *et al.*, 2006), and small losses (<0.05 kg P/ha/year, with concentrations of 0.18–0.30 mg P/l) have been measured through this pathway in sheep-grazed pastures (Ridley *et al.*, 2003). Recent work has also suggested that in intensively managed sheep-grazed pastures with flood irrigation, up to 8 kg P/ha/year may be lost due to outwash – the loss of irrigation water out the end of an irrigation bay (McDowell, unpublished data).

Environmentally and agriculturally significant loss of N may occur through overland flow, interflow or deep drainage. Loss of N from sheep-grazed land has been estimated at <1–19 kg N/ha/year from overland flow and interflow, with concentrations between 0.7 and 10.8 mg N/l (McColl and Gibson, 1979; Costin, 1980; Smith, 1987; Nelson *et al.*, 1996; Ridley *et al.*, 2001; Melland, 2003; Ridley *et al.*, 2003; Parfitt *et al.*, 2007). Estimated losses through deep drainage range from <1 to 50 kg N/ha/year, with concentrations generally between 2 and 25 mg N/l (Cuttle *et al.*, 1992; Ruz-Jerez *et al.*, 1995; Magesan *et al.*, 1996; Ridley *et al.*, 2001, 2003; Parfitt *et al.*, 2007).

As with dairy pastures, the sources of P and N in water flowing from sheep pastures may include soil, plants, animal excreta or fertilizers (McDowell *et al.*, 2007). While losses from sheep farms are sometimes reported to be less than those from cattle farms (Lambert *et al.*, 1985; Di and Cameron, 2002), this mainly reflects the relatively extensive nature of most sheep farms (lower stocking rates, soil fertility, inputs of fertilizer and supplementary feed; Haygarth *et al.*, 1998a). However, less urine volumes from sheep, compared with cattle, may also decrease nitrate loss in deep drainage (Di and Cameron, 2002). Sheep have a relatively small footprint and exert slightly less hoof pressure than cattle (Willatt and Pullar, 1983), which may result in less damage to surface soil structure. Sheep dung is potentially as damaging to water quality as cattle or deer dung (McDowell, 2006). When sheep are confined in feedlots in large numbers, the wastes can present similar problems for water quality as the wastes from cattle or pigs (Ham and DeSutter, 2000; Rosen *et al.*, 2004).

On sheep-grazed land with poor ground cover, much of the waterborne P and N loss can be in particulate form via soil erosion (Costin, 1980; Elliot and Carlson, 2004). Where ground cover is close to complete, most of the P and N loss is in soluble form (McDowell *et al.*, 2003). The soluble P is often predominantly inorganic 'molybdate-reactive' phosphate (McDowell *et al.*, 2003). It adsorbs readily to soil particles, so P concentrations are decreased when overland flow encounters sediment or bare soil, or when water flows through soil as interflow or drainage (Haygarth *et al.*, 1998b). Soluble N, on the other hand, may contain significant proportions of both organic and inorganic (ammonium and nitrate) N (Nelson *et al.*, 1996; Elliot and Carlson, 2004; Robertson and Nash, 2007). Loss of N through deep drainage is invariably measured as nitrate, most of the ammonium being retained in the upper soil through fixation and adsorption to soil particles (Cameron *et al.*, 2002). Loss of soluble organic N in drainage is also possible (Vinther *et al.*, 2006), but this does not seem to have been investigated in sheep pastures.

Transport of nutrients in water at paddock scale is highly variable in space and time (Sharpley *et al.*, 1999) as a consequence of variability in water flows and in concentrations of mobile nutrients. In general, factors that increase either water flows or mobilization of nutrients into water will increase the loss of nutrients. In sheep systems, P and N mobilization into water can be increased by dung and urine deposition, grazing, fertilizer application, inclusion of legumes in the pasture, cultivation of pasture and the formation of sheep camping areas (Di and Cameron, 2002; McDowell *et al.*, 2003; McDowell, 2006; Robertson and Nash, 2007). Climatic effects on P and N cycling processes can also have large effects on P and N mobilization (Di and Cameron, 2002; Robertson and Nash, 2007).

The P and N concentrations in overland flow, interflow and drainage from sheep-grazed land in the studies mentioned in the preceding discussion usually exceeded water quality standards for unpolluted aquatic ecosystems. Standards vary in different regions, but maximum concentrations set in New Zealand and Australia are 0.01 mg P/l and 0.44 mg N/l (e.g. ANZECC, 2002). Sheep farms, therefore, have the potential to contribute to eutrophication if nutrient concentrations are not reduced between the paddock and the receiving waterway (e.g. by dilution, microbial or plant uptake, sedimentation or soil adsorption). Unfortunately, this is an area where current understanding is very poor. Relating nutrient losses from farms to impact in waterways is further complicated because the impact will vary according to biotic, chemical and physical conditions in the waterway (McDowell *et al.*, 2004).

Nutrient concentrations in waterways can also be increased by direct deposition of urine and dung. Figure 6.3 shows sheep with direct access to a creek in the North Island of New Zealand. Direct nutrient deposition into waterways by sheep has not been measured, but is potentially as important as the nutrients entering waterways in surface runoff from sheep-grazed fields. For example, a single urination and defaecation by a sheep every day for 1 year could deposit 0.4–1.1 kg N and 0.11–0.25 kg P (calculated from data presented by Haynes and Williams, 1993 and McDowell, 2006). Direct deposition of nutrients into watercourses may also occur during aerial application of fertilizer. Sharpley and Syers (1979) reported a 30-fold increase in stream P concentrations after aerial application of superphosphate. However, recent work on more accurate aerial application and the improvements possible in fertilizer-use efficiency and pasture production has been done (Morton and Roberts, 1999; Murray and Yule, 2007a,b; Murray *et al.*, 2007).

Although losses of C, sediment and faecal coliforms have been reported from sheep-grazed systems in concentrations that may affect water quality (Nelson *et al.*, 1996; McDowell, 2006), there is insufficient information relating to sheep farming to be able to assess the prevalence of such losses.

Fig. 6.3. Sheep with direct access to a creek in the North Island of New Zealand. Note erosion onside bank of stream. Poplars can be seen in the middle distance. These will have been planted to minimize erosion.

A wide array of pesticides may be used in sheep farming, but information on the extent of use of particular compounds is generally not available. Many of the products used as anthelmintics, ectoparasiticides, herbicides, fungicides and insecticides have the potential to impair water quality (ANZECC, 2002; APVMA, 2006), but few instances of this occurrence have been documented for sheep-farmed land. An exception is the point-source contamination of several rivers in the UK with sheep dip chemicals (e.g. Virtue and Clayton, 1997; Hooda *et al.*, 2000). Some chemicals are no longer used in sheep dips in most countries (e.g. dieldrin, DDT, As), but may continue to be a source of contamination for many years where historical dip sites are connected to waterways (Dewar *et al.*, 2004; Environment ACT, 2007).

Atmosphere

As with other ruminant livestock systems, sheep farming has an effect on the gaseous composition of the atmosphere. Sheep produce the greenhouse gas CH_4, contribute to NH_3 emissions and influence the greenhouse effect by affecting CO_2, N_2O and NO emissions from grazed forage.

Methane

Lockyer (1997) found under grazing akin to rotational grazing of pasture in the southern UK that when intake was unlimited, CH_4 production by ewes was

25–30 g/day and lambs 15–18 g/day. However, as feed intake and quality declined, so did CH_4 production. The resulting estimate of mean CH_4 production, when stock were not feed-limited, was 23.1 g/day and lambs 15 g/day. Griffith *et al.* (2002) found similar data for sheep in New South Wales, Australia.

Murray *et al.* (2001) investigated CH_4 emissions from sheep grazing pasture with differing N input management in the southern UK. CH_4 emissions per unit body weight from pasture fertilized with 70 or 270 kg N/ha or with a high proportion of white clover did not differ. Because other studies have found differences in emissions with feed quality, the pasture quality in Murray *et al.*'s (2001) study may have been similar, especially as they were all predominantly ryegrass.

Pelchen and Peters (1998) surveyed the literature in an attempt to calculate CH_4 emissions based on descriptive variables of different rations. Their approach clearly has merit, as the range in feed quality consumed by sheep is great. They concluded that the different factors that influence CH_4 emissions should be considered. They noted that, on average, CH_4 emissions from sheep were 7.2% of gross energy intake, or 22 g/day. In particular, with increasing digestibility of the ration, emissions increased up to 72%, whereas with crude fibre content they increased steeply up to 18% then progressively declined. Not surprisingly, CH_4 emissions increased with increasing energy density up to 10.5 kJ ME/g DM (good-quality pasture) then sharply declined. Further, emissions increased with live weight and level of feeding, i.e. the greater the intake above maintenance the greater the CH_4 emissions. Pelchen and Peters' (1998) findings suggest that CH_4 emissions from sheep fed on high-quality pasture at progressively higher intakes from maintenance will have increased compared with emissions from sheep grazed on poorer-quality pasture at or below maintenance feeding. Feed quality was shown to differ seasonally in a study of seasonal CH_4 emissions from Romney ewes grazing ryegrass white clover pastures in New Zealand (Ulyatt *et al.*, 2002). However, seasonal differences in CH_4 emissions from the ewes were small, with the greatest being in late spring when feed quality was lowest.

Degradation of Mongolian steppe by heavy sheep grazing has been previously noted. Such degradation can result in a change in plant community structure and a decline in sheep productivity (Wang *et al.*, 2007), resulting in increased CH_4 production during the growing season per unit of product produced and per stock unit. Decreased grazing pressure and concentrate supplementation was found to decrease CH_4 production per unit of product produced. However, total CH_4 production per stock unit increased (Wang *et al.*, 2007).

All the above studies measured CH_4 emissions from individual sheep. Considerable variation in CH_4 emissions between animals has been observed in the above and other studies (Lassey *et al.*, 1997; Judd *et al.*, 1999). Micrometeorological methods have, among other advantages, the ability to average across larger flock sizes. Judd *et al.* (1999) measured both individual and micrometeorological CH_4 emissions from 6-month-old 40 kg wethers grazing highly digestible abundant pasture at 20 wethers/ha (14 su/ha) on the west coast of the North Island of New Zealand. They found mean emissions from individual animals to be 19.5 g/day. Micrometeorological measurements showed net flock emission rates of 46 mg/m^2/day (168 kg/ha/year and breath emission rate 39 ± 9.6 mg/m^2/day).

However, Judd *et al.* (1999) reported CH_4 emissions from sheep grazing high-quality lowland pastures of 155 kg/ha/year.

Measurements of CH_4 emissions, whether micrometeorological or individual animals, have been at small time and spatial scales. Using such data to estimate emissions at greater scales is not a straightforward exercise but requires rigorous uncertainty analysis. An example of such an approach is that carried out by Kelliher *et al.* (2007).

A clear pattern of CH_4 emissions measured or estimated from relevant data at the paddock scale with different levels of management intensity has not been recorded. It is therefore not clear across the wide range of sheep-grazed ecosystems that reliable management approaches to decrease CH_4 emissions exist in absolutes terms. It may be possible to decrease CH_4 emissions as a proportion of product produced by ensuring a smaller proportion of feed is used for maintenance relative to growth and production.

Nitrous Oxide

Both atmospheric N_2O and NO are involved in photochemical reactions that destroy ozone. Furthermore, N_2O contributes to what is known as the greenhouse effect (Bouwman, 1990). In soil, nitrification is the main source of NO, while denitrification is the main source for most N_2O (de Klein *et al.*, 2001). These processes depend on available soil N. Clearly, additions of available N as urine or fertilizer (Cole *et al.*, 1997) can strongly influence gaseous emissions, suggesting potential for greater emissions per unit area as stocking rates and soil fertility increase. De Klein *et al.* (2001) presented a review of N_2O emissions from agricultural soils in New Zealand. Although not specifically focused on sheep or the world, the reviews' relevance is that New Zealand's high sheep density and intensive and extensive sheep farming systems enable an understanding of global implications.

Soil oxygen has been regarded as the main factor in N_2O emissions (Luo *et al.*, 1999). However, de Klein *et al.* (2001) noted that soil oxygen supply is regulated by soil water content, resulting in peak N_2O emissions occurring after rainfall or irrigation, conditions that generally coincide with intensive sheep farming systems. Furthermore, greater emissions have been observed from compacted than from uncompacted soil (McTaggart *et al.*, 1997), conditions also accentuated by intensive sheep farming systems. Emissions from sheep- and cattle-grazed pastures, however, are less than from dairy cow-grazed pastures (Carran *et al.*, 1995), presumably due to their propensity for less compaction (Drewry *et al.*, 2000).

In Australia, Wang *et al.* (1997) indicated that nitrification in semi-arid and arid N-limited systems was a more important source of N_2O than denitrification; however, total losses may be small under these dry and low-producing environments. Under summer dry, windy, Mediterranean conditions, when NO and N_2O gaseous losses from sheep urine N would be expected to be maximal, Bronson *et al.* (1999) found them to be negligible. In New Zealand, Kelliher *et al.* (2002) determined an area-integrated N_2O emission rate overnight of 24 ± 5 ng N/m^2/s in a paddock stocked at ten 70 kg ewes/ha in autumn.

Validated mechanistic models enable an integration of research, describe the mechanisms involved and help provide predictive tools. Saggar *et al.* (2007)

modified the denitrification-decomposition model of Li *et al.* (1992a,b; 2001) to represent grazed pasture systems. N_2O emissions were simulated on dairy and sheep farms in New Zealand (Saggar *et al.*, 2007). Total measured emission from the sheep farm was 3.8 kg N_2O-N/ha/year, while the simulated emission was 5.23 kg N_2O-N/ha/year. Both simulated and measured emissions were significantly less than from the dairy farms. These were attributed to different emission factors for dairy and sheep-grazed systems. Both dairy and sheep farms were on highly productive pastures producing approximately 16 t DM/ha/year with the sheep farm stocked at 15–22 su/ha and fertilizer urea-N input to the sheep farm of 90 kg N/ha/year.

Mitigation of N_2O emissions is problematic as other environmentally deleterious forms of N loss could be increased. Jarvis *et al.* (1996) suggested, in the first instance, increasing N efficiency. Although liming and the use of nitrification inhibitors can decrease N_2O emissions (de Klein *et al.*, 2001), the modelling work of Saggar *et al.* (2007) suggests N_2O emissions can be decreased by changes in management. For example, decreases can occur by applying less N, especially when there are no significant changes in pasture production. Because the timing and amount of each N application affect emissions, making certain the soil is not too wet and N applications do not exceed the pasture's ability to uptake applied N ensures decreased emissions. As a consequence, frequent smaller amounts of applied N result in decreased emissions. Decreases in stocking rate also decrease emissions and set stocking emitted less than a rotational grazing system (Saggar *et al.*, 2007).

Ammonia

Although NH_3 is not a gas that contributes to global warming, it does contribute to acid rainfall (Younie and Baars, 2005) and the atmospheric concentration of N that can be transported to N-limited ecosystems from pastoral farming areas (Parfitt *et al.*, 2006). Increased soil N in N-limited forest can significantly alter the recipient ecosystems, for example, by decreasing biodiversity (Clark *et al.*, 2007). Under warm, moist conditions, urine N (primarily urea) is rapidly hydrolysed on contact with soil and could be lost via NH_3 volatilization (Whitehead and Raistrick, 1991). For example, Bronson *et al.* (1999) found 38% of N loss was as volatilized NH_3 from sheep urine N in summer on sandy soil in West Australia. This is consistent with other studies (Whitehead and Raistrick, 1992; Thompson and Fillery, 1997, 1998). Unfortunately, there are no management options to decrease NH_3 loss from urine other than decreased stocking rates (Bronson *et al.*, 1999).

Carbon dioxide

There is limited research specifically targeted at the contribution of sheep farming to atmospheric CO_2. Nevertheless, as noted in the soil C section, declines in soil C have been observed under intensive sheep grazing systems. This loss of soil C has contributed to the increasing atmospheric CO_2 content. Preliminary C budgets for New Zealand by Tate *et al.* (2000) suggest that improved grassland may have a large negative C balance (i.e. C was being lost) relative to all land covers considered. The results suggest intensification leads to greater CO_2 emissions. However, Tate *et al.* (2000) noted that the negative balance must be treated with caution given the uncertainties in the assessment of soil respiration and NPP and the limited data set. In contrast, Cao *et al.* (2004) found light grazing in an Alpine meadow on the Tibetan plateau at 2.6 sheep/ha produced CO_2 efflux (includes measurements of biomass and soil respiration) about 33% greater than with twice the stocking rate. However, their study did not include CO_2 measurements of all system components.

The Mongolian steppe is a large inland area of semi-arid and arid grassland grazed by sheep, horses and cattle. An eddy covariance study of net ecosystem CO_2 exchange (NEE) over moderately sheep-grazed Mongolian steppe was C neutral (Li *et al.*, 2005). This is consistent with the estimates made by Tate *et al.* (2000) in New Zealand from unimproved and tussock grasslands.

Climate change

Because sheep farming contributes to greenhouse gas emissions it becomes a contributor to climate change. Clearly, a discussion of climate change is not appropriate here, but sheep farming's contribution to a deterioration in air quality should be noted. With elevated CO_2 concentrations and a non-limiting soil N supply, increased atmospheric CO_2 may enhance forage quality (Lilley *et al.*, 2001). However, where soil N is limited, which occurs over significant areas where sheep are grazed (native rangelands, semi-natural grasslands), forage digestibility may be decreased (Körner, 2002). Furthermore, with increased atmospheric CO_2 concentrations and climate change are implications for weed invasions (Smith *et al.*, 2000; Watkinson and Ormerod, 2001), the spread of woody plants into former semi-arid and arid grasslands (Polley *et al.*, 1997; Scholes and Archer, 1997), and other changes in species composition in semi-natural grasslands (Duckworth *et al.*, 2000; Teyssonneyre *et al.*, 2002). Morgan (2005) suggests management changes will be required as a consequence of the predicted significant effects of increases in atmospheric CO_2 and climate change.

Above-ground biodiversity

Biodiversity, a key descriptor of ecosystem condition (Aguiar, 2005), is defined as the sum of total biotic variation from gene to landscape. Dorrough *et al.* (2007) and Chapin *et al.* (2000) considered that land-use change associated with agriculture might be the greatest threat to global biodiversity. Sheep farming, a global activity within large areas of sensitive and highly modified environments, may be a contributory cause. Clearly, a highly modified environment where an indigenous forest has been replaced by exotic pasture species results in the death and displacement of many animal species. Moreover, the same can be said for sheep grazing of natural landscapes (Borrelli and Cibils, 2005).

In UK temperate grasslands, a bell-shaped curve describes the effect of grazing pressures on plant species diversity (Grime, 1979). Very high or low grazing pressure causes few species, whereas an intermediate level results in maximum species diversity. For example, in a natural rangeland in Scotland, Evans *et al.* (2006) found maximum biodiversity occurred at low-intensity livestock grazing, compared with excluding livestock completely. They suggested maintaining the open character of moorland habitats would benefit key upland species generally and could be achieved by reducing sheep grazing pressure and introducing low-intensity mixed livestock grazing throughout the uplands. A similar story has been found in Australia (Kemp *et al.*, 2003; Dorrough *et al.*, 2007), and Patagonia (Borrelli and Cibils, 2005; Oliva *et al.*, 2005).

Van Wieren (1991) suggested low grazing pressure maximizes biodiversity of invertebrates because the structural diversity of the vegetation is greater as a consequence of a lower frequency of defoliation and trampling but with a supply of dung and carrion still available. However, there can be a decrease in invertebrate species diversity with fertilizer (Morris, 1990b). Furthermore, mowing for hay or silage causes a decrease in invertebrate diversity (Morris, 1990a; Kirby, 1992) because of the abrupt loss of flowers and seeds.

Bird species can also be affected by grazing. Harding *et al.* (1994) reported that grazing can influence bird species by: (i) changing habitat structure and composition which can alter food resources; (ii) increasing invertebrate numbers by providing dung; (iii) destroying nesting sites through trampling; and (iv) increasing the amount of carrion available (Milne, 1997).

Mammals and other vertebrates can also be affected, even to the point of extinction, by altering competition for food and breeding sites; changing the structure of the vegetation and plant species present; influencing the transmission of disease; and affecting other species higher up the food chain through impacts on prey abundance (Milne, 1997; Lunney, 2001).

Sheep grazing has facilitated the invasion of exotic plant species. D'Antonio and Vitousek (1992), reviewing literature on the ecosystem effects of biological invasion by exotic grasses, found that the pattern of invasion is common. They considered biological invasions have caused more species extinction than climate change or atmospheric composition change. They stressed that biological invasions change not only the compositional attributes of biodiversity but also the functional attributes of biodiversity.

Clearly, sheep grazing can cause severe damage to above-ground ecosystems. However, with some landscapes the inclusion of moderate grazing may enhance biodiversity and ecosystem function. General simplistic judgements that sheep farming is harmful to above-ground ecosystems are not warranted; however, it seems that intensive highly productive systems over large areas may pose a serious threat to ecosystem function in the long term. Research to clarify this is needed.

Mitigation Measures

As intensity of sheep farming increases the effect on the environment becomes greater. For example, increases in nutrient loss to waterways, greenhouse gas emissions, deleterious effects of veterinary products and pesticides occur, as do declines in biodiversity and the suggestion of decreased ecosystem function. Many of these environmental problems potentially arising from sheep farming can be at least partly mitigated through management. This requires management strategies to achieve: (i) maximal ground cover; (ii) minimal soil compaction; and

(iii) efficient nutrient utilization. The achievement and the consequences of these management elements are closely interrelated.

Ground cover includes live plants and dead plant litter (standing or on the soil surface), both of which decline as the intensity of grazing increases (Rauzi and Hanson, 1966; Naeth *et al.*, 1991; Mapfumo *et al.*, 2002). Soil compaction (increased bulk density, decreased pore space) may also occur with increasing grazing pressure (Rauzi and Hanson, 1966; Proffitt *et al.*, 1993; McCaskill and Cayley, 2000). Ground cover can therefore be maximized and compaction minimized through grazing management, as well as by management of animal containment areas and animal and vehicle traffic.

Near-complete ground cover and the absence of soil compaction are necessary for optimal infiltration of rainfall and avoidance of excessive surface runoff, contaminant loss and erosion (Johnston, 1962; Rauzi and Hanson, 1966; Proffitt *et al.*, 1993). To maximize the hydrological benefits in water-limited environments, it may be as important to manage for litter accumulation as for live plant cover (Branson, 1984; Naeth *et al.*, 1991). With current sheep grazing methods, however, ground cover per se is rarely considered (Pratley and Godyn, 1991; Saul, 2006).

In situations where the soil is saturated – a seasonal occurrence in most sheep farming systems – a high degree of ground cover will protect against loss of soil and attached nutrients, organic matter and agricultural chemicals to waterways. Loss of dissolved forms of these contaminants (through surface runoff or leaching), however, can remain a potential threat to water quality. Acute ('incidental') contaminant losses may be minimized by avoiding application of fertilizer and pesticides: (i) immediately before heavy rain is expected; (ii) close to waterways; and (iii) at excessive rates.

That applied nutrients are used efficiently by the plants is a prerequisite for the mitigation of nutrient losses to the environment. Efficient nutrient management involves balancing nutrient inputs (from fertilizer, legumes or manure) with exports in harvested products, while keeping other exports to a minimum. In practice, however, nutrient inputs to sheep production systems will probably exceed exports because of the small removal of nutrients in meat and wool, except where pastures contain no sown legume and remain unfertilized (Smoliak *et al.*, 1972; McCaskill and Cayley, 2000). For example, P, S, Ca and N can accumulate in the soil even with small inputs (4 kg P/ha/year) of superphosphate (McCaskill and Cayley, 2000). Despite such accumulation, continued inputs of nutrients (usually P and N) are required to maintain productivity (Cayley *et al.*, 1999) due to the chemical and biological immobilization of plant-available forms of the nutrients, and transfer of nutrients to sheep camps. The main strategy available to farmers for promoting efficient nutrient use is to ensure nutrient inputs are adequate but not excessive for plant requirements, and to minimize other constraints to plant growth such as soil structure, acidity or water stress.

Unfortunately, even efficient nutrient use and control of erosion and incidental nutrient losses will not necessarily prevent nutrient movement from paddocks to waterways. This is because a significant part of the dissolved nutrient loss from grasslands may be due to the general nutrient enrichment of the soil-plant system through years of nutrient inputs (Nash *et al.*, 2005; Robertson and Nash, 2007). While nutrient losses from these background sources can be decreased by minimization of excessive infiltration-excess runoff, losses from saturation-excess runoff cannot, with current knowledge, be effectively controlled.

The formation of camping areas in set-stocked sheep grazing systems may be of particular environmental concern if connected to a waterway. The soils in these areas typically have poor ground cover, are compacted and have large accumulations of dung and total and soluble nutrients (McCaskill and Cayley, 2000). These features make it likely that sheep camps can contribute disproportionately to sediment, P and N losses to waterways, and N losses to the atmosphere (McTaggart *et al.*, 1997a; Saggar *et al.*, 2007). Figure 6.4 shows soil, dung and organic matter washed from a sheep camp near Ararat in Victoria, Australia, after a heavy rainfall event (Fig. 6.4a). The outwash is seen collected in a fence line (Fig. 6.4b) and after spillage into a waterway can be seen in a farm dam (Fig. 6.4c). The development of camping areas can be difficult to prevent, but they may be able to be minimized through choice of fencing configurations, location of watering points,

Fig. 6.4. Soil, dung and organic matter washed from a sheep camp near Ararat in Victoria, Australia after a heavy rainfall event (a). The outwash is seen collected in a fence line (b) and after spillage into a waterway can be seen in a farm dam (c). (Photographs courtesy of Reto Zollinger and used with permission.)

proximity of neighbouring stock or by adoption of rotational grazing (Haynes and Williams, 1993; Sargeant, 2003; Saul, 2006).

The concept of critical source areas to describe those (usually) small parts of the landscape that have a high potential for both nutrient mobilization and overland flow generation is a fairly recent advance in our understanding of nutrient movement from agricultural catchments. Directing mitigation measures at critical source areas instead of whole fields may lead to better mitigation results (Heathwaite *et al.*, 2005). However, while research in this direction is progressing (Gburek *et al.*, 2002; Heathwaite *et al.*, 2005; Hughes *et al.*, 2005; Srinivasan and McDowell, 2007), the identification of critical source areas is often difficult without substantial financial and other resources (Heathwaite *et al.*, 2005; Hughes *et al.*, 2005), and other approaches may in practice be as successful (McDowell *et al.*, 2005).

Riparian buffer strips vegetated with grass, and sometimes also trees, are widely advocated as means of decreasing sediment and nutrient movement to waterways (Dorioz *et al.*, 2006). Such buffers may be effective where erosion is an important process, but in pastures with good ground cover, where most of the nutrient loss is in dissolved form, buffers are unlikely to decrease nutrient losses (Nash and Murdoch, 1997; Dougherty *et al.*, 2004). The long-term viability of buffer strips is questionable, as buffers that are initially nutrient sinks may eventually become nutrient sources (Bedard-Haughn *et al.*, 2004; Dorioz *et al.*, 2006) or be bypassed when flow converges.

In addition to minimizing nutrient movement from grazing land, watercourses also need to be protected from direct deposition of nutrients, organic matter and sediment. In effect, this means restricting the access of sheep to waterways, and avoiding high-risk fertilization methods such as aerial application. Nevertheless, attempts at improving the accuracy of aerial application are being made (Murray and Yule, 2007a,b)

Clearly, the importance of minimizing the amount of nutrient applied relative to that which is essential cannot be over-emphasized. The use of models that, for example, help improve the precision of fertilizer use and can predict nutrient loss (McDowell *et al.*, 2005; Srinivasan and McDowell, 2007), is becoming more important. Modelling provides an integrative quantitative means of describing and predicting processes and outcomes for a particular system. Various nutrient budgeting models to estimate contaminant loss from farmland have been developed (McDowell *et al.*, 2005; Srinivasan and McDowell, 2007). A good example is that of McDowell *et al.* (2005), who under New Zealand pastoral conditions, introduced into the nutrient budgeting computer model Overseer a model to estimate P loss from pastoral systems to surface waters. The model quantifies background and incidental P loss (Haygarth and Jarvis, 1999; McDowell *et al.*, 2004) under a set of site, transport and management factors. The parameters are mechanistically relevant and not too numerous to deter evaluating farming systems. Further extending the use of models, some companies and agencies have, for example, used the USDA P index in conjunction with farm mapping to isolate high-risk areas of P loss (Hart and Quin, 2003).

Ideal scenarios would be to improve environmental outcomes without sacrificing production and profitability (Jackson *et al.*, 2007). However, some environments may be too sensitive or have a high societal value, which makes them unsuitable for sustainable farming. Management changes that achieve one or more of the three aims mentioned above may require destocking entirely. This is occurring, despite continued debate as to its merits, in the higher elevations of New Zealand's South Island high country. Conversion to organic or other farming systems that exclude many of the inputs that clearly degrade the environment is another management approach that has been proposed.

While the literature on organic sheep farming is sparse, some findings for other organic farming systems have applicability to sheep systems. Younie and Baars (2005) presented a review of organic grassland farming, and Condron *et al.* (2000) reviewed the literature to compare soil and environmental quality under organic and conventional farming systems in New Zealand. MacKerron *et al.* (2007) considered that organic systems do not have a single policy. Consequently, a simple definition of it is difficult. Nevertheless, according to Condron *et al.* (2000, p. 444), organic farming characteristics include:

> protecting the long-term fertility and quality of the soil; providing nutrients in natural and organic fertilizers; nitrogen self-sufficiency through legumes; weed, disease, and pest control through crop rotations, natural predators, diversity, organic manuring, and limited biological and chemical intervention; extensive management of livestock; and, minimizing the impact on the wider environment.

A popular belief exists that under organic farming soil biological activity is enhanced. Parfitt *et al.* (2005), however, found no difference in soil microbial biomass pools or soil C/N and N/P ratios or earthworm numbers between organic and conventional sheep farming. Moreover, microbiological and microfaunal data from organically farmed sites were found to be on the same soil N dependent trend lines as conventionally farmed sites.

Although it is assumed organic systems will have lower N and P losses than conventional systems, it is the quantity of N rather than its source that determines potential losses to the wider environment (Watson and Younie, 1995; Condron *et al.*, 2000). Many organic systems can have soil test P concentrations equal or in excess of conventional systems, which means both can lose a similar amount of P from soil. Nevertheless, organic systems tend to have less soil N as a consequence of lower N inputs. However, Younie and Armstrong (1996) observed little species diversity in an intensively managed and highly productive organically managed sward over 9 years from sowing with a mix of species. Whether organically managed or not, the outcomes from high-fertility, intensively managed pastures can be the same. Furthermore, to achieve acceptable levels of animal performance, especially finishing, requires high-quality and highly productive pastures, which in turn require high-fertility. To maintain species diversity in the Netherlands, Younie and Baars (2005) observed that N input must be limited to 50–150 kg N/ha/year. One advantage for organic systems is that they are commonly required to use only reactive phosphate rock (RPR). Under wet conditions the risk of P loss to water ways can be decreased in the first 60 days after fertilizer application if RPR is applied as a substitute for a water-soluble fertilizer like superphosphate (McDowell *et al.*, 2003). However, if applied during drier periods P losses are similar.

The negative effect of sheep farming on above-ground biodiversity is a significant issue. It seems that production and biodiversity may be incompatible not only in intensive highly productive systems over large areas but also in sensitive landscapes such as arid and semi-arid rangelands within old landscapes in a retrogressive development stage (Walker *et al.*, 1983; Braunack and Walker, 1985; Walker *et al.*, 2001). Although further research is required to determine the relationship between agrobiodiversity and ecosystem function clearly (Andrén *et al.*, 1999; Ritz *et al.*, 2003; Jackson *et al.*, 2007), segregation or integration of different sward types or selected levels of biodiversity could be considered at landscape, farm or paddock scales (Wilkins, 2002). A precautionary approach in high-fertility, low-biodiversity and nutrient-leaky landscapes may include interspersed areas that are not managed as intensively but have as their management objective to enhance selected biodiversity indices that ensure long-term stability of a broad range of ecosystem services. This approach may increase the probability of long-term resilience of key ecosystem functions. Figure 6.5 shows a North Island of New Zealand hill country farmscape with a mix of pasture, poplar trees for erosion control and native forest on more sensitive sections of the landscape. Nevertheless, Dorrough *et al.* (2007) in southern Australia discounted what seemed superficially a workable scenario of intensification of selected areas to make other areas available for increased biodiversity. This was because intermediate levels of grazing often have little impact on plant diversity when nutrient levels are low:

> [I]ncreasing productivity via fertilizer application is likely to require intensification on even more land and could come at the cost of biodiversity. In contrast, improving grazing management across broad scales is likely to result in enhanced profitability and could also benefit native vegetation. Extensive management may be necessary to maintain biodiversity and prevent further long-term degradation of the resource base.
>
> (Dorrough *et al.*, 2007, p. 222)

Fig. 6.5. North Island of New Zealand hill country farmscape with a mix of pasture, poplar trees for erosion control and native forest on more sensitive sections of the landscape.

Although often considered a 'blunt instrument', decreased stocking rates are a management approach to mitigate sheep grazing impact. An example of decreased stocking rate and improved grazing management has occurred in Patagonian grasslands (Borrelli and Cibils, 2005). Borrelli and Cibils (2005) noted that sheep farming is often considered to have a negative impact on the environment. In Patagonia, this impact was a consequence of continuous overgrazing on traditional farms. Oliva *et al.* (1998) suggested that there is no evidence that carefully managed sheep flocks cause significant environmental damage. Furthermore, continued sheep farming in Patagonia may be compatible with rangeland conservation under judicious grazing. However, the issues are not simple, and solutions can raise further difficult questions (McIntosh *et al.*, 1999).

An encouraging trend may come from consumers who want better quality product produced under improved environmental standards. This may be reflected in Hodgson *et al.*'s (2005) comment: 'The industry will think more in terms of quality not quantity, and value not volume.'

Administrative Body Policy Implications

Clearly, there are conflicts between sheep production and environmental health. Is there a constructive role for public administrative bodies to deal with this conflict? There are many examples of top-down government meddling in agriculture that have failed (Erickson, 2006); there are other examples of beneficial involvement (Sheldrick, 1997). Although significant research still remains to be conducted, there are measures that can currently be taken to remedy environmental harm – but some consider progress is slow. Johns (2007) argues that politics is part of the solution. Whether there is political drive for change or not there are three possible approaches that administrative bodies can take (Ministry for the Environment (MFE), 1997): voluntary, economic and regulatory. These involve encouragement and education, financial incentives and required or compulsory methods.

An example of a voluntary approach is the educational effort in New Zealand to protect and enhance the country's 'clean-green' brand by promoting environmentally friendly farm practices to farmers. Economic incentives can involve direct payments to farmers for the preservation of wildlife habitats and flora (Sheldrick, 1997). There are strong arguments for nature to be included as a capital asset in production activities (Dasgupta, 2007) and consequently, in the long run, it is important to include ecological implications in the economics of sheep farming systems. Regulatory mechanisms can, for example, involve legal restrictions on fertilizer application rates such as nitrogen.

The increasing intensity of resource use on the planet will continue as the world's population increases and there is more money to be made by increasing farm production. Redman (1999) integrated archaeological and historical records in a conceptual model of the interactions between natural and social systems. He identified three phases in human systems: expansion, intensification and abandonment. Aguiar (2005) reviewed and discussed biodiversity in grasslands and explained Redman's (1999) model by noting that intensification results in the appropriation of more resources to support expansion and new functions of a society. The resulting pressures on ecosystems result in a structure and function beyond their resilience, leading to their collapse. A science conference in 2007 titled 'Tipping Points in the Biosphere: Agriculture, Water and Resilience' is indicative of the increasing concern scientists have about links between agriculture and biosphere tipping points. Clearly, administrative bodies will need to be proactively engaged in strategic approaches to facilitate mitigation of environmental degradation by sheep farming and restorative measures.

Conclusions and Future Research

Sheep farming is practised globally. Its environmental impact ranges from beneficial to benign to highly destructive. It can enhance and diminish above and below biodiversity. Under sheep grazing soil physical properties are generally diminished. Soil C concentrations can be enhanced or diminished. Soil nutrient concentrations can increase to beneficial levels but can also exceed the ability of the ecosystem to retain them. Damaging amounts of N and P can be lost to waterways and greenhouse gases emitted directly or indirectly to the atmosphere. Harmful contaminants can be introduced through fertilizers, pesticides and pharmaceuticals. These adverse environmental impacts become greater as sheep farming intensity increases, with more sensitive environments being more obviously impacted.

Changes in management can mitigate many negative environmental impacts by enabling: (i) maximal ground cover; (ii) minimal soil compaction; and (iii) efficient nutrient utilization. These requirements become more difficult as intensity of production increases. Furthermore, when one or more of these aspects are not achieved it appears biodiversity may be diminished. A significant research need is the link between biodiversity and the range of ecosystem functions, for example, achieving less N and P loss to waters while improving or maintaining profitability and increasing biodiversity. Administrative bodies have an important role in safe guarding the environment. This need not be mutually exclusive from production requirements or farmer's financial returns. The long-term viability of sheep farming will ultimately depend on the resilience and ability of the ecosystem to provide many more services than the production of sheep products.

References

Agritech (2007) Available at: http://www.agritech.org.nz/sheep.shtml

Aguiar, M.R. (2005) Biodiversity in grasslands: current changes and future scenarios. In: Reynolds, S.G. and Frame, J. (eds) *Grasslands: Developments Opportunities Perspectives*, Science Publishers in association with the Food and Agriculture Organisation of the United Nations (FAO), Enfield, New Hampshire, pp. 261–281.

Almås, A.R., Bakken, L.R. and Mulder, J. (2004) Changes in tolerance of soil microbial communities in Zn and Cd contaminated soils. *Soil Biology and Biochemistry* 36, 805–813.

Anderson, G. (1980) Assessing organic phosphorus in soils. In: Khasaweh, F., Sample, E. and

Kamprath, E.E. (eds) *The Role of Phosphorus in Agriculture*. ASA, Madison, Wisconsin, pp. 441–431.

Andrén, O., Brussard, L. and Clarholm, M. (1999) Soil organism influence on ecosystem-level processes – bypassing the ecological hierarchy? *Applied Soil Ecology* 11, 177–188.

ANZECC (2002) *Australian and New Zealand Guidelines for Fresh and Marine Water Quality*. Volume 1, The guidelines (Chapters 1–7). Australian and New Zealand Environment and Conservation Council and the Agriculture and Resource Management Council of Australia and New Zealand, Canberra.

APVMA (2006) *The Reconsideration of Approvals of Selected Sheep Ectoparasiticide Products and Their Associated Labels*. Preliminary review findings. Volume 1 of 2. Australian Pesticides and Veterinary Medicines Authority, Canberra.

Australian Wool Innovation (2006) Wool facts. Australia's wool industry. Available at: http://www.woolinnovation.com.au/attachments/Education/21053_AWI_WoolFacts.pdf

Australian Wool Innovation (2007) Wool facts. Australia's wool industry. Available at: http://www.woolinnovation.com.au/attachments/Education/21053_AWI_WoolFacts.pdf

Ball, P.R. and Keeney, D.R. (1983) Nitrogen losses from urine-affected areas of a New Zealand pasture, under contrasting seasonal conditions. In: Smith, J.A. and Hays, V.W. (eds) *Proceedings of the XIV International Grassland Congress*, Lexington, Kentucky, 15–24 June, 1981. Westview Press, Boulder, Colorado, pp. 341–344.

Bardgett, R.D. (2005) *The Biology of Soil: A Community and Ecosystem Approach*. Oxford University Press, Oxford.

Bardgett, R.D., Frankland, J.C. and Whittaker, J.B. (1993) The effects of agricultural management on the soil biota of some upland grasslands. *Agriculture, Ecosystems and Environment* 45, 25–45.

Bardgett, R.D., Leemans, D.K., Cook, R. and Hobbs, P.J. (1997) Seasonality of the soil biota of grazed and ungrazed hill grasslands. *Soil Biology and Biochemistry* 29, 1285–1294.

Bardgett, R.D., Jones, A.C., Jones, D.L., Kemmitt, S.J., Cook, R. and Hobbs, P.J. (2001) Soil microbial community patterns related to the history and intensity of grazing in sub-montane ecosystems. *Soil Biology and Biochemistry* 33, 1653–1664.

Bargh, B.J. (1978) Output of water, suspended sediment, and phosphorus and nitrogen forms from a small agricultural catchment. *New Zealand Journal of Agricultural Research* 21, 29–38.

Barrow, N.J. and Lambourne, L.J. (1962) Partition of excreted nitrogen, sulphur, and phosphorus between the faeces and urine of sheep being fed pasture. *Australian Journal of Agricultural Research* 13, 461–471.

Bedard-Haughn, A., Tate, K.W. and van Kessel, C. (2004) Using nitrogen-15 to quantify vegetative buffer effectiveness for sequestering nitrogen in runoff. *Journal of Environmental Quality* 33, 2252–2262.

Beeton, R.J.S., Buckley, K.I., Jones, G.J., Morgan, D., Reichelt, R.E. and Trewin, D. (2006) *Australia State of the Environment 2006*. Independent report to the Australian Government Minister for the Environment and Heritage. Department of the Environment and Heritage, Canberra.

Betteridge, K., Mackay, A.D., Shepherd, T.G., Barker, D.J., Budding, P.J., Devantier, B.P. and Costall, D.A. (1999) Effect of cattle and sheep treading on surface configuration of a sedimentary hill soil. *Australian Journal of Soil Research* 37, 743–760.

Black, A.L. (1988) Toxic effects of cadmium in animals and man. In: Simpson, J. and Curnow, W.J. (eds) *Cadmium Accumulations in Australian Agriculture*, Bureau of Rural Resources, Canberra, ACT, pp. 161–166

Bligh, M.W. (1990) A comparative study of the nitrogen cycling in a brigalow woodland and a derived green panic pasture. MSc thesis, The University of Queensland, Brisbane, Australia.

Borrelli, P. and Cibils, A.F. (2005) Rural depopulation and grassland management in Patagonia. In: Reynolds, S.G. and Frame, J. (eds) *Grasslands: Developments Opportunities Perspectives*. Science Publishers in association with the Food and Agriculture Organisation of the United Nations (FAO), Enfield, New Hampshire, pp. 461–487.

Boul, H.L. (1994) DDT residues in the environment – a review with a New Zealand perspective. *New Zealand Journal of Agricultural Research* 38, 257–277.

Bouwman, L.A. (1990) *Soils and the Green House Effect*. Wiley, Chichester, UK.

Branson, F.A. (1984) Evaluation of impacts of grazing intensity and specialized grazing systems on watershed characteristics and responses. In: *Developing Strategies for Rangeland Management*. Westview Press, Boulder, Colorado.

Braunack, M.V. and Walker, J. (1985) Recovery of some surface soil properties of ecological interest after sheep grazing in a semi-arid woodland. *Australian Journal of Ecology* 10, 451–460.

Bronson, K.F., Sparling, G.P. and Fillery, I.R.P. (1999) Short-term N dynamics following application of N-15-labeled urine to a sandy soil in summer. *Soil Biology and Biochemistry* 31, 1049–1057.

Cameron, K.C., Di, H.J. and Condron, L.M. (2002) Nutrient and pesticide transfer from agricultural soils to water in New Zealand. In: Haygarth, P.M. and Jarvis, S.C. (eds) *Agriculture, Hydrology and Water Quality*. CAB International, Wallingford, UK, pp. 373–393.

Cao, G.M., Tang, Y.H., Mo, W.H., Wang, Y.A., Li, Y.N. and Zhao, X.Q. (2004) Grazing intensity alters soil respiration in an alpine meadow on the Tibetan plateau. *Soil Biology and Biochemistry* 36, 237–243.

Carran, R.A., Ball, P.R., Theobald, P.W. and Collins, M.E.G. (1982) Soil nitrogen balances in urine-affected areas under two moisture regimes in Southland. *Zealand Journal of Experimental Agriculture* 10, 377–381.

Carran, R.A., Theobald, P.W. and Evans, J.P. (1995) Emission of nitrous oxide from some grazed pasture soils in New Zealand. *Australian Journal of Soil Research* 33, 341–352.

Cayley, J.W.D., Kearney, G.A., Saul, G.R. and Lescun, C.L. (1999) The long-term influence of superphosphate and stocking rate on the production of spring-lambing Merino sheep in the high rainfall zone of southern Australia. *Australian Journal of Agricultural Research* 50, 1179–1190.

Chapin III, F.S., Zavaleta, E.S., Eviner, V.T., Naylor, R.L., Vitousek, P.M., Reynolds, H.L., Hooper, D.U., Lavorel, S., Sala, O.E. and Hobbie, S.E. (2000) Consequences of changing biodiversity. *Nature* 405, 234–242.

Chittleborough, D.J., Smettem, K.R.J., Cotsaris, E. and Leaney, F.W. (1992) Seasonal changes in pathways of dissolved organic carbon through a hillslope soil (Xeralf) with contrasting texture. *Australian Journal of Soil Research* 30, 465–476.

Clark, C.M., Cleland, E.E., Collins, S.L., Fargione, J.E., Gough, L., Gross, K.L., Pennings, S.C., Suding, K.N. and Grace, J.B. (2007) Environmental and plant community determinants of species loss following nitrogen enrichment. *Ecology Letters* 10, 1–12.

Clark, H., de Klein, C. and Newton, P. (2001) *Potential Management Practices and Technologies to Reduce Nitrous Oxide, Methane and Carbon Dioxide Emissions from New Zealand Agriculture*. Prepared for the Ministry of Agriculture and Forestry. September 2001.

Climo, W.J. and Richardson, M.A. (1984) Factors affecting the susceptibility of 3 soils in the Manawatu to stock treading. *New Zealand Journal of Agricultural Science* 27, 247–253.

Cole, C.V., Duxbury, J., Freney, J., Heinemeyer, O., Minami, K., Mosier, A., Paustian, K., Rosenberg, N., Sampson, N., Sauerbeck, D. and Zhao, Q. (1997) Global estimates of potential mitigation of greenhouse gas emissions by agriculture. *Nutrient Cycling in Agroecosystems* 49, 221–228.

Condron, L.M. and Goh, K.M. (1989) Effects of long-term phosphatic fertilizer applications on amounts and forms of phosphorus in soils under irrigated pasture in New Zealand. *Journal of Soil Science* 40, 383–395.

Condron, L.M., Cameron, K.C., Di, H.J., Clough, T.J., Forbes, E.A., McLaren, R.G. and Silva, R.G. (2000) A comparison of soil and environmental quality under organic and conventional farming systems in New Zealand. *New Zealand Journal of Agricultural Research* 43, 443–466.

Condron, L.M., Sinaj, S., McDowell, R.W., Dudler-Guela, J., Scott, J.T. and Metherell, A.K. (2006) Influence of long-term irrigation on the distribution and availability of soil phosphorus under permanent pasture. *Australian Journal of Soil Research* 44, 127–133.

Connolly, R.D., Freebairn, D.M. and Bell, M.J. (1998) Change in soil infiltration associated with leys in south-eastern Queensland. *Australian Journal of Soil Research* 36, 1057–1072.

Costin, A.B. (1980) Runoff and soil and nutrient losses from an improved pasture at Ginninderra, Southern Tablelands, New South Wales. *Australian Journal of Agricultural Research* 31, 533–546.

Cottle, D.J. (1991) The sheep industry. In: Cottle, D.J. (ed.) *Australian Sheep and Wool Handbook*. Inkata Press, Melbourne, Australia, pp. 3–18.

Cox, J.W., Varcoe, J., Chittleborough, D.J. and van Leeuwen, J. (2005) Using gypsum to reduce phosphorus in runoff from subcatchments in South Australia. *Journal of Environmental Quality* 34, 2118–2128.

Cuttle, S.P., Hallard, M., Daniel, G. and Scurlock, R.V. (1992) Nitrate leaching from sheep-grazed grass/clover and fertilised grass pastures. *Journal of Agricultural Science* 119, 335–343.

D'Antonio, C.M. and Vitousek, P.M. (1992) Biological invasions by exotic grasses, the grass/fire cycle, and global change. *Annual Review of Ecology and Systematics* 23, 63–87.

Dasgupta, P. (2007) Nature and the economy. *Journal of Applied Ecology* 44, 475–487.

Dawe, S. (1991) Intensive systems. In: Cottle, D.J. (ed.) *Australian Sheep and Wool Handbook*. Inkata Press, Melbourne, Australia, pp. 380–401.

Decau, M.L., Simon, J.C. and Jacquet, A. (2004) Nitrate leaching under grassland as affected by mineral nitrogen fertilization and cattle urine. *Journal of Environmental Quality* 33, 637–644.

defra (2004) Review of key issues related to economic instruments to reduce the environmental impact of pesticide use. Risk and Policy Analysts for the Department for the Environment, Food and Rural Affairs. Available at: http://www.defra.gov.uk/environment/chemicals/pesticides

Degens, B.P., Schipper, L.A., Sparling, G.P. and Duncan, L.C. (2001) Is the microbial community in a soil with reduced catabolic diversity less resistant to stress or disturbance? *Soil Biology and Biochemistry* 33, 1143–1153.

de Haan, C., Steinfeld, H. and Blackburn, H. (1997) *Livestock and the Environment: Finding a Balance*. European Commission Directorate General for Development, Commission of the European Community, Brussels.

de Klein, C.A.M., Sherlock, R.R., Cameron, K.C. and van der Weerden, T.J. (2001) Nitrous oxide emissions from agricultural soils in New Zealand – a review of current knowledge and directions for future research. *Journal of the Royal Society of New Zealand* 31, 543–574.

Dewar, D., Palmer, G.T., Northcott, G.L. and Wilkins, A. (2004) The fate and extent of contaminants from historical sheep dip sites. Supersoil 2004: Program and Abstracts for the 3rd Australian New Zealand Soils Conference. Available at: www.regional.org.au/au/asssi/

Di, H.J. and Cameron, K.C. (2002) Nitrate leaching in temperate agroecosystems: sources, factors and mitigating strategies. *Nutrient Cycling in Agroecosystems* 64, 237–256.

Diamond, J. (2005) *Collapse: How Societies Choose to Fail or Survive*. Penguin, London.

Doak, B.W. (1952) Some chemical changes in the nitrogenous constituents of urine when voided on pasture. *Journal of Agricultural Science* 42, 162–171.

Dorioz, J.M., Wang, D., Poulenard, J. and Trévisan, D. (2006) The effect of grass buffer strips on phosphorus dynamics – A critical review and synthesis as a basis for application in agricultural landscapes in France. *Agriculture, Ecosystems and Environment* 117, 4–21.

Dorrough, J., Moll, J. and Crosthwaite, J. (2007) Can intensification of temperate Australian livestock production systems save land for native biodiversity? *Agriculture, Ecosystems and Environment* 121, 222–232.

Dougherty, W.J., Fleming, N.K., Cox, J.W. and Chittleborough, D.J. (2004) Phosphorus transfer in surface runoff from intensive pasture systems at various scales: a review. *Journal of Environmental Quality* 33, 1973–1988.

Drewry, J.J. (2006) Natural recovery of soil physical properties from treading damage of pastoral. *Agriculture Ecosystems and Environment* 114, 159–169.

Drewry, J.J., Littlejohn, R.P. and Paton, R.J. (2000) A survey of physical properties on sheep and dairy farms in southern New

Zealand. *New Zealand Journal of Agricultural Research* 43, 251–258.

Duckworth, J.C., Bunce, R.G.H. and Malloch, A.J.C. (2000) Vegetation gradients in Atlantic Europe: the use of existing phytosociological data in preliminary investigations on the potential effects of climate change on British vegetation. *Global Ecology and Biogeography* 9, 187–199.

Edmond, D.B. (1964) Some effects of sheep treading on ten pasture species. *New Zealand Journal of Agricultural Research* 7, 1–16.

Eldridge, D.J. (1998) Trampling of microphytic crusts on calcareous soils, and its impact on erosion under rain-impacted flow. *Catena* 33, 221–239.

Elliot, A.H. and Carlson, W.T. (2004) Effects of sheep grazing episodes on sediment and nutrient loss in overland flow. *Australian Journal of Soil Research* 42, 213–220.

Elliot, A.H., Tian, Y.Q., Rutherford, J.C. and Carlson, W.T. (2002) Effect of cattle treading on interrill erosion from hill pasture: modelling concepts and analysis of rainfall simulator data. *Australian Journal of Soil Research* 40, 963–976.

Environment ACT (2007) Contaminated sites. Towards a solution. Available at: http://www.environment.act.gov.au/businessandindustry/contaminatedsites

Erickson, C.L. (2006) Intensification, political economy, and the farming community: in defense of a bottom-up perspective of the past. In: Marcus, J. and Stanish, C. (eds) *Agricultural Strategies*, Cotsen Institute of Archaelogy, University of California, Los Angeles, California, pp. 334–363.

Evans, D.M., Redpath, S.M., Elston, D.A., Evans, S.A., Mitchell, R.J. and Dennis, P. (2006) To graze or not to graze? Sheep, voles, forestry and nature conservation in the British Uplands. *Journal of Applied Ecology* 43, 499–505.

FAO (2007) The global livestock production and health atlas. Animal production and health division. Available at: http://www.fao.org/ag/aga/glipha/index.jsp

Floate, M.J.S. (1970a) Mineralization of nitrogen and phosphorus from organic materials of plant and animal origin and its significance in the nutrient cycle in grazed upland and hill soils. *Journal of the British Grassland Society* 25, 295–302.

Floate, M.J.S. (1970b) Decomposition of organic materials from hill soils and pastures II. Comparative studies on the mineralisation of carbon, nitrogen and phosphorus from plant materials and sheep faeces. *Soil Biology and Biochemistry* 2, 173–185.

Forster, P., Ramaswamy, V., Artaxo, P., Berntsen, T., Betts, R., Fahey, D.W., Haywood, J., Lean, J., Lowe, D.C., Myhre, G., Nganga, J., Prinn, R., Raga, G., Schulz, M. and Van Dorland, R. (2007) Changes in atmospheric constituents and in radiative forcing. In: Solomon, S., Qin, D., Manning, M., Chen, Z., Marquis, M., Averyt, K.B., Tignor, M. and Miller, H.L. (eds) *Climate Change 2007: The Physical Science Basis.* Contribution of Working Group I to the Fourth Assessment Report of the Intergovernmental Panel on Climate Change, Cambridge University Press, Cambridge/New York.

Fraser, P.M. and Piercy, J.E. (1996) *Effects of Summer Irrigation on the Seasonal Activity, Population Size, Composition and Biomass of Lumbricid Earthworms in a Long-Term Irrigation Trial at Winchmore, New Zealand.* ASSSI and NZSSS National Soils Conference, July 1996. pp 89–90.

Frossard, E., Condron, L.M., Oberson, A., Sinaj, S. and Fardeau, J.C. (2000) Processes governing phosphorus availability in temperate soils. *Journal of Environmental Quality* 29, 15–23.

Gburek, W.J., Drungil, C.C., Srinivasan, M.S., Needelman, B.A. and Woodward, D.E. (2002) Variable-source-area controls on phosphorus transport: bridging the gap between research and design. *Journal of Soil and Water Conservation* 57, 534–543.

Gillingham, A.G. and During, C. (1973) Pasture production and transfer of fertility within a long-established hill pasture. *New Zealand Journal of Experimental Agriculture* 1, 227–232.

Gillingham, A.G., Syers, J.K. and Gregg, P.E.H. (1984) Nutrient cycling and management in grazed pastures. *New Zealand Agricultural Science* 18, 115–119.

Greene, R.S.B., Kinnell, P.I.A. and Wood, J.T. (1994) Role of plant cover and stock trampling on runoff and soil erosion from semi-arid wooded rangelands. *Australian Journal of Soil Research* 32, 953–973.

Greenwood, K.L. and McKenzie, B.M. (2001) Grazing effects on soil physical properties and the consequences for pastures: a review. *Australian Journal of Experimental Agriculture* 41, 1231–1250.

Greenwood, P.B. and McNamara, R.M. (1992) An analysis of the physical condition of two intensively grazed Southland soils. *Proceedings of the New Zealand Grassland Association* 54, 71–75.

Griffith, D.W.T., Leuning, R., Denmead, O.T. and Jamie, I.M. (2002) Air-land exchanges of CO_2, CH_4 and N_2O measured by FTIR spectrometry and micrometeorological techniques. *Atmospheric Environment* 36, 1833–1842.

Grime, J.P. (1979) *Plant Strategies and Vegetation Processes*. Wiley, Chichester, UK.

Guo, L.B. and Gifford, R.M. (2002) Soil carbon stocks and land use change: a meta analysis. *Global Change Biology* 8, 345–360.

Hairsine, P.B., Moran, C.J. and Rose, C.W. (1992) Recent developments regarding the influence of soil surface characteristics on overland flow and erosion. *Australian Journal of Soil Research* 30, 249–264.

Ham, J.M. and DeSutter, T.M. (2000) Toward site-specific design standards for animal-waste lagoons: protecting ground water quality. *Journal of Environmental Quality* 29, 1721–1732.

Harding, N.J., Green, R.E. and Summers, R.W. (1994) *The Effect of Future Changes in Land Use on Upland Birds in Britain*. Royal Society for the Protection of Birds, Edinburgh, UK.

Harrod, T.R. and Theurer, F.D. (2002) Sediment. In: Haygarth, P.M. and Jarvis, S.C. (eds) *Agriculture, Hydrology and Water Quality*. CAB International, Wallingford, UK, pp. 155–170.

Hart, M.R. and Quin, B.F. (2003) Farm environment mapping – a farmer-friendly risk-management route to environmental protection. In: Currie, L.D. and Hanly, J.A. (eds) *Tools for Nutrient and Pollutant Management: Applications to Agriculture and Environmental Quality*, Occasional report No. 17. Fertilizer and Lime Research Centre, Massey University, Palmerston North, New Zealand, pp. 299–311.

Hatch, D., Goulding, K. and Murphy, D. (2002) Nitrogen. In: Haygarth, P.M. and Jarvis, S.C. (eds) *Agriculture, Hydrology and Water Quality*. CAB International, Wallingford, UK, pp. 7–27.

Haygarth, P.M. and Jarvis, S.C. (1999) Transfer of phosphorus from agricultural soils. *Advances in Agronomy* 66, 195–249.

Haygarth, P.M., Chapman, P.J., Jarvis, S.C. and Smith, R.V. (1998a) Phosphorus budgets for two contrasting grassland farming systems in the UK. *Soil Use and Management* 14, 160–167.

Haygarth, P.M., Hepworth, L. and Jarvis, S.C. (1998b) Forms of phosphorus transfer in hydrological pathways from soil under grazed grassland. *European Journal of Soil Science* 49, 65–72.

Haynes, R.J. and Williams, P.H. (1993) Nutrient cycling and soil fertility in the grazed pasture ecosystem. *Advances in Agronomy* 49, 119–199.

He, Z.L., Wilson, M.J., Campbell, C.O., Edwards, A.C. and Chapman, S.J. (1995) Distribution of phosphorus in soil aggregate fractions and its significance with regard to phosphorus transport in agricultural runoff. *Water Air and Soil Pollution* 83, 69–84.

Heathwaite, A.L., Quinn, P.F. and Hewett, C.J.M. (2005) Modelling and managing critical source areas of diffuse pollution from agricultural land using flow connectivity simulation. *Journal of Hydrology* 304, 446–461.

Henderson, D.C. (2000) Carbon storage in grazed prairie grasslands of Alberta. MS thesis, University of Alberta, Edmonton, Canada.

Hicks, D.M. and Griffiths, G.A. (1992) Sediment load. In: Mosley, P.A. (ed.) *Waters of New Zealand*, New Zealand Hydrological Society, Wellington, pp. 229–248.

Hodgson, J., Cameron, K., Clark, D., Condron, L., Fraser, T., Hedley, M., Holmes, C., Kemp, P., Lucas, R., Moot, D., Morris, S., Nicholas, P., Shadbolt, N., Sheath, G., Valentine, I., Waghorn, G. and Woodfield, D. (2005) New Zealand's pastoral industries: efficient use of grassland resources. In: Reynolds, S.G. and Frame, J. (eds) *Grasslands: Developments Opportunities Perspectives*, Science Publishers in association with the Food and Agriculture Organisation of the United Nations (FAO), Enfield, New Hampshire, pp. 181–205.

Hoglund, J.H. (1985) Grazing intensity and soil nitrogen accumulation. *Proceedings of the New Zealand Grassland Association* 46, 65–69.

Hooda, P.S., Edwards, A.C., Anderson, H.A. and Miller, A. (2000) A review of water quality concerns in livestock farming areas. *Science of the Total Environment* 250, 143–167.

Hughes, K.J., Magette, W.L. and Kurz, I. (2005) Identifying critical source areas for phosphorus loss in Ireland using field and catchment scale ranking schemes. *Journal of Hydrology* 304, 430–445.

Jackson, L.E., Pascual, U. and Hodgkin, T. (2007) Utilizing and conserving agrobiodiversity in agricultural landscapes. *Agriculture Ecosystems and Environment* 121, 196–210.

Jarvis, S.C., Wilkins, R.J. and Pain, B.F. (1996) Opportunities for reducing the environmental impact of dairy farming managements: a systems approach. *Grass and Forage Science* 51, 21–31.

Johns, D. (2007) Like it or not, politics is the solution. *Conservation Biology* 21, 287–288.

Johnston, A. (1962) Effects of grazing intensity and cover on the water-intake rate of fescue grassland. *Journal of Range Management* 15, 79–87.

Judd, M.J., Kellier, F.M., Ulyatt, M.J., Lassey, K.R., Tate, K.R., Shelton, D., Harvey, M.J. and Walker, C.F. (1999) Net methane emissions from grazing sheep. *Global Change Biology* 5, 647–657.

Karlen, D.L., Mausbach, J.W., Cline, R.G., Harris, R.F. and Schuman, G.E. (1997) Soil quality: a concept, definition, and framework for evaluation (a guest editorial). *Soil Science Society of America Journal* 61, 4–10.

Karlen, D.L., Andrews, S.S. and Doran, J.W. (2001) Soil quality: current concepts and applications. *Advances in Agronomy* 74, 1–40.

Kelliher, F.M., Reisinger, A.R., Martin, R.J., Harvey, M.J., Price, S.J. and Sherlock, R.R. (2002) Measuring nitrous oxide emission rate from grazed pasture using Fourier-transform infrared spectroscopy in the nocturnal boundary layer. *Agricultural and Forest Meteorology* 111, 29–38.

Kelliher, F.M., Dymond, J.R., Arnold, G.C., Clark, H. and Rys, G. (2007) Estimating the uncertainty of methane emissions from New Zealand's ruminant animals. *Agricultural and Forest Meteorology* 143, 146–150.

Kemp, D.R., King, W.M., Gilmour, A.R., Lodge, G.M., Murphy, S.R., Quigley, P.E., Sanford, P. and Andrew, M.H. (2003) SGS biodiversity theme: impact of plant biodiversity on the productivity and stability of grazing systems across southern Australia. *Australian Journal of Experimental Agriculture* 43, 961–975.

King, L.K. and Hutchison, K.J. (1976) The effects of sheep stocking intensity on the abundance and distribution of mesofauna in pastures. *Journal of Applied Ecology* 13, 41–55.

King, L.K., Hutchinson, K.J. and Greenslade, P. (1976) The effects of sheep numbers on associations of Collembola in sown pastures. *Journal of Applied Ecology* 13, 731–739.

Kirby, P. (1992) *Habitat Management for Invertebrates: A Practical Handbook*. Royal Society for the Protection of Birds, Joint Nature Conservation Committee and Natural Power, UK.

Körner, C. (2002) Grasslands in a CO_2 enriched world. *Grassland Science in Europe* 7, 611–624.

Lakzian, A., Murphy, P., Turner, A., Beynon, J.L. and Giller, K.E. (2002) *Rhizobium leguminosarum* bv. viciae populations in soils with increasing heavy metal contamination: abundance, plasmid profiles, diversity and metal tolerance. *Soil Biology and Biochemistry* 34, 519–529.

Lambert, M.G., Renton, S.W. and Grant, D.A. (1982) Nitrogen balance studies in some North Island hill pastures. In: Gandar, P.W. (ed.) *Nitrogen Balances in New Zealand Ecosystems*, Department of Scientific and Industrial Research, Palmerston North, New Zealand, pp. 35–39.

Lambert, M.G., Devantier, B.P., Nes, P. and Penny, P.E. (1985) Losses of nitrogen, phosphorus, and sediment in runoff from hill country under different fertiliser and grazing management regimes. *New Zealand Journal of Agricultural Research* 28, 371–379.

Lambert, M.G., Clark, D.A., Mackay, A.D. and Costall, D.A. (2000) Effects of fertiliser application on nutrient status and organic matter content of hill soils. *New Zealand Journal of Agricultural Research* 43, 127–138.

Lang, R.D. and McCaffrey, L.A.H. (1984) Ground cover – its effects on soil loss from grazed runoff plots, Gunnedah. *Journal of Soil Conservation, New South Wales* 40, 56–61.

Lassey, K.R., Ulyatt, M.J., Martin, R.J., Walker, C.F. and Shelton, I.D. (1997) Methane emissions measured directly from grazing livestock in New Zealand. *Atmospheric Environment* 31, 2905–2914.

Ledgard, S.F. (2001) Nitrogen cycling in low input legume-based agriculture, with emphasis on legume/grass pastures. *Plant and Soil* 228, 43–59.

Leinweber, P., Turner, B.L. and Meissner, R. (2002) Phosphorus. In: Haygarth, P.M. and Jarvis, S.C. (eds) *Agriculture, Hydrology and Water Quality*. CAB International, Wallingford, UK, pp. 29–55.

Letey, J., Sojka, R.E., Upchurch, D.R., Cassel, D.K., Olson, K.R., Payne, W.A., Petrie, S.E., Price, G.H., Reginato, R.J., Scott, H.D., Smethurst, P.J. and Triplett, G.B. (2003) Deficiencies in the soil quality concept and its application. *Journal of Soil and Water Conservation* 58, 180–187.

Leytem, A.B., Mikkelsen, R.L. and Gilliam, J.W. (2002) Sorption of organic phosphorus compounds in Atlantic coastal plain soils. *Soil Science* 167, 652–658.

Li, C., Frolking, S. and Frolking, T.A. (1992a) A model of nitrous oxide evolution from soil driven by rainfall events: 1. Model structure and sensitivity. *Journal of Geophysical Research* 97, 9759–9776.

Li, C., Frolking, S. and Frolking, T.A. (1992b) Amodel of nitrous oxide evolution from soil driven by rainfall events: 2. Model applications. *Journal of Geophysical Research* 97, 9777–9783.

Li, C., Zhuang, Y., Cao, M., Crill, P., Dai, Z., Frolking, S., Moore, B.L., Salas, W., Sonf, W. and Wang, X.C. (2001) Comparing a national inventory of N_2O Emissions from arable lands in China developed with a process-based agroecosystem model to the IPCC methodology. *Nutrient Cycling in Agricultural Ecosystems* 60, 159–175.

Li, S.-G., Asanuma, J., Eugster, W., Kotani, A., Liu, J.-J., Urano, T., Oikawa, T., Davaa, G., Oyunbaatar, D. and Sugita, M. (2005) Net ecosystem carbon dioxide exchange over grazed steppe in central Mongolia. *Global Change Biology* 11, 1941–1955.

Lilburne, L., Sparling, G. and Schipper, L. (2004) Soil quality monitoring in New Zealand: development of an interpretative framework. *Agriculture Ecosystems and Environment* 104, 535–544.

Lilley, J.M., Bolger, T.P., Peoples, M.B. and Gifford, R.M. (2001) Nutritive value and the nitrogen dynamics of Trifolium subterraneum and Phalaris aquatica under warmer, high CO_2 conditions. *New Phytologist* 150, 385–395.

Lockyer, D.R. (1997) Methane emissions from grazing sheep and calves. *Agriculture Ecosystems and Environment* 66, 11–18.

Loganathan, P., Hedley, M.J., Grace, N.D., Lee, J., Cronin, S.J., Bolan, N.S. and Zanders, J.M. (2003) Fertiliser contaminants in New Zealand grazed pasture with special reference to cadmium and fluorine: a review. *Australian Journal of Soil Research* 41, 501–532.

Lunney, D. (2001) Causes of the extinction of native mammals of the western division of New South Wales: an ecological interpretation of the nineteenth century historical record. *Rangeland Journal* 23, 44–70.

Luo, J., Tillman, R.W. and Ball, P.R. (1999a) Factors regulating denitrification in a soil under pasture. *Soil Biology and Biochemistry* 31, 913–927.

MacKerron, D.K.L., Duncan, J.M., Hillman, J.R., MacKay, G.R., Robinson, D.J., Trudgill, D.L. and Wheatley, R.J. (2007) Organic farming: science and belief. Annual Report 1998–1999.Scottish Crop Research Institute. pp. 60–72. Available at: http://www.scri.sari.ac.uk/publications/annualreports

MacLoed, C.J. and Moller, H. (2006) Intensification and diversification of New Zealand agriculture since 1960: an evaluation of current indicators of land use change. *Agriculture Ecosystems and Environment* 115, 201–218.

Magesan, G.N., White, R.E. and Scotter, D.R. (1996) Nitrate leaching from a drained, sheep-grazed pasture. I. Experimental results and environmental implications. *Australian Journal of Soil Research* 34, 55–67.

Mapfumo, E., Naeth, M.A., Baron, V.S., Dick, A.C. and Chanasyk, D.S. (2002) Grazing impacts on litter and roots: perennial versus annual grasses. *Journal of Range Management* 55, 16–22.

Mathews, P.N.P., Hodgson, J. and White, J.G.H. (1999) Livestock farming systems in New Zealand. In: White, J. and Hodgson, J. (eds) *New Zealand Pasture and Crop Science*. Oxford University Press, Auckland, New Zealand, pp. 133–152.

McCallum, M.H., Kirkegaard, J.A., Green, T.W., Cresswell, H.P., Davies, S.L., Angus, J.F. and Peoples, M.B. (2004) Improved subsoil macroporosity following perennial pastures. *Australian Journal of Experimental Agriculture* 44, 299–307.

McCaskill, M.R. and Cayley, J.W.D. (2000) Soil audit of a long-term phosphate experiment in south-western Victoria: total phosphorus, sulfur, nitrogen, and major cations. *Australian Journal of Agricultural Research* 51, 737–748.

McCaskill, M. and Quigley, P. (2006) Fertilising pastures. In: Nie, Z. and Saul, G. (eds) *Greener Pastures for South West Victoria*, 2nd edn. Victorian Department of Primary Industries, Hamilton, Victoria, Australia, pp. 43–52.

McCaskill, M.R., Ridley, A.M., Okom, A., White, R.E., Andrew, M.H., Michalk, D.L., Melland, A., Johnson, W.H. and Murphy, S.R. (2003) SGS nutrient theme: environmental assessment of nutrient application to extensive pastures in the high rainfall zone of southern Australia. *Australian Journal of Experimental Agriculture* 43, 927–944.

McColl, R.H.S. and Gibson, A.R. (1979) Downslope movement of nutrients in hill pasture, Taita, New Zealand. 3. Amounts involved and management implications. *New Zealand Journal of Agricultural Research* 22, 279–286.

McDowell, R.W. (2006) Contaminant losses in overland flow from cattle, deer and sheep dung. *Water, Air and Soil Pollution* 174, 211–222.

McDowell, R.W. and Catto, W. (2005) Alternative fertilisers and management to decrease incidental phosphorus loss. *Environmental Chemistry Letters* 2, 169–174.

McDowell, R.W. and Monaghan, R.M. (2002) The potential for phosphorus loss in relation to nitrogen fertiliser application and cultivation. *New Zealand Journal of Agricultural Research* 45, 245–253.

McDowell, R.W. and Stewart, I. (2006) The phosphorus composition of contrasting soils in pastoral, native and forest management in Otago, New Zealand: sequential extraction and ^{31}P NMR. *Geoderma* 130, 176–189.

McDowell, R.W., Monaghan, R.M. and Carey, P.L. (2003) Potential phosphorus losses in overland flow from pastoral soils receiving long-term applications of either superphosphate or reactive phosphate rock. *New Zealand Journal of Agricultural Research* 46, 329–337.

McDowell, R.W., Drewry, J.J. and Paton, R.J. (2004) Effects of deer grazing and fence-line pacing on water and soil quality. *Soil Use and Management* 20, 302–307.

McDowell, R.W., Monaghan, R.M. and Wheeler, D. (2005) Modelling phosphorus losses from pastoral farming systems in New Zealand. *New Zealand Journal of Agricultural Research* 48, 131–141.

McDowell, R.W., Scott, J.T. and Condron, L.M. (2006) Influence of aggregate size on phosphorus loss and ryegrass yield in a soil cultivated intermittently. *Soil Use and Management* 22, 224–226.

McDowell, R.W., Nash, D.M. and Robertson, F. (2007) Sources of phosphorus lost from a grazed pasture receiving simulated rainfall. *Journal of Environmental Quality* 36, 1281–1288.

McIntosh, P.D., Gibson, R.S., Saggar, S., Yeates, G.W. and McGimpsey, P. (1999) Effect of contrasting farm management on vegetation and biochemical, chemical, and biological condition of moist steepland soils of the South Island high country, New Zealand. *Australian Journal of Soil Research* 37, 847–865.

McKellar, Q.A. (1997) Ecotoxicology and residues of anthelmintic compounds. *Veterinary Parasitology* 72, 413–435.

McKellar, Q.A. (2006) The health of the sheep industry and the medicines to maintain it. *Small Ruminant Research* 62, 7–12.

McTaggart, I.P., Douglas, J.T., Clayton, H., Smith, K.A., Jarvis, S.C. and Pain, B.F. (1997) Nitrous oxide emissions from slurry and mineral nitrogen fertilizer applied to grassland. In: Jarvis, S.C. and Pain, B.F. (eds) *Gaseous Nitrogen Emissions from Grasslands*. CAB International, Wallingford, UK, pp. 201–209.

Melland, A.R. (2003) Pathways and processes of phosphorus loss from pastures grazed by sheep. PhD thesis, The University of Melbourne, Australia.

Metherell, A.K. (2003) Management effects on soil carbon storage in New Zealand pastures. *Proceedings of the New Zealand Grasslands Association Conference* 65, 259–264.

Metherell, A.K., Late Stewart, D.P.C., Carey, P.L. and Moss, R.A. (2002) Long-term irrigation improves soil quality, but decreases soil carbon sequestration. *Soil Quality and Sustainable Land Management Conference* (3–5 April 2002), Massey University, New Zealand, pp. 55–61.

Milne, J. (1997) Grassland management and biodiversity. In: Sheldrick, R.D. (ed.) *Grassland Management in the Environmentally Sensitive Areas*. Proceedings of the British Grassland Society Conference, University of Lancaster, 23–25 September 1997. British Grassland Society, Arrowhead Books, Reading, UK, pp. 25–32.

Ministry for the Environment (1997) *The State of New Zealand's Environment*. Ministry for the Environment, Wellington.

Morgan, J.A. (2005) Rising atmospheric CO2 and global climate change: management implications for grazing lands. In: Reynolds, S.G. and Frame, J. (eds) *Grasslands: Developments Opportunities Perspectives*, Science Publishers in association with the Food and Agriculture Organisation of the United Nations (FAO), Enfield, New Hampshire, pp. 235–260.

Morris, M.G. (1990a) The effects of management on the invertebrate community of calcareous grassland. In: Hillier, S.H., Walton, D.W.H. and Wells, D.A. (eds) *Calcarious Grasslands – Ecology and Management*. Bluntisham Books, Blunisham, UK, pp. 128–133.

Morris, M.G. (1990b) The Hemiptera of two sown calcareous grasslands. III Comparisons with the Auchenorhyncha faunas of other grasslands. *Journal of Applied Ecology* 27, 394–409.

Morton, J.D. and Roberts, A.H.C. (1999) *Fertiliser Use on New Zealand Sheep and Beef Farms: The Principles and Practice of Soil Fertility and Fertiliser Use on New Zealand Sheep and Beef Farms*. New Zealand Fertiliser Manufacturers' Research Association, Auckland, New Zealand.

Murphy, S.R., Lodge, G.M. and Harden, S. (2004) Surface soil water dynamics in pastures in northern New South Wales. 2. Surface runoff. *Australian Journal of Experimental Agriculture* 44, 283–298.

Murray, P.J., Gill, E., Balsdon, S.L. and Jarvis, S.C. (2001) A comparison of methane emissions from sheep grazing pastures with differing management intensities. *Nutrient Cycling in Agroecosystems* 60, 93–97.

Murray, R.I. and Yule, I.J. (2007a) Developing variable rate application technology: scenario development and agronomic evaluation. *New Zealand Journal of Agricultural Research* 50, 53–63.

Murray, R.I. and Yule, I.J. (2007b) Developing variable rate application technology: economic impact for farm owners and topdressing operators. *New Zealand Journal of Agricultural Research* 50, 65–72.

Murray, R.I., Yule, I.J. and Gillingham, A.G. (2007) Developing variable rate application technology: modelling annual pasture production on hill country. *New Zealand Journal of Agricultural Research* 50, 41–52.

Naeth, M.A., Bailey, A.W., Pluth, D.J., Chanasyk, D.S. and Hardin, R.T. (1991) Grazing impacts on litter and soil organic matter in mixed prairie and fescue grassland ecosystems in Alberta. *Journal of Range Management* 44, 7–12.

Nash, D. and Murdoch, C. (1997) Phosphorus in runoff from a fertile dairy pasture. *Australian Journal of Soil Research* 35, 419–429.

Nash, D., Clemow, L., Hannah, M., Barlow, K. and Gangaiya, P. (2005) Modelling phosphorus exports from rain-fed and irrigated pastures in southern Australia. *Australian Journal of Soil Research* 43, 745–755.

Nelson, P.N., Cotsaris, E. and Oades, J.M. (1996) Nitrogen, phosphorus, and organic carbon in streams draining two grazed catchments. *Journal of Environmental Quality* 25, 1221–1229.

Newman, R.H. and Tate, K.R. (1980) Soil phosphorus characterisation by ^{31}P nuclear magnetic resonance. *Communications in Soil Science and Plant Analysis* 11, 835–842.

Nguyen, M.L. and Goh, K.M. (1990) Accumulation of soil sulphur fractions in grazed pastures receiving long-term superphosphate applications. *New Zealand Journal of Agricultural Research* 33, 111–128.

Oliva, G., Cibils, A., Borrelli, P. and Humano, G. (1998) Stable states in relation to grazing in Patagonia: a 10 year experimental trial. *Journal of Arid Environments* 40, 113–131.

Oliva, G., Collantes, M. and Humano, G. (2005) Demography of grazed tussock grass populations in Patagonia. *Rangeland Ecology and Management* 58, 466–473.

Parfitt, R.L., Yeates, G.W., Ross, D.J., MacKay, A.D. and Budding, P.J. (2005) Relationships between soil biota, nitrogen and phosphorus availability, and pasture growth under organic and conventional management. *Applied Soil Ecology* 28, 1–13.

Parfitt, R.L., Schipper, L.A., Baisden, W.T. and Elliot, A.H. (2006) Nitrogen inputs and outputs for New Zealand in 2001 at national and regional scales. *Biogeochemistry* 80, 71–88.

Parfitt, R.L., Mackay, A.D., Yeates, G.W. and Ross, D.J. (2007) Movement of N, P, and C from soils to waters in hill country pastures. In: Currie, L.D. and Yates, L.J. (eds) *Designing Sustainable Farms: Critical Aspects of Soil and Water Management*. Occasional report No. 20. Fertilizer and Lime Research Centre, Massey University, Palmerston North, New Zealand, pp.177–184.

Pelchen, A. and Peters, K.J. (1998) Methane emissions from sheep. *Small Ruminant Research* 27, 137–150.

Polley, H.W., Mayeux, H.S., Johnson, H.B. and Tischler, C.R. (1997) Viewpoint: atmospheric CO_2, soil water and shrub/grass ratios on rangelands. *Journal of Range Management* 50, 278–284.

Povirk, K. (1999) Carbon and nitrogen dynamics of an alpine grassland: effects of grazing history and experimental warming on CO_2 flux and soil properties. MS thesis, University of Wyoming, Laramie, Wyoming.

Pratley, J.E. and Godyn, D.L. (1991) Pasture management. In: Cottle, D.J. (ed.) *Australian Sheep and Wool Handbook*. Inkata Press, Melbourne, Australia, pp. 267–285.

Proffitt, A.P.B., Bendotti, S., Howell, M.R. and Eastham, J. (1993) The effect of sheep trampling and grazing on soil physical properties and pasture growth for a red-brown earth. *Australian Journal of Agricultural Research* 44, 317–331.

Quinn, H., Fitzpatrick, L., Whatley, J. and Fry, P. (2005) *Farm Monitor Project*. Summary of results 2004–2005. Department of Primary Industries, Melbourne, Australia.

Rauzi, F. and Hanson, C.L. (1966) Water intake and runoff as affected by intensity of grazing. *Journal of Range Management* 19, 351–356.

Redman, C. (1999) *Human Impact on Ancient Environments*. University of Arizona Press, Tucson, Arizona.

Ridley, A.M., White, R.E., Helyar, K.R., Morrison, G.R., Heng, L.K. and Fisher, R. (2001) Nitrate leaching loss under annual and perennial pastures with and without lime on a duplex (texture contrast) soil in humid southeastern Australia. *European Journal of Soil Science* 52, 237–252.

Ridley, A.M., Christy, B.P., White, R.E., McLean, T. and Green, R. (2003) North-east Victoria SGS national experiment site: water and nutrient losses from grazing systems on contrasting soil types and levels of inputs. *Australian Journal of Experimental Agriculture* 43, 799–815.

Ritz, K., McHugh, M. and Harris, J. (2003) Biological diversity and function in soils: contemporary perspectives and implications in relation to the formulation of effective indicators. In: Francaviglia, R. (ed.) *Proceedings of OECD Expert Meeting on Soil Erosion and Soil Biodiversity Indicators – Rome March 2003*. MIPAF-MINAMB. Available at: http://www.oecd.org/agr/env/indicators.htm

Robertson, F.A. and Nash, D.M. (2008) Phosphorus and nitrogen in soils, plants, and overland flow from sheep-grazed pastures fertilised with different rates of superphosphate. *Agriculture, Ecosystems and Environment* 126, 195–208.

Rosen, M.R., Reeves, R.R., Green, S., Clothier, B. and Ironside, N. (2004) Prediction of groundwater nitrate contamination after closure of an unlined sheep feedlot. *Vadose Zone Journal* 3, 990–1006.

Russell, J.S. (1960) Soil fertility changes in the long-term experimental plots at Kybybolite, South Australia. I. Changes in pH, total nitrogen, organic carbon, and bulk density. *Australian Journal of Agricultural Research* 11, 902–926.

Ruz-Jerez, B.E., White, R.E. and Ball, P.R. (1995) A comparison of nitrate leaching under clover-based pastures and nitrogen-fertilized grass grazed by sheep. *Journal of Agricultural Science* 125, 361–369.

Saggar, S., Giltrap, D.L., Li, C. and Tate, K.R. (2007) Modelling nitrous oxide emissions from grazed grasslands in New Zealand. *Agriculture Ecosystems and Environment* 119, 205–216.

Sakadevan, K., MacKay, A.D. and Hedley, M.J. (1993) Influence of sheep excreta on pasture uptake and leaching losses of sulfur, nitrogen and potassium from grazed pastures. *Australian Journal of Soil Research* 31, 151–162.

Sankaran, M. and Augustine, D.J. (2004) Large herbivores suppress decomposer abundance in a semiarid grazing ecosystem. *Ecology* 85, 1052–1061.

Sargeant, K. (2003) Effect of grazing management and fertiliser regime on the spatial distribution of available phosphorus, nitrogen and potassium in soil. Honours thesis, La Trobe University, Melbourne, Australia.

Sarmah, A.K., Muller, K. and Ahmad, R. (2004) Fate and behaviour of pesticides in the agroecosystem – a review with a New Zealand perspective. *Australian Journal of Soil Research* 42, 125–154.

Sarmah, A.K., Meyer, M.T. and Boxall, A.B.A. (2006) A global perspective on the use, sales, exposure pathways, occurrence, fate and effects of veterinary antibiotics (VAs) in the environment. *Chemosphere* 65, 725–179.

Sauerbeck, D. (1992) Conditions controlling the bioavailability of trace elements and heavy metals derived from phosphate fertilizers in soils. *Proceedings of the 4th International Imphos Conference on Phosphorus, Life and Environment, Casablanca, 1992*. pp 419–448.

Saul, G. (2006) Grazing pastures. In: Nie, Z. and Saul, G. (eds) *Greener Pastures for South West Victoria*, 2nd edn. Victorian Department of Primary Industries, Hamilton, Victoria, Australia, pp. 53–60.

Schipper, L.A., Baisden, W.T., Parfitt, R.L., Ross, C. and Claydon, J.J. (2007) Large losses of soil C and N from soil profiles under pasture in New Zealand during the past 20 years. *Global Change Biology* 13, 1138–1144.

Scholes, R.J. and Archer, S.R. (1997) Tree-grass interactions in savannas. *Annual Review of Ecology and Systematics* 28, 517–544.

Schuman, G.E., Janzen, H.H. and Herrick, J.E. (2002) Soil carbon dynamics and potential carbon sequestration by rangelands. *Environmental Pollution* 116, 391–396.

Scott, J.T., Condron, L.M., McDowell, R.W. (2005) Soil aggregate dynamics, particulate organic matter and phosphate under dryland and irrigated pasture. In: Jarvis, S.C., Murray, P.J. and Roker, J.A. (eds) *Optimisation of Nutrient Cycling and Soil Quality for Sustainable Grasslands*. Proceedings of a satellite workshop of the XXth International Grasslands Congress, July 2005, Oxford. p. 102.

Sears, P.D. (1950) Soil fertility and pasture growth. *Grass and Forage Science* 5, 267–280.

Sharpley, A.N. and Syers, J.K. (1976) Phosphorus transport in surface runoff as influenced by fertiliser and grazing cattle. *New Zealand Journal of Science* 19, 277–282.

Sharpley, A.N. and Syers, J.K. (1979) Effect of aerial topdressing with superphosphate on the loss of phosphate from a pasture catchment. *New Zealand Journal of Agricultural Research* 22, 273–277.

Sharpley, A.N., Gburek, W.J., Folmar, G. and Pionke, H.B. (1999) Sources of phosphorus exported from an agricultural watershed in Pennsylvania. *Agricultural Water Management* 41, 77–89.

Sheldrick, R.D. (1997) Grassland management in the environmentally sensitive areas. *Proceedings of the British Grassland Society Conference, University of Lancaster, 23–25 September 1997*. British Grassland Society, Arrowhead Books, Reading, UK.

Sherlock, R.R. and Goh, K.M. (1984) Dynamics of ammonia volatilization from simulated urine patches and aqueous urea applied to pasture. I. Field experiments. *Fertilizer Research* 5, 181–195.

Silva, R.G., Cameron, K.C., Di, H.J. and Hendry, T. (1999) A lysimeter study of the impact of cow urine, dairy shed effluent, and nitrogen fertiliser on nitrate leaching. *Australian Journal of Soil Research* 37, 357–369.

Six, J., Guggenberger, G., Paustian, K., Haumaier, L., Elliott, E.T. and Zech, W. (2001) Sources and composition of soil organic matter fractions between and within soil aggregates. *European Journal of Soil Science* 52, 607–618.

Six, J., Conant, R.T., Paul, E.A. and Paustian, K. (2002) Stabilization mechanisms of soil organic matter: implications for C-saturation of soils. *Plant and Soil* 241, 155–176.

Smith, C.M. (1987) Sediment, phosphorus, and nitrogen in channelised surface runoff from a New Zealand pastoral catchment. *New Zealand Journal of Marine and Freshwater Research* 21, 627–639.

Smith, S.D., Huxman, T.E., Zitzer, S.F., Charlet, T.N., Housman, D.C., Coleman, J.S., Fenstermaker, L.K., Seeman, J.R. and Nowak, R.S. (2000) Elevated CO_2 increases productivity and invasive species success in an arid ecosystem. *Nature* 408, 79–82.

Smoliak, S., Dormaar, J.F. and Johnston, A. (1972) Long-term grazing effects on Stipa-Bouteloua prairie soils. *Journal of Range Management* 25, 246–250.

Soane, B.D. (1970) The effects of traffic and implements on soil compaction. *Journal and Proceedings of the Institution of Agricultural Engineers* 25, 115–126.

Sparling, G.P. and Schipper, L.A. (2002) Soil quality at a national scale in New Zealand. *Journal of Environmental Quality* 31, 1848–1857.

Sparling, G. and Schipper, L. (2004) Soil quality monitoring in New Zealand: trends and issues arising from a broad-scale survey. *Agriculture, Ecosystems and Environment* 104, 545–552.

Sparling, G.P., Schipper, L.A., Bettjeman, W. and Hill, R. (2004) Soil quality monitoring in New Zealand: practical lessons from a 6 year trial. *Agriculture, Ecosystems and Environment* 104, 523–534.

Sparling, G.P., Wheeler, D., Vesely, E.-T. and Schipper, L.A. (2006) What is soil organic matter worth? *Journal of Environmental Quality* 35, 548–557.

Squires, V.R. (1991) Extensive grazing systems. In: Cottle, D.J. (ed.) *Australian Sheep and Wool Handbook*. Inkata Press, Melbourne, Australia, pp. 369–379.

Srinivasan, M.S. and McDowell, R.W. (2007) Hydrological approaches to the delineation of critical-source areas of runoff. *New Zealand Journal of Agricultural Research* 50, 249–265.

Stevens, D.P., Cox, J.W. and Chittleborough, D.J. (1999) Pathways of phosphorus, nitrogen, and carbon movement over and through texturally differentiated soils, South Australia. *Australian Journal of Soil Research* 37, 679–693.

Stewart, D.P.C. and Metherell, A.K. (1999a) Carbon (^{13}C) uptake and allocation in pasture plants following field pulse labelling. *Plant and Soil* 210, 61–73.

Stewart, D.P.C. and Metherell, A.K. (1999b) Long-term influence of management practices on the soil carbon of pastoral ecosystems. In: Currie, L.D., Hedley, M.J., Horne, D.J. and Loganathan, P. (eds) *Best Soil Management Practices for Production, Occasional Report No. 12*, Fertilizer and Lime Research Centre, Massey University, Palmerston North, New Zealand pp. 317–326.

Stewart, J.W.B. and Tiessen, H. (1987) Dynamics of soil organic phosphorus. *Biogeochemistry* 4, 41–60.

Tate, K.R., Parshotam, A. and Ross, D.J. (1995) Soil carbon storage and turnover in temperate forests and grasslands-a New Zealand perspective. *Journal of Biogeography* 22, 695–700.

Tate, K.R., Scott, N.A., Parshotam, A., Brown, L., Wilde, R.H., Giltrap, D.J., Trustrum, N.A., Gomez, B. and Ross, D.J. (2000) A multi-scale analysis of a terrestrial carbon budget: is New Zealand a source or sink of carbon? *Agriculture, Ecosystems and Environment* 82, 229–246.

Teyssonneyre, F., Picon-Cochard, C., Falcimagne, R. and Soussana, J.F. (2002) Effects of elevated CO_2 and cutting frequency on plant community structure in a temperate grassland. *Global Change Biology* 8, 1034–1046.

Thompson, R.B. and Fillery, I.R.P. (1997) Transformation in soil and turnover to wheat of nitrogen from components of grazed pasture in the south of Western Australia. *Australian Journal of Agricultural Research* 48, 1033–1047.

Thompson, R.B. and Fillery, I.R.P. (1998) Fate of urea nitrogen in sheep urine applied to soil at different times of the year in the pasture-wheat rotation in south Western Australia. *Australian Journal of Agricultural Research* 49, 495–510.

Thomsen, I.K., Schjønning, P. and Chirstensen, B.T. (2003) Mineralisation of N-15-labelled sheep manure in soils of different texture and water contents. *Biology and Fertility of Soils* 37, 295–301.

Tolls, J. (2001) Sorption of veterinary pharmaceuticals in soils: a review. *Environmental Science and Technology* 35, 3397–3406.

Toor, G.S., Condron, L.M., Cade-Menun, B.J., Di, H.J. and Cameron, K.C. (2005) Preferential phosphorus leaching from an irrigated grassland soil. *European Journal of Soil Science* 56, 155–167.

Turner, B.L., Mahieu, N. and Condron, L.M. (2003) The phosphorus composition of temperate pasture soils determined by NaOH-EDTA extraction and solution ^{31}P NMR spectroscopy. *Organic Geochemistry* 34, 1199–1210.

Ulyatt, M.J., Lassey, K.R., Shelton, I.D. and Walker, C.F. (2002) Seasonal variation in methane emission from dairy cows and breeding ewes. *New Zealand Journal of Agricultural Research* 45, 217–226.

van Wieren, S.E. (1991) The management of populations of large mammals. In: Spellerberg, F.B., Goldsmith, C.B. and Morris, M.G. (eds) *The Scientific Management of Temperate Communities for Conservation.* Blackwells, Oxford, pp. 103–128.

Veterinary Medicines Directorate (2007) Sheep dips. Available at: http://www.vmd.gov.uk/general/sheepdip/sheepdips.htm

Victorian Catchment Management Council (2002) *The Health of Our Catchments – A Victorian Report Card 2002.* Victorian Catchment Management Council, Melbourne, Australia.

Vinten, A.J.A., Douglas, J.T., Lewis, D.R., Aitken, M.N. and Fenlon, D.R. (2004) Relative risk of surface water pollution by *E. coli* derived from faeces of grazing animals compared to slurry application. *Soil Use and Management* 20, 13–22.

Vinther, F.P., Hansen, E.M. and Eriksen, J. (2006) Leaching of soil organic carbon and nitrogen in sandy soils after cultivating grass-clover swards. *Biology and Fertility of Soils* 43, 12–19.

Virtue, W.A. and Clayton, J.W. (1997) Sheep dip chemicals and water pollution: UK fluxes to the North Sea, Land Ocean Interaction Study (LOIS) Rivers Basins Research, the first two years. *Science of the Total Environment* 194–195, 207–217.

Walker, J.W., Thompson, C.H. and Jehne, W. (1983) Soil weathering stage, vegetation succession, and canopy dieback. *Pacific Science* 37, 471–481.

Walker, J., Thompson, C.H., Reddell, P. and Rapport, D.J. (2001) The importance of landscape age in influencing landscape health. *Ecosystem Health* 7, 7–14.

Walsingham, J.M. (1976) Effect of sheep grazing on the invertebrate population of agricultural grassland. *Proceedings of the Royal Society of Dublin* 11, 297–304.

Wang, C.J., Wang, S.P., Zhou, H. and Glindemann, T. (2007) Effects of forage composition and growing season on methane emission from sheep in the Inner Mongolia steppe of China. *Ecological Research* 22, 41–48.

Wang, Y.P., Meyer, C.P. and Galbally, I.E. (1997) Comparisons of field measurements of carbon diooxide and nitrous oxide fluxes with model simulations for a legume pasture in south east Australia. *Journal of Geophysical Research* 102, 28013–28024.

Ward, R.C. (1984) On the response to precipitation of headwater streams in humid areas. *Journal of Hydrology* 74, 171–189.

Wardle, D.A. and Bardgett, R.D. (2004) Human-induced changes in large herbivorous mammal density: the consequences for decomposers. *Frontiers in Ecology and the Environment* 2, 145–153.

Watkinson, A.R. and Ormerod, S.J. (2001) Grasslands, grazing and biodiversity: editors' introduction. *Journal of Applied Ecology* 38, 233–237.

Watson, C.A. and Younie, D. (1995) Nitrogen balances in organically and conventionally managed beef production systems. In: Pollott, G.E. (ed.) *Grasslands into the 21st Century: Challenges and Opportunities.* British Grassland Society, Reading, UK, pp. 197–199.

White, R.E., Helyar, K.R., Ridley, A.M., Chen, D., Heng, L.K., Evans, J., Fisher, R., Hirth, J.R., Mele, P.M., Morrison, G.R., Cresswell, H.P., Paydar, Z., Dunin, F.X., Dove, H. and Simpson, R.J. (2000) Soil factors affecting the sustainability and productivity of perennial and annual pastures in the high rainfall zone of south-eastern Australia. *Australian Journal of Experimental Agriculture* 40, 267–283.

White, R.E., Christy, B.P., Ridley, A.M., Okom, A.E., Murphy, S.R., Johnston, W.H., Michalk, D.L., Sanford, P., McCaskill, M.R., Johnson, I.R., Garden, D.L., Hall, D.J.M. and Andrew, M.H. (2003) SGS water theme: influence of soil, pasture type and management on water use in grazing systems across the high rainfall zone of southern Australia. *Australian Journal of Experimental Agriculture* 43, 907–926.

Whitehead, D.C. (2000) *Nutrient Elements in Grassland: Soil-Plant-Animal Relationships.* CAB International, Wallingford, UK.

Whitehead, D.C. and Raistrick, N. (1991) Effects of some environmental factors on ammonia volatilization from simulated livestock urine applied to soil. *Biology and Fertility of Soils* 11, 279–284.

Whitehead, D.C. and Raistrick, N. (1992) Effects of plant material on ammonia volatilization from simulated livestock urine applied to soil. *Biology and Fertility of Soils* 13, 92–95.

Wilkins, R.J. (2002) Environmental sustainability of wool production. *Wool Technology and Sheep Breeding* 50, 705–723.

Willatt, S.T. and Pullar, D.M. (1983) Changes in soil properties under grazed pastures. *Australian Journal Soil Research* 22, 343–348.

Wright, I. (2005) Meat and milk production from grass-based systems. In: Reynolds, S.G. and Frame, J. (eds) *Grasslands: Developments Opportunities Perspectives*, Science Publishers in association with the Food and Agriculture Organization of the United Nations (FAO), Enfield, New Hampshire, pp. 161–179.

Wright, R. (2004) *A Short History of Progress.* Text Publishing, Melbourne, Australia.

Yeates, G.W., Skipp, R.A., Gray, R.A.J., Chen, L.Y. and Waghorn, T.S. (2007) Impact on soil fauna of sheep faeces containing a range of parasite control agents. *Applied Soil Ecology* 35, 380–389.

Younie, D. and Armstrong, G. (1996) Botanical and invertebrate diversity in organic and intensively fertilized grassland. In: Isart, J. and Llerena, J. (eds) *Resource Use in Organic Farming*, Proceedings of the first Workshop of the European Network for Scientific Coordination in Organic Farming. Bonn, Germany, December 1995, pp. 35–44.

Younie, D. and Baars, T. (2005) Organic grassland: principles, practices and potential. In: Reynolds, S.G. and Frame, J. (eds) *Grasslands: Developments Opportunities Perspectives*, Science Publishers in association with the Food and Agriculture Organization of the United Nations (FAO), Enfield, New Hampshire, pp. 207–232.

Zhou, Z.Y., Sun, O.J., Huang, J.H., Li, L.H., Liu, P. and Han, X.G. (2007) Soil carbon and nitrogen stores and storage potential as affected by land use in an agro-pastoral ecotone of northern China. *Biogeochemistry* 82, 127.

7 Pressures on Beef Grazing in Mixed Production Farming

A.N. Sharpley and C. West

Department of Crop, Soil and Environmental Sciences, University of Arkansas, Arkansas, USA

Introduction

Grazing of beef cattle is of major economic importance to agricultural production systems worldwide. With the advent of new technologies, mechanization, increased chemical use and government incentives, agricultural production has more than doubled and become concentrated on less agricultural land and on fewer, but larger, farms (Evans *et al.*, 1996). Also, farming systems have become more specialized, with crop and animal operations efficiently coexisting but in separate regions of the country, as seen by the switch from crop- to poultry- and swine-based systems in several important agricultural states of the USA, such as Alabama, Arkansas, Delaware, Maryland, North Carolina and Oklahoma (Kellogg *et al.*, 2000; Lanyon, 2000).

As a consequence of the spatial separation of crop and animal production systems, fertilizer is imported to areas of grain production. The grain and harvested nitrogen (N) and phosphorus (P) are then transported to areas of animal production, where most is excreted in manure (70%) because of inefficient nutrient utilization by poultry and swine. Since the late 1980s in the USA, cattle, poultry and swine numbers have increased 10–30%, while the number of farms on which they were reared has decreased 40–70% (Gardner, 1998). This intensification has been driven by a greater demand for animal products and an improved profitability associated with economies of scale. This has led to a large-scale, one-way transfer of nutrients from grain- to animal-producing areas and dramatically broadened the emphasis of nutrient management strategies from field and watershed scales to national scales (Lanyon, 2005; Sharpley *et al.*, 2005).

One outcome of the shift in agricultural production systems is the coexistence of confined animal feeding operations (CAFO), with beef grazing or pastoral farming in several areas of the USA. Pastoral farming is typically practised in soils that are too erosive or otherwise unsuitable for grain production, and therefore, offer limited income potential. Integration with poultry and swine feeding operations provided not only additional farm income but a ready source of available nutrients in manure to fertilize pastures at a much lower cost than with commercial mineral fertilizers. However, the rapid growth of CAFOs, coupled with the continued impairment of water quality associated with nutrient enrichment, has necessitated local and regional reduction strategies for nutrient loss (US Department of Agriculture and US Environmental Protection Agency, 1999; National Research Council, 2000; US Department of Agriculture – Natural Resources Conservation Service, 2003). For example, harmful algal blooms in the Chesapeake Bay, Neuse River and hypoxia in the northern Gulf of Mexico have induced point and non-point source reduction strategies in contributing drainage basins. One overarching consequence to agriculture has been the implementation of comprehensive nutrient management planning targeted to CAFOs (US Department of Agriculture – Natural Resources Conservation Service, 2003; US Environmental Protection Agency, 2004). In many cases, this has resulted in a decrease of manure-based nutrients applied to cropland and to pastures.

For watersheds designated as impaired, either by local water quality problems or regional total maximum daily loads (TMDL) development, nutrient management planning has become P-based (US Environmental Protection Agency, 1998, 2002). Where beef grazing has been integrated with CAFOs, this means that on-farm manure can only be applied at rates equivalent to harvested crop removal. Not only does this usually create an on-farm manure nutrient excess, but graziers often have to supplement N not applied in manure by purchasing costly N fertilizer. As the price of N fertilizers is directly linked to that of natural gas (methane), fertilizer costs can fluctuate. For example, between 2005 and 2007, fertilizer N has risen 50% in price. Manufacturing 1 mg of anhydrous NH_3 fertilizer requires 950 m^3 of natural gas (http://www.noble.org/Ag/Soils/NitrogenPrices/Index.htm). This cost comprises most of the costs associated with manufacturing anhydrous NH_3. At US$90/1000 m^3, the natural gas used to manufacture 1 mg of anhydrous NH_3 fertilizer costs US$92. If the price rises to US$250/1000 m^3, the cost of natural gas used in manufacturing that tonne of anhydrous NH_3 rises to US$260, an increase to the manufacturer of US$168. Early 2006 anhydrous ammonia cost about US$240–255 mg in Nebraska, while current price quotes (spring 2007) are around US$460 mg (Bennett, 2007).

In extreme cases, litigation has led to strategies to target and remediate sources of P, such as in the Eucha-Spavinaw Watershed in north-west Arkansas and north-east Oklahoma, which collects and supplies water to the metropolitan area of Tulsa, Oklahoma. In 2003, the City of Tulsa and Tulsa Metropolitan Utility Authority reached an agreement with several poultry integrators and the City of Decatur, Arkansas, waste water treatment plant to address allegations that excess agricultural P runoff from pastures fertilized with poultry litter, as well as waste water discharge from Decatur, were the cause of prolific algal growth and subsequent taste and odour problems in drinking water. The agreement stated that contract poultry producers in the watershed could not land-apply poultry litter until a new P-based nutrient management protocol was developed for use in the entire watershed (DeLaune *et al.*, 2007). Clearly pressure external to beef grazing operations and pastoral farming has put severe economic pressure on graziers.

This chapter describes the integration of farming systems in terms of grazing management impacts on nutrient losses, and how pasture management may aid the economically and environmentally sound coexistence of these systems. This includes an evaluation of a farm system where manure from CAFOs is applied to grazed and harvested grasslands and how management of all system components can influence the potential for nutrient loss.

Impacts of CAFO-Beef Grazing Systems on Nutrient Loss

The combination of CAFO and beef grazing systems can present unique challenges to managing nutrient losses. While the CAFO requires land application of generated manure, usually as N and P source for pastures, the effect of treading on soil compaction by cattle can enhance nutrient loss potential.

Two mechanisms of soil damage occur in pastures. Fine-textured soils become plastic when wetted sufficiently and soil pore space is decreased through compaction. In waterlogged or saturated soils, livestock hooves cause soil to be laterally displaced, disrupting pore continuity, which in turn decreases pore function (Kuykendall *et al.*, 1999; Natural Resources, Agriculture, and Engineering Service, 2006). Half the runoff coming from pastures that are used to over-winter livestock occurs during the dormant season when compaction is highest (Owens *et al.*, 1982). Most sediment losses occur late winter and early spring when soils are wet, loosened at hoof prints and often exposed due to close grazing and plant loss (Owens *et al.*, 1982). Once compacted, pasture soils tend to remain so until rested from livestock traffic. Frost action, shrink-swell of some clays, plant root growth and decay, and earthworm and other burrowing insect activity decrease soil compaction with time. Mechanical soil aerators can be used to decrease compaction; however, their use has been limited due to the expense involved and the brief time which the loosening lasts (Pote *et al.*, 2003).

Nutrient budgeting

The adoption of comprehensive nutrient management planning for CAFOs requires an inventory of the nutrient budget of the farm, which includes the amounts of N and P brought on to the farm in feed, output in forage and animal produce, soil reserves and forage needs, in an effort to minimize nutrient losses (US Department of Agriculture – Natural Resources Conservation Service, 2003). Nutrient flows on a typical CAFO (poultry broilers)–beef farm system are presented in Fig. 7.1, illustrating the challenges facing nutrient budgeting of these systems. In general, N and P inputs far exceed outputs at a farm level. While there is the potential for N use as a fertilizer for forage production, P can rapidly accumulate (West and Waller, 2007).

Estimates of annual flows and balance of N and P for pathways shown in Fig. 7.1 are presented in Table 7.1 for a representative CAFO (poultry broiler–beef operation in north-west Arkansas). For this example, only 15% of the imported N is exported in animal produce and for P, 12% (Table 7.1). Fifteen per cent of the N and 17% of P is recycled back to the pasture through ungrazed vegetation

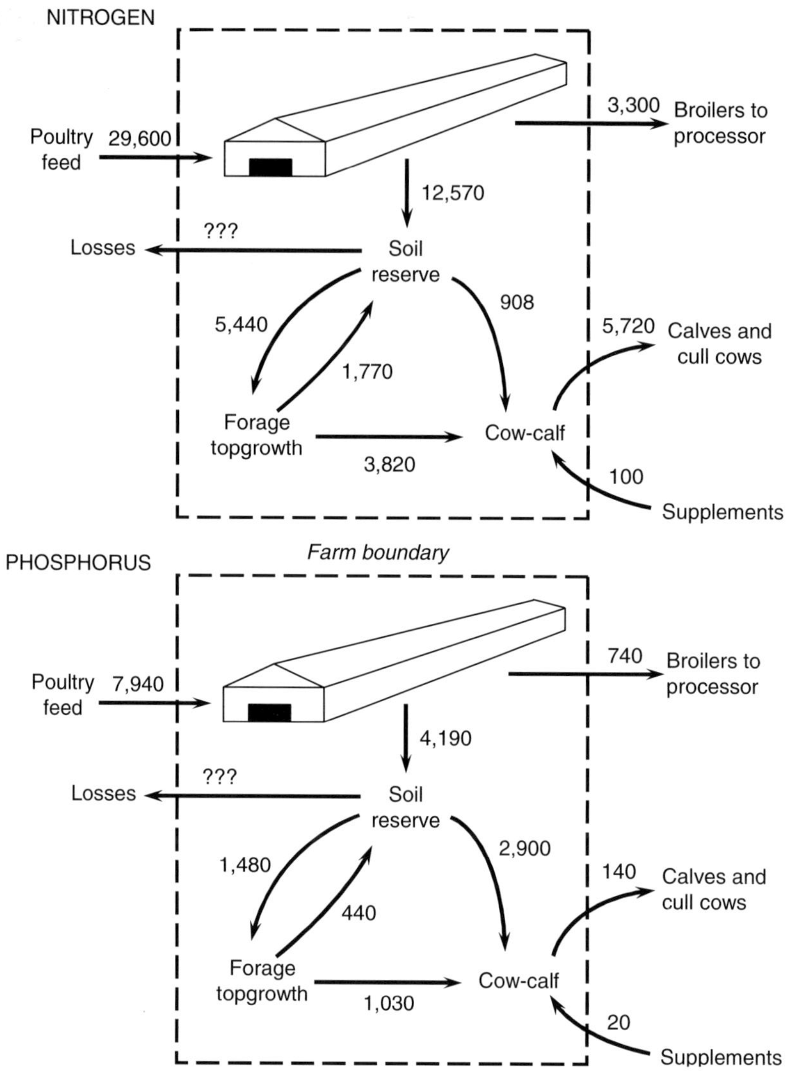

Fig. 7.1. Farm-scale N and P budget for a theoretical 80 ha farm in north-west Arkansas with three broiler houses and beef cows. Values are total N and P in kg/year. The cow-calf pool includes heifers and bulls. (Data adapted from West and Waller, 2007).

and cattle excreta. The remaining N and P (about 70%) is unaccounted for with the farm system (i.e. 263 kg N and 72 kg P/ha/year). This scenario illustrates the potential for N and P to accumulate within a CAFO–beef grazing system and while the litter-based N can be used to maintain forage production, excess soil P can limit on-farm utilization of the litter-based N. Management options to minimize this P build-up include exporting poultry litter and harvesting forage crops for off-farm sale; however, this requires use of off-farm mineral N or N_2-fixing legumes (West and Waller, 2007).

Table 7.1. Annual N and P balance and flow through components of a beef grazing–CAFO system (poultry operation in north-west Arkansas). (Adapted from West and Waller, 2007.)

	kg/ha/year	
Farm component[a]	Nitrogen	Phosphorus
Poultry P balance[b]		
N and P import in feed	370	100
N and P export in poultry	40	10
Litter (manure) produced	3750	3750
N and P recovered in litter and applied to pastures	100	52
Cattle/forage N and P balance[c]		
Forage dry matter produced	5770	5770
N and P uptake into topgowth[d]	127	19
Forage N and P consumed by cattle at 0.7 grazing utilization	89	13
Ungrazed forage N and P returned to soil	38	6
Supplement N and P consumed by cattle	1	<1
N and P excreted by cattle on pasture	81	11
N and P exported in cattle live weight: weaned calves and cull cows	10	2
Whole-farm nutrient balance		
Total N and P import in feed and supplement	371	101
Total N and P export in poultry and beef	51	12
Excess N and P (import–export)	320	89
N and P returned to pasture as ungrazed forage and cattle excreta	119	17
Unaccounted for N and P	201	72

[a]Eighty-ha farm in forage (bermudagrass, tall fescue, white clover and some annuals).
[b]Three poultry houses with five broiler flocks per year, producing 734,700 kg of bird live weight.
[c]Eighty beef cows, 72 calves, 12 heifers and 3 bulls. Farm is self-sufficient in feed production for the cattle except winter energy supplement for cows and heifers and mineral supplement. No hay is imported or exported, and no phosphatic fertilizer is imported.
[d]Topgrowth uptake is 2.20% N and 0.32% P.

Nitrogen

Nitrate (NO_3) leaching from intensively grazed pastures occurs where precipitation exceeds evapotranspiration and results from the high levels of N fertilization and uneven recycling of N in urine and dung (Ball and Ryden, 1984). In fact, NO_3 leached from below-grazed grasslands is affected by grazing density and soil type (Garwood and Ryden, 1986; Steenvoorden *et al.*, 1986; Cuttle and Scholfield, 1994). Increasing grazing density can promote NO_3 leaching because the N consumed by the animal is largely returned to the pasture as urea in urine patches to only a very small part of the grazed sward (Jarvis *et al.*, 1989). For example, cattle urine spots impact about a 60 cm diameter area and N concentrations under urine spots are equivalent to a 700 kg/ha fertilizer N application, which exceeds the capacity of the soil to assimilate N and with water contained in the urine, leaching occurs (Stout *et al.*, 1997, 2000).

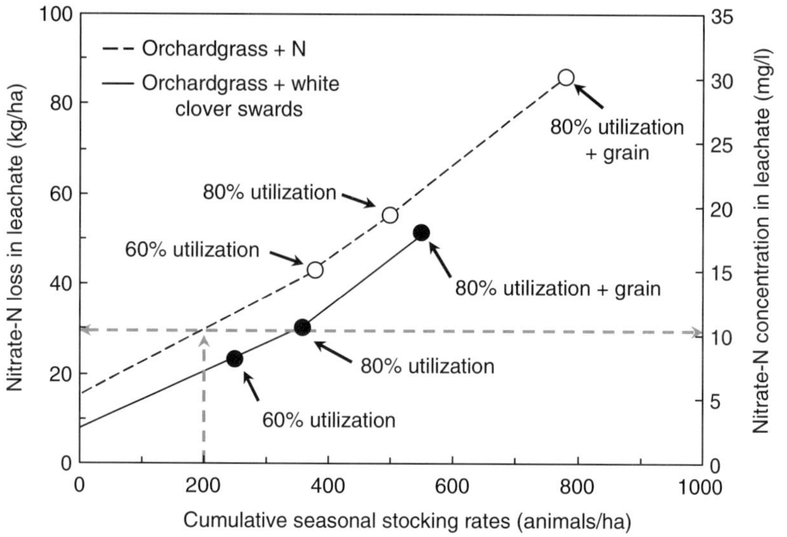

Fig. 7.2. Relationship between stocking density and nitrate-N in leachate beneath N-fertilized orchardgrass and orchardgrass/white clover swards.

N loss from urine patches is affected by animal size, animal type, forage quality and growth stage, and climatic factors, while N loss from the pasture as a whole can be affected by soil hydrologic properties, fertilization rate and pasture species composition (Whitehead, 1995). In the North-east USA, Stout *et al.* (1997) showed that about 25% of the N excreted in cattle urine can leach below the root zone and can contribute to a sizeable amount of NO_3 leached when projected over the entire pasture (Fig. 7.2). A relatively low cumulative seasonal stocking rate of 200 cows-days/ha (590 kg mature Holstein dairy cows) would result in 10 mg/l NO_3–N in leachate beneath a fertilized cocksfoot (*Dactylis glymerata* L.) pasture (Fig. 7.2).

Phosphorus

High concentrations of soil P can occur in grazed pastures where manure is deposited (Gerrish *et al.*, 1995; Sharpley *et al.*, 1998). For example, West *et al.* (1989) found Bray-I soil P ranged from 3 to 9 mg/kg in Kentucky 31 tall fescue [*Lolium arundinaceum* (Schreb.) Darbysh.] permanent pasture. After 5 years grazing (5 beef steers/ha), Bray-I P ranged from 10 to 150 mg/kg, with the higher soil P values extending only 10–20 m from the water source (West *et al.*, 1989). This, along with soil compaction, increases the potential for P loss from grazed pastures in runoff or drainage waters (Breeuwsma *et al.*, 1995; Nelson *et al.*, 1996; Sharpley and Syers, 1976). For example, Owens *et al.* (1997) found that decreasing grazing density and duration dramatically decreased runoff and erosion from a pastured watershed in Ohio (Table 7.2).

Table 7.2. The effect of grazing on runoff and erosion from a pasture watershed in Ohio. (Adapted from Owens *et al.*, 1997.)

	Period	Runoff		Erosion
Pasture management	Years	mm	%[a]	kg/ha/year
Continuous grazing with hay feed	12 (1974–1986)	120	11	2259
Rotational grazing in summer	3 (1986–1989)	14	1	146
No grazing	5 (1989–1994)	6	1	9

[a]Per cent of annual precipitation as runoff.

Clearly, increased runoff and erosion with grazing will enhance the potential for P loss. Olness *et al.* (1975) found that P losses were greater from continuously (4.6 kg P/ha/year) than rotationally grazed pastures (1.3 kg P/ha/year). In fact, P losses with continuous grazing were greater than from ungrazed lucerne (2.5 kg P/ha/year) and wheat (2.9 kg P/ha/year; Olness *et al.*, 1975). Owens *et al.* (1997) showed that when management is changed, the impacts of the previous grazing event or scheme were not long-lasting, changing within a year. Similarly, Sharpley and Syers (1976, 1979) found a rapid increase in the concentration of dissolved and particulate P of surface runoff and tile drainage following grazing. Also, organic P from manure may be more readily leached through a soil profile to shallow aquifers (Haygarth and Jarvis, 1996; Chardon *et al.*, 1997).

Pathogens

Cattle are susceptible to a variety of bacterial and protozoan pathogens which are also pathogenic to humans. For example, the protozoan parasites *Giardia* spp. and *Cryptosporidium parvum* and the enterotoxigenic/enteropathogenic *Escherichia coli* cause scours in calves and diarrhoea in humans (Aiello, 1998). These pathogens are excreted in large numbers by symptomatic animals; asymptomatic animals may shed at lower levels (Garber *et al.*, 1994). Healthy cattle may also serve as passive carriers of human pathogens. For example, the enterohemorrhagic *E. coli* O157, *Campylobacter* spp. and *Listeria monocytogenes* have been documented in cattle feces (Shere *et al.*, 1998; Wesley *et al.*, 2000). There is limited, and often conflicting, information regarding the survival of specific pathogens in pasture environments. Laboratory studies indicate that faecal bacteria (*E. coli* O157 and *Campylobacter* spp.) die rapidly in soils. However, watershed studies suggest that faecal bacteria may persist along river/stream banks and in channels. Increased concentrations of faecal bacteria are frequently observed in rivers/streams during storms owing to microbial resuspension during periods of turbulent flow (McDonald *et al.*, 1982).

Managing Nutrients Through the Grazing Animal

Three major factors limit the sustainability of land application of animal manure to perennial grassland: (i) the mismatch between N to P ratio of the manure (2:1; Edwards, 1996) and the ratio of their uptake by forage plants (5:1 to 10:1; Robinson, 1996); (ii) the impracticality of incorporating manure into the surface soil on the sloping, rocky lands typical of many perennial grasslands; and (iii) the very low export of P in the marketed products of the grazing animals. West and Waller (2007) estimated that only 3.5% of the P from poultry manure applied to representative pastures in Arkansas, USA, is recovered by beef cull cow and calf body mass and marketed off the farm. Therefore, multi-year applications of manures from confined poultry and swine feeding operations to surrounding grasslands result in massive build-up of soil P levels at the soil surface. The readily hydrolysable and soluble forms of P in such manures render this P vulnerable to runoff losses.

Managing nitrogen

Nutrient management planning programmes in high-runoff-risk watersheds restrict or disallow manure applications to grassland soils showing high surface-P build-up. This forces cattle producers to forego animal manures as a low-cost source of N for growing forages, causing a decline in livestock carrying capacity. Options for preventing this loss in pasture production include the purchase of expensive synthetic N fertilizer or incorporating legumes in the forage mixtures. Using Arkansas, USA, forage production data (Huneycutt *et al.*, 1988; 1200 mm rainfall, warm temperate zone), we estimate that applying 4.5 Mg/ha of poultry litter per year to a tall fescue/bermudagrass mixture can sustain beef cattle at 1.0–1.25 cow + calf/ha/year (including replacement heifers and bulls) at a cost of US$45–90/ha depending on transport distance. Replacing the manure with synthetic N from urea would require 100–120 kg/ha of N to maintain the carrying capacity, at a current cost of US$110–130/ha, causing a net decrease in profitability.

Legumes such as white clover (*Trifolium repens* L.) are compatible with pasture grasses in humid zones, and can fix atmospheric N_2 via symbiotic association with *Bradyrhizobium* bacteria in root nodules. Established stands of temperate legumes in a mixture with perennial grasses can fix from 90–285 kg N/ha annually (West and Mallarino, 1996), with amounts increasing as percentage legume in the mixture increases (Mallarino *et al.*, 1990a). A portion of the N fixed by legumes is transferred to associated grass via decomposition of nodules, roots and ungrazed leaves and stems. In addition, legume N in the form of protein is consumed by livestock and largely excreted in the urine as urea. Part of the excreted N is mineralized and taken up by the companion grass. Mallarino *et al.* (1990b) measured an annual average of 41 kg/ha of fixed N transferred from white clover and recovered in tall fescue forage using an ^{15}N-tracer technique. Even though grazing animals can act as an intermediary for transferring fixed N to grass, leakage of N via ammonia volatilization and nitrate leaching from urine limits the efficacy of this pathway.

Legumes are limited in the degree to which they can replace fertilizer N because of their poor competitiveness against aggressive grasses and low tolerance of drought, insects, diseases and grazing. In the Arkansas scenario above where poultry-litter application ceases, white clover would be expected to comprise around 25–30% of the grass-legume mixture and support no more than 1.0 cow + calf/ha, with high risk of legume stand loss due to summer drought. Pasture carrying capacity would be moderately decreased by replacing 4.5 Mg/ha of poultry manure with legumes, but would provide a higher quality forage capable of producing greater profit per head (Hoveland, 1986). Successful use of legumes as an alternative N source depends on fine-tuned management practices, such as maintaining favourable soil pH, replenishment of the soil seed-bank to promote continual recruitment of new legume seedlings, preventing overgrazing to maintain legume plant vigour and N_2 fixation rate and avoiding insufficient grazing of the grass component to prevent excessive shading of the legumes by grass.

Managing phosphorus

Another challenge facing managers of excessively high-P soils in vulnerable watersheds is to draw down soil P to levels considered low risk for P loss in runoff. As grazing beef animals excrete >90% of the P they consume and retain so little in their body mass (<2 kg in a 225 kg calf and <8 kg in a cull cow marketed off the pasture), forage management may shift from all-grazing to harvesting and removing some or all the herbage as hay or silage. This is considered a remediation effort since fairly high applications of N are needed to drive plant growth and P uptake. Annual declines in soil-test P may be a small fraction of the relatively labile P pool in the soil, leading to slow improvements in the potential P runoff. Robinson (1996) summarized annual P removal rates with bermudagrass (*Cynodon dactylon* (L.) Pers.) hay of 26 kg/ha for a common variety to an extreme of 82 kg/ha on swine-slurry treated fields for 'Coastal'. Coblentz *et al.* (2004) reported an annual P removal rate with common bermudagrass on soils with a history of poultry-litter application of 43 kg/ha when receiving 112 kg/ha/year of inorganic N in western Arkansas. Mehlich-3 extractable soil P declined by 48 mg/kg over 2 years.

Forages harvested from high-P sites can be fed back to animals in identifiable low-P, low runoff-risk areas on the same farm; however, that does not mitigate the farm-scale P accumulation problem, and soil adsorption capacity at the low-P site will eventually become saturated with P. Export of the harvested forage from the watershed as a cash crop to buyers demanding feed of high nutritional quality, such as dairy and horse producers, offers the best opportunity to draw down soil P to sustainable levels while making a profit.

Rotational stocking is often suggested as a practice to improve nutrient recycling on pastures by promoting a more uniform redistribution of excreta. High-density grazing in relatively small paddocks with long rest periods for plant recovery results in more uniform defoliation of the pasture compared with continuous stocking (Sollenberger and Newman, 2007). Such behavior would theoretically result in a more uniform distribution of excreted N and P, which would decrease their build-up and potential loss in frequented (e.g. camping) sites, such as at water and shade (Peterson and Gerrish, 1996). However, in US CAFO-beef grazing systems, spatial distribution of cattle excreta appears to be driven more by stocking density, location of shade and distance of travel to water, independently of rotational versus continuous stocking. Mathews *et al.* (1994) observed no difference in uniformity of excreta deposition in continuous, 3-paddock and 15-paddock rotations, but macronutrients consistently accumulated most in the third of pasture area closest to camping areas. Providing readily accessible water supply in each paddock and away from other shade and feeding sites is recommended to maximize recycling of excreted nutrients for pasture regrowth (Peterson and Gerrish, 1996).

Managing forage species

Rotational grazing practices consistently increase pasture-carrying capacity over continuous grazing and defoliation. Increased carrying capacity translates into greater output of animal live weight gain or milk per unit land area for a given level of resource input, as long as per-animal production rate is not compromised. Aiken (1998) reported a 39% increase in carrying capacity and 44% increase in weight gain per hectare with steers grazing wheat (*Triticum aestivum* L.) and annual ryegrass (*L. multiflorum* Lam.) during spring in an 11-paddock rotation. Hoveland *et al.* (1997) observed a 37% increase in weight gain per hectare in rotational over continuous stocking with tall fescue/bermudagrass, which was explained entirely by an increase in carrying capacity. Multi-paddock rotation at high stock density enhances carrying capacity by increasing the utilization rate of forage by the animal and by reducing defoliation stress on the plant, thereby increasing plant growth and nutrient uptake from the soil. The enhanced export of P in the marketing of more animal product would be too small, however, to affect the farm-scale P balance. The rationale for intensifying the degree of rotation where the nutrient management plan calls for cessation of manure application is that increased per-hectare output of animal product generates additional income to offset the cost of the more expensive N from synthetic fertilizer, or a similar per-hectare output can be realized from a lower level of N fertilizer input.

Diversifying the types of forages on a farm can more thoroughly exploit changing growing conditions throughout the year for maximizing nutrient uptake and recycling. Farms in warm-temperate and subtropical conditions rely heavily on warm-season forages for grazing and hay production. Bermudagrass is the predominant grass used for that purpose in the southern USA; however, this grass leaves a 5- to 7-month production gap during which temperatures are too cold for growth (West and Waller, 2007). Winter annual grasses and legumes can be autumn-planted and grazed during the winter and early spring, and/or allowed to accumulate growth in spring for a harvest of hay or silage. The winter crops would take up N and P during a time of year when these nutrients are most subject to leaching and runoff losses. Winter uptake rates of P in Mississippi, USA, amounted to 20, 17 and 17 kg/ha for annual ryegrass, wheat and crimson clover, respectively, when fertilized with poultry litter (calculated from Pederson *et al.*,

2002). These amounts add significantly to the range of P uptake rates of 26–82 kg/ha during summer growth of bermudagrass cited by Robinson (1996).

Tall fescue is the predominant perennial temperate forage grass in the humid USA owing to its high yield and adaptation to widely variable soil and climatic conditions and grazing managements (West and Waller, 2007). Animal productivity and health are depressed because of ergot-alkaloid toxins produced by a fungal endophyte (*Neotyphodium coenophialum*) in the grass. The toxins exacerbate heat stress in cattle in hot, humid environments and reduce blood flow to the extremities in cold weather, causing animals to seek shade or stand in ponds for relief. Endophyte-free cultivars of tall fescue lack such toxins, but do not persist well under the combined stresses of drought and heavy grazing pressure. New cultivars contain endophytes specifically selected for lack of ergot-alkaloid production, but which retain the benefits of drought and grazing tolerance for host grass persistence (Bouton *et al.*, 2002).

Such cultivars could potentially improve the efficiency of nutrient recycling in pastures and decrease nutrient losses by causing animals to spend a greater proportion of their time grazing and less time in shade and water (Schomberg *et al.*, 2000). Parish *et al.* (2003) reported that steers grazing tall fescue with a non-toxic endophyte spent less time idling and standing, used less water and consumed more forage than steers grazing toxic fescue, indicating the potential for better redistribution of excreted nutrients when using non-toxic endophytes. Steer-calf weaning weight was increased by 15% by grazing cow-calf pairs on tall fescue infected with a non-toxic endophyte compared with the wild-type toxic endophyte, plus the cows maintained higher body weight on the new endophyte (Watson *et al.*, 2004). Greater live weight gain indicates that non-toxic endophytes can enhance retention of all ingested nutrients in cattle, including P. Moreover, converting fescue pastures from toxic to non-toxic types increases overall conversion efficiency of all resource inputs and profitability of enterprise (Stuedemann and Seman, 2005).

Managing for soil carbon

The effect of grazing on soil organic C (SOC) was reviewed by Schnabel *et al.* (2000), who summarized the effect of various grazing management practices on SOC and animal production (Table 7.3). While the effect of some specific practices such as utilizing C4 grasses have predictable effects on SOC, the effect of the overall practice of adopting rotational stocking on SOC was found to be unclear. Recent unpublished research on cattle grazing effects on SOC in the Southern Piedmont region of the USA has been summarized in abstract form (Lovell *et al.*, 1997). During the first 3 years of steers grazing 'Coastal' bermudagrass, SOC increased at a rate of 1.5–1.8 Mg/ha/year. Soil organic C under bermudagrass that was harvested as hay or left unharvested for conservation increased at a rate of only 0.3–0.4 Mg/ha/year. The higher rates of soil C accretion under grazing were due to return of the forage-derived but rumen-undigested C to the soil that readily became part of the SOC pool. Following 15–19 years of cattle grazing 'Tifton 44' or Coastal bermudagrass, SOC to a depth of 20 cm averaged 36.7 Mg/ha (Franzluebbers *et al.*, 1998). Three paired fields that

Table 7.3. Implications of changes in grassland management on potential carbon sequestration and animal production. (Adapted from Schnabel *et al.*, 2000.)

	Influence	
Factor	C sequestration	Animal production
C4 grasses replacing C3 grasses	*Increase*; higher C/N; more SOC accumulation	Beef cattle – possible increased production, but increased management costs Dairy cattle – decreased production on lactating animals
Replacing endophyte-infected tall fescue with non-infected fescue or with non-toxic endophyte	*Decrease*; lower microbial degradation in soil	Increased weight gains, also suitable for dairy production
Adopting intensive grazing management	*Unknown*; increased biomass; increased SOC; increased forage quality; increased degradation	Increased animal gains and milk production
N fertilization – hayland	*Increase*; higher biomass production; higher SOM	Increased animal production per unit land area
N fertilization – intensive pastures	*Unknown*; increased biomass; increased SOC; increased forage quality; increased degradation	Increased animal production per unit land area
P fertilization	*Decrease*; increased legume in swards; lower C/N; less SOC accumulation	Increased animal production per unit land area

were hayed, instead of grazed, contained 31.1 Mg SOC/ha. Most of the difference in SOC between grazed and hayed bermudagrass occurred in the surface 5 cm. Carbon in surface residue was also greater ($P < 0.01$) under grazed (1.8 Mg/ha) than under hayed (1.2 Mg/ha) bermudagrass.

The effect of grazing versus haying on SOC in the Netherlands was mixed. Soil organic C in the surface 25 cm averaged 8.9 Mg/ha greater under grazing than under haying at the end of 3 years of comparison (Hassink and Neetson, 1991). Little response in SOC to fertilizer addition (250–700 kg N/ha/year) was observed. In a later study at the same location, the effect of grazing and haying on SOC was small and inconsistent (Hassink, 1994).

Intensity of grazing management may influence SOC. Where soil water supply does not constrain yield, SOC levels would be expected to be greater under management-intensive rotational stocking > intensive continuous stocking > extensive continuous stocking. In contrast, where a water deficit limits production, as in western rangelands, long-term intensive grazing may damage the stand with a concomitant loss of SOC (Dormaar and Smoliak, 1985; Hoglund, 1985).

Fertilizer and lime additions and use of improved seed make intensively managed pastures more productive than extensively managed pastures.

Spatial redistribution of plant residue and excreta from foraging areas to camping areas can alter SOC. In a 15-year-old pasture in New Zealand, SOC in camping areas ranged from 37 to 84 g/kg, but 34 to 72 g/kg in non-camping areas (Nguyen and Goh, 1992). In Georgia, SOC was also most concentrated near permanent shade and water sources (i.e. camping area) in 7- to 15-year-old tall fescue pastures. Although some literature exists on redistribution of a few different nutrients in pastures (West *et al.*, 1989; Wilkinson *et al.*, 1989; Sagger *et al.*, 1990; Peterson and Gerrish, 1996), much more research is needed to understand SOC redistribution within pastures and its potential to sequester C.

Best Management Practices for Mixed Grazing and CAFO Farming

Pasture management

Vegetative cover is widely accepted as one of the main factors that can be managed to minimize runoff and erosion potential. By intercepting raindrop impact, vegetation decreases the energy imparted to surface soil. Pastoral systems, forage type and season of growth can affect both canopy cover and structure. In fact, Self-Davis *et al.* (2003) showed that 'Kentucky-31' tall fescue decreased runoff volume 50% in spring, summer and fall compared with runoff from 'Alamo' switchgrass (*Panicum virgatum* L.), caucasian bluestem (*Bothriochloa caucasia* (Trin.) C.E. Hubbard), 'Greenfield' bermudagrass and 'Pete' eastern gamagrass (*Tripsacum dactyloides* (L.) L.). Infiltration was 17% greater with tall fescue than the other grasses (Self-Davis *et al.*, 2003). Although nutrient loss was not determined in this study, the decrease in runoff would translate into a lower potential for N and P loss. Further, tall fescue produces a major part of its biomass (up to 50%) in April and May in the south-east climate of the USA. As there is a propensity for rainfall-generated runoff to occur in early spring, the early biomass production of tall fescue is likely to further decrease P loss, when compared with warm-season grasses (bermudagrass, bluestem, gamagrass and switchgrass), which obtain heavy canopy cover starting later in May and June.

Periodic aeration of pastures has the potential to decrease runoff and associated nutrient loss via increased infiltration of rainfall, as well as extending forage production (Chen *et al.*, 2001; Shah *et al.*, 2004; Franklin *et al.*, 2005). Franklin *et al.* (2007) found that tall fescue/bermudagrass hay fields (0.8 ha), which had been fertilized with poultry litter for several years, decreased surface runoff volume and dissolved P loss by 35% on well-drained soils in Georgia. In contrast, aeration of poorly drained soils actually increased surface runoff volume and P loss because the presence of hydrologically impeding soil morphological features (depth to Bt horizon) was closer to the soil surface than for well-drained soils. For soils that remain wet near the surface, compaction by tractor traffic during manure application can negate any benefit of aeration (Franklin *et al.*, 2007).

Riparian buffers

Grassed riparian buffers at least 10 m wide on either side of the stream can prevent overland sediment flow and decrease sediment-borne nutrient loading (Welsh, 1992; Lowrance *et al.*, 1994). Excluding grazing from this zone keeps livestock from concentrating their time in the shade along the buffer strip and creating a nutrient hot spot close to the stream. Periodic grazing or harvesting at 10 cm residual stubble height will remove accumulated nutrients and maintain stand density in grassed riparian buffers (Clary and Webster, 1989).

The principle means by which a riparian buffer can improve water quality is by decreasing sediment and sediment-borne nutrient loads. It is much less efficient in removing soluble P from surface runoff or ground water (Corps of Engineers, 1991; Lowrance *et al.*, 1994). In some examples cited, neither grass nor forested riparian buffers were very effective in removing soluble P. Nitrogen removal is dependent in large measure (up to 80%) upon denitrification. In order for this to occur, the soils in the riparian buffer need to be poorly drained, highly organic and anaerobic most of the time (Corps of Engineers, 1991). If not, N removal by the riparian buffer, forested or grass may be as little as 4% of the total N exported from the contributing watershed if the buffer has a very narrow area that supports denitrification (Lowrance *et al.*, 1994). Grassed buffers, where denitrification could occur, were effective in removing about 40–60% of various forms of N. This was accomplished within 8 m of the land side entry point (Lowrance *et al.*, 1994). Riparian areas subject to overflow may actually release accumulated nutrients sequestered in sediment and litter to overlying floodwaters (Corps of Engineers, 1991).

Riparian shade can also attract grazing cattle and exacerbate P loss in stream flow. Byers *et al.* (2005) found that 0.3 kg/ha/year dissolved P and 1.2 kg/ha/year total P were exported in stream flow over 3 years from a 14.2 ha tall fescue and bermudagrass watershed grazed by 20 cows in Eatonton, Georgia, with 1.8 ha of non-riparian shade. In comparison, 0.6 kg/ha/year dissolved P and 4.6 kg/ha/year total P were exported in stream flow from a 17.5 ha watershed with only 0.6 ha of non-riparian shade. Byers *et al.* (2005) concluded that as both watersheds had similar areas of unfenced riparian shade (0.5 ha), the greater area of non-riparian shade attracted cattle away from the stream, resulting in a 50% decrease in dissolved and 74% decrease in total P export.

Livestock exclusion from streams

Livestock that defaecate and urinate into stream and near-stream areas can potentially contribute significant loads of nutrients over time. For instance, an average dairy cow can defaecate up to 15 times and urinate 12 times daily, with a single defaecation containing an average of 6.6 g of N and 1.8 mg of P (Whitehead, 1995; James, 2005). While urine does not contain significant concentrations of P (no pun intended), mean concentrations of N in urine range from 8 to 10 g/l, with an approximate N load of 27 g in each Holstein urination (Whitehead, 1995).

By observing four pastures where cattle had access to streams over four intervals during the spring and summer of 2003 in the Cannonsville Watershed, southcentral New York, James *et al.* (2007) were able to estimate faecal P contributions to streams. On average, approximately 30% of all faecal deposits expected from a herd were observed to fall on land within 40 m of a stream, and 7% fell directly into streams. Although amenities in pasture, such as waterers, feeders, salt, and shade, located away from the stream did affect where cattle congregated, the stream was a consistent draw. Using spatial databases of streams, pasture boundaries and livestock characteristics (i.e. number of cattle, and time in pasture) for 90% of the farms in the Cannonsville watershed, approximately 3600 kg of manure P was estimated as deposited directly into streams, with 7650 kg deposited in pasture near streams (<10 m) from the 11,000 cattle in the watershed. Thus, at 12% of the agriculturally derived P loading, cattle excreta contributed a significant amount of P to stream water draining this watershed (Scott *et al.*, 1998).

Recent efforts to exclude cattle from streams as part of the Conservation Reserve Enhancement Program (CREP) were estimated to have resulted in a 32% decrease in P loadings to streams within the Cannonsville watersheds. Thus, livestock exclusionary programmes like CREP and stream bank fencing decrease nutrient loading by mitigating excreta deposition. Clearly, grazing management and placement of stream bank fencing effectively minimizes watershed export of P. For instance, herd size, grazing density and duration could all be used to prioritize sites for stream bank fencing installation. In addition, field observations, such as those by James *et al.* (2007), show that installation of alternative watering sources does not necessarily preclude continued use of streams as a preferred water source.

Stream bank fences must be properly maintained and replaced to ensure effective exclusion, especially after flooding. Also, general wear and tear on the fences by livestock must be continuously monitored. Even though numerous federal, state, county and non-governmental entities currently subsidize stream bank fencing, farmer participation in these programmes is mixed. These programmes may include stipulations concerning reimbursement, maintenance and upkeep that cause farmers to balk. In addition, riparian exclusion may result in various secondary effects that are not subsidized. Farmers often complain about the loss of productive pasture land. Because riparian areas serve as watering sources to livestock as well as sanctuaries from oppressive heat, an alternative to these services must be implemented. Finally, the layout of paddocks may require investments in infrastructure such as stream crossings and bridges.

Some key management practices that can help control the environmental impact of grazing are listed in Table 7.4 and involve:

1. Using on-farm manure to meet the P requirements of the pasture and hay crops, and providing other nutrients through purchased fertilizers to correct crop deficiencies.
2. Growing legumes as the pasture N source to supplement N added in manure. This will decrease the potential for N leaching, provide a lower-cost alternative to commercial fertilizer as a source of N, and increase marketable per-animal production resulting from high forage quality.
3. Feeding digestible energy supplements to improve N retention in the animal (Ledgard *et al.*, 1999). This will decrease N excretion in urine and consequently decrease N leaching, provided stocking rates are not increased.

Table 7.4. Best management practices for beef grazing component within integrated CAFO systems.

Best management practice
Nutrient management for pastures
Solid/liquid separation and treatment of swine slurry
Manure treatment that increases N/P ratio
Frequency and timing of manure application
Fertilize pastures for optimal N and P balance for meeting forage needs
Develop off-farm markets for manure
Forage species and harvest managements that reduce runoff volume
Maintain vigorous legume content to sustain low-cost pasture production
Reduce soil compaction to decrease runoff and N and P loss
Periodic aeration of pastures
Controlled rotational grazing to spread around the livestock impact
Lower grazing densities
Improved loafing areas for over-wintering cattle
Supplemental forage and feed for balanced dietary energy and N
Exclusion of cattle from streams
Alternative forages to tall fescue containing toxic endophyte
Creation of riparian buffers
Alternative water and shade sources
Off-site remediation
Grassed drainage and waterways
Natural and constructed wetlands

4. For CAFOs with liquid manures, separating slurry solids (rich in P) and liquids (rich in N) to more precisely target those nutrients to specific fields and thus optimize their N and P balance. Additionally, the solids can be mixed with P-sequestering by-products, thereby decreasing their soluble P concentrations.
5. Maintaining proper grazing intensity and soil fertility to sustain vigorous pasture swards and prevent soil erosion and compaction.
6. Harvesting hay from high soil fertility fields and near-stream margins for feeding on low fertility sites with low runoff potential or to sell off-farm as high-value forage. This will bring the system or individual fields closer to being nutrient-balanced.
7. Diversifying forages, especially for off-season growth and year-round nutrient uptake, to decrease the potential for soil nutrient build-up.
8. Replacing tall fescue and its toxic endophyte with a variety that is endophyte-free or with a non-toxic endophyte to decrease animal loafing in shade and water and increase marketable animal production.
9. Using rotational stocking to increase the uniformity of grazing, decrease travel to loafing areas and to increase per-area animal productivity.
10. Periodic aeration of pastures to decrease surface compaction, increase infiltration and decrease surface runoff and erosion of nutrients.
11. Installing stream bank fencing and off-stream watering systems to decrease animal access to streams and direct defaecation and urination into streams, and

planting to provide shade and shelter away from streams and decrease stream bank degradation.
12. Installing off-site conservation measures such as grassed waterways, small impoundments and wetlands to impede runoff from pastures and decrease nutrient export.

Conclusions

Grazing is an animal production system whose main purpose is to produce low-cost, high-value animal products on land unsuitable for crop farming, by efficiently cycling grassland nutrients and energy through ruminant animals. As with confinement production systems, grazing will have negative impacts on the environment, and care must be taken to ensure that these impacts are within limits acceptable to society. In contrast to confinement systems, however, grazing managers do not have the same degree of control over nutrient and energy flows. Consequently, if environmental impacts of grazing systems are to be minimized, careful management practices must be invoked.

Some critical questions need to be answered in enhancing the economic and environmental sustainability of mixed farming systems, which integrate pastoral grazing with CAFOs. If manure is exported from the farm, can low-P input systems provide sufficient P for beef cattle to develop and reproduce, while at the same time optimizing forage production? Will the high cost of N fertilizer reduce its use and decrease farm productivity owing to chronic N deficiency? Can grazing be sufficiently and economically controlled on pastures near sensitive water bodies to reduce nutrient losses even when receiving manures? Can intensive management and improved genetics of legumes be implemented for high persistence and N-fixing ability to sufficiently provide N on P-enriched CAFO farms where manure application is restricted?

References

Aiello, S.E. (1998) *The Merck Veterinary Manual.* 8th edn. Merck, Whitehouse Station, New Jersey, 2305 p.

Aiken, G.E. (1998) Steer performance and nutritive value for continuously and rotationally stocked bermudagrass sod-seeded with wheat and ryegrass. *Journal of Production Agriculture* 11, 185–190.

Ball, P.R. and Ryden, J.C. (1984) Nitrogen relationships in intensively managed temperate grasslands. *Plant and Soil* 76, 23–33.

Bennett, D. (2007) *Fuel, Fertilizer Price, Supply Now Uncertain.* Delta Farm Press, Overland Park, KS. Available at: http://deltafarmpress.com/mag/farming_ fuel_fertilizer_price/

Bouton, J.H., Latch, G.C.M., Hill, N.S., Hoveland, C.S., McCann, M.A., Watson, R.H., Parish, J.A., Hawkins, L.L. and Thompson, F.N. (2002) Reinfection of tall fescue cultivars with non-ergot alkaloid-producing endophytes. *Agronomy Journal* 94, 567–574.

Breeuwsma, A., Reijerink, J.G.A. and Schoumans, O.F. (1995) Impact of manure on accumulation and leaching of phosphate in areas on intensive livestock farming. In: Steele, K. (ed.) *Animal Waste and the Land–Water Interface.* Lewis Publishers, New York, pp. 239–249.

Byers, H.L., Cabrera, M.L., Matthews, M.K., Franklin, D.H., Andrae, J.G., Radcliffe, D.E.,

McCann, M.A., Kuykendall, H.A., Hoveland, C.S. and Calvert, V.H. (2005) Phosphorus, sediment, and *Escherichia coli* loads in unfenced streams of the Georgia Piedmont, USA. *Journal of Environmental Quality* 34, 2293–2300.

Chardon, W.J., Oenema, O., del Castillo, P., Vriesema, R., Japenga, J. and Blaauw, D. (1997) Organic phosphorus in solutions and leachates from soils treated with animal manure. *Journal of Environmental Quality* 26, 372–378.

Chen, Y., Zhang, Q. and Petkau, D.S. (2001) Evaluation of different techniques for liquid manure application on grassland. *Applied Engineering in Agriculture* 17, 489–496.

Clary, W.P. and B.F. Webster (1989) Managing grazing of riparian areas in the intermountain region. USDA, Forest Service. Intermountain Research Station. *General Technical Report INT-263*, p.11.

Coblentz, W.K., Daniels, M.B., Gunsaulis, J.L., Turner, J.E., Scarbrough, D.A., Humphry, J.B., Coffey, K.P., Moore, Jr., P.A., Teague, K.A. and Speight, J.D. (2004) Effects of nitrogen fertilization on phosphorus uptake in bermudagrass forage grown on high soil-test phosphorus sites. *The Professional Animal Scientist* 20, 146–154.

Corps of Engineers. (1991) New England Division, US Army. Buffer Strips for Riparian Zone Management. January, 1991, 56 p.

Cuttle S.P. and Scholfield, D. (1994) *Management Options to Limit Nitrate Leaching from Grassland.* Transactions of the 15th World Congress of Soil Science. Acapulco, Mexico, pp. 138–150.

DeLaune, P.B., Haggard, B.E., Daniel, T.C., Chaubey, I. and Cochran, M.J. (2007) The Eucha/Spavinaw phosphorus index: a court mandated index for litter management. *Journal of Soil and Water Conservation* 61, 96–105.

Dormaar, J.F. and Smoliak, S. (1985) Recovery of vegetative cover and soil organic matter during revegetation of abandoned farmland in a semi-arid climate. *Journal of Range Management* 51, 487–491.

Edwards, D.R. (1996) Recycling livestock manure on pastures. In: Joost, R.E. and Roberts, C.A. (eds) *Nutrient Cycling in Forage Systems.* Potash and Phosphate Institute, Norcross, Georgia, pp. 45–64.

Evans, R., Cuffman-Neff, L.C. and Nehring, R. (1996) Increases in agricultural productivity, 1948–1993. In: *Updates on Agricultural Resources and Environmental Indicators No. 6.* US Department of Agriculture-Economic Research Service. US Government Printing Office, Washington, DC, 85 p.

Franklin, D.H., Cabrera, M.L. and Calvert, V.H. (2005) Fertilizer source and soil aeration effects on runoff volume and quality. *Soil Science Society America Journal* 70, 84–89.

Franklin, D.H., Cabrera, M.L., West, L.T., Calvert, V.H. and Rema, J.A. (2007) Aerating grasslands: effects on runoff and phosphorus losses from applied broiler litter. *Journal of Environmental Quality* 36, 208–215.

Franzluebbers, A.J., Hons, F.M. and Zuberer, D.A. (1998) *In situ* and potential CO_2 evolution from a Fluventic Ustochrept in southcentral Texas as affected by tillage and cropping intensity. *Soil Tillage Research* 47, 303–308.

Garber, L.P., Salman, M.D., Hurd, H.S., Keefe, T. and Schlater, J.L. (1994) Potential risk factors for *Cryptosporidium* infection in dairy calves. *Journal of American Veterinary Medical Association* 205, 86–91.

Gardner, G. (1998) Recycling organic wastes. In: Brown, L., Flavin, C. and French, H. (eds) *State of the World.* W.W. Norton, New York, pp. 96–112.

Garwood, E.A. and Ryden, J.C. (1986) Nitrate loss through leaching and surface runoff from grassland: effects of water supply, soil type and management. In: van der Meer, H.G., Ryden, J.C. and Ennik, G.C. (eds) *Nitrogen Fluxes in Intensive Grassland Systems.* Martinus Nijoff Publishers, Dordrecht, the Netherlands, pp. 90–113.

Gerrish, J.R., Peterson, P.R. and Brown, J.R. (1995) Grazing management affects soil phosphorus and potassium levels. In: *American Forage and Grassland Council.* Proceedings, March 1995, Lexington, Kentucky, pp. 174–179.

Hassink, J. (1994) Effects of soil texture and grassland management on soil organic C and N and rates of C and N mineralization. *Soil Biology and Biochemistry* 26, 1221–1231.

Hassink, J. and Neetson, J.J. (1991) Effects of grassland management on the amounts of soil organic N and C. *Netherlands Journal of Agricultural Science* 39, 225–236.

Haygarth, P.M. and Jarvis, S.C. (1996) Pathways and forms of phosphorus losses from grazed grassland hillslopes. In: Anderson, M.G. and Brooks, S.M. (eds) *Advances in Hillslope Processes*. Wiley, Chichester, UK, pp. 283–294.

Hoglund, J.H. (1985) Grazing intensity and soil nitrogen accumulation. *Proceedings of the New Zealand Grassland Association* 46, 65–69.

Hoveland, C.S. (1986) Beef-forage systems for the southeastern United States. *Journal of Animal Science* 63, 978–985.

Hoveland, C.S., McCann, M.A. and Hill, N.S. (1997) Rotational vs. continuous stocking of beef cows and calves on mixed endophyte-free tall fescue-bermudagrass pasture. *Journal of Production Agriculture* 10, 245–250.

Huneycutt, H.J., West, C.P. and Phillips, J.M. (1988) Responses of bermudagrass, tall fescue and fescue-clover to poultry fertilizer and commercial fertilizer. *Arkansas Agricultural Experiment Station Bulletin* 913.

James, E.E. (2005) *Factors Influencing the Adoption and Non-adoption of the Conservation Reserve Enhancement Program in the Cannonsville Watershed, New York*. MSc dissertation thesis, Department of Agricultural Economics and Rural Sociology, The Pennsylvania State University, University Park, Pennsylvania, 174 p.

James, E.E., Kleinman, P.J.A., Veith, T., Stedman, R. and Sharpley, A.N. (2007) Phosphorus contributions from pastured dairy cattle to streams. *Journal of Soil and Water Conservation* 62, 40–47.

Jarvis, S.C., Hatch, D.J. and Roberts, D.H. (1989) The effects of grassland management on nitrogen losses from grazed swards through ammonia volatilization, the relationship to excretal N returns from cattle. *Journal of Agricultural Science (Cambridge)* 112, 205–216.

Kellogg, R.L., Lander, C.H., Moffitt, D.C. and Gollehon, N. (2000) Manure nutrients relative to the capacity of cropland and pastureland to assimilate nutrients: spatial and temporal trends for the United States. *Resource Assessment and Strategic Planning Working Paper 98-1*, USDA, Natural Resources Conservation Service and Economic Research Service, General Services Administration, National Forms and Publication Center, Fort Worth, Texas, 140 p. Available at: http://www.nrcs.usda.gov/technical/land/pubs/manntr.pdf

Kuykendall, H.A., Cabrera, M.L., Hoveland, C.S. and McCann, M.A. (1999) Stocking methods effects on nutrient runoff from pastures fertilized with broiler litter. *Journal of Environmental Quality* 28, 1886–1890.

Lanyon, L.E. (2000) Nutrient management: regional issues affecting the Chesapeake Bay. In: Sharpley, A.N. (ed.) *Agriculture and Phosphorus Management: The Chesapeake Bay*. CRC Press, Boca Raton, Florida, pp. 145–158.

Lanyon, L.E. (2005) Phosphorus, animal nutrition and feeding: overview. In: Sims, J.T. and Sharpley, A.N. (eds) *Phosphorus; Agriculture and the Environment*. American Society Of Agronomy Monograph. American Society of Agronomy, Madison, Wisconsin, pp. 561–586.

Ledgard, S.F., Penno, J.W. and Sprosen, M.S. (1999) Nitrogen inputs and losses from clover/grass pastures grazed by dairy cows, as affected by nitrogen fertilizer application. *Journal of Agricultural Science (Cambridge)* 132 Part 2, 215–225.

Lovell, A.D., Wilkinson, S.R., Stuedemann, J.A., Seman, D.H. and Franzluebbers, A.J. (1997) Broiler litter and grazing pressure impacts on soil organic C and N pools. Agronomy Abstracts, American Society of Agronomy, Madison, Wisconsin, p. 217.

Lowrance, R., Altier, L., Newbold, D., Schnabel, R., Groffman, P., Denver, J., Correll, D., Gilliam, W., Robinson, J., Brinsfield, R., Staver, K., Lucas, W. and Todd, A. (1994) *Riparian Forest Buffer Systems in the Chesapeake Bay Watersheds – Potential Uses*. US Department of Agriculture – Agricultural Research Service. Southeast Watershed Research Laboratory, Tifton, Georgia, 110 p.

Mallarino, A.P., Wedin, W.F., Goyenola, R.S., Perdomo, C.H. and West, C.P. (1990a) Legume species and proportion effects on symbiotic nitrogen fixation in legume-grass mixtures. *Agronomy Journal* 82, 785–789.

Mallarino, A.P., Wedin, W.F., Perdomo, C.H., Goyenola, R.S. and West, C.P. (1990b) Nitrogen transfer from white clover, red clover, and birdsfoot trefoil to associated grass. *Agronomy Journal* 82, 790–795.

Mathews, B.W., Sollenberger, L.E., Nair, V.D. and Staples, C.R. (1994) Impact of grazing management on soil nitrogen, phosphorus, potassium, and sulfur distribution. *Journal of Environmental Quality* 23, 1006–1013.

McDonald, A., Kay, D. and Jenkins, A. (1982) Generation of fecal and total coliform surges by stream flow manipulation in the absence of normal hydrometeorological stimuli. *Applied Environmental Microbiology* 44, 292–300.

National Research Council. (2000) *Watershed Management for Potable Water Supply: Assessing the New York City strategy*. National Academy Press, Washington, DC, 549 p.

Natural Resources, Agriculture, and Engineering Service. (2006) *Forage Production for Pasture Based Livestock Production*. In: Rayburn, E.B. (ed.) NRAES-172. Cooperative Extension, Cornell University, Ithaca, New York.

Nelson, P.N., Costaris, E. and Oades, J.M. (1996) Nitrogen phosphorus and organic carbon in streams draining two grazed catchments. *Journal of Environmental Quality* 25, 1221–1229.

Nguyen, M.L. and Goh, K.M. (1992) Status and distribution of soil sulphur fractions, total nitrogen and organic carbon in camp and non-camp of grazed pastures supplied with long-term superphosphate. *Biology and Fertility of Soils* 14, 181–190.

Olness, A.E., Smith, S.J., Rhoades, E.D. and Menzel, R.G. (1975) Nutrient and sediment discharge from agricultural watersheds in Oklahoma. *Journal of Environmental Quality* 4, 331–336.

Owens, L.B., Edwards, W.M. and Van Keuren, R.W. (1997) Runoff and sediment losses resulting from winter feeding on pastures. *Journal of Soil and Water Conservation* 52, 194–197.

Owens, L.B., Van Keuren, R.W. and Edwards, W.M. (1982) Environmental effects of a medium-fertility 12-month pasture program: hydrology and soil loss. *Journal of Environmental Quality* 11, 236–240.

Parish, J.A., McCann, M.A., Watson, R.H., Paiva, N.N., Hoveland, C.S., Parks, A.H., Upchurch, B.L., Hill, N.S. and Bouton, J.H. (2003) Use of nonergot alkaloid-producing endophytes for alleviating tall fescue toxicosis in stocker cattle. *Journal of Animal Science* 81, 2856–2868.

Pederson, G.A., Brink, G.E. and Fairbrother, T.E. (2002) Nutrient uptake in plant parts of sixteen forages fertilized with poultry litter: nitrogen, phosphorus, potassium, copper, and zinc. *Agronomy Journal* 94, 895–904.

Peterson, P.R. and Gerrish, J.R. (1996) Grazing systems and spatial distribution of nutrients in pastures: livestock management considerations. In: Joost, R.E. and Roberts, C.A. (eds) *Nutrient Cycling in Forage Systems*, Potash and Phosphate Institute, Norcross, Georgia and Foundation Agronomic Research, Manhattan, Kansas, pp. 203–212.

Pote, D.H., Kingery, W.L., Aiken, G.E., Han, F.X., Moore, Jr. P.A. and Buddington, K. (2003) Water quality effects of incorporating poultry litter into perennial grassland soils. *Journal of Environmental Quality* 32, 2392–2398.

Robinson, D.L. (1996) Fertilization and nutrient utilization in harvested forage systems – Southern forage crops. In: Joost, R.E. and Roberts, C.A. (eds) *Nutrient Cycling in Forage Systems*. Potash and Phosphate Institute, Norcross, Georgia, pp. 65–92.

Sagger, S., Mackay, A.D., Hedley, M.J., Lambert, M.G. and Clark, D.A. (1990) A nutrient-transfer model to explain the fate of phosphorus and sulphur in grazed hill-country pasture. *Agriculture, Ecosystems and Environment* 30, 295–315.

Schnabel, R.R., Franzlubbers, A.J., Stout, W.L., Sanderson, M.A. and Steudemann, J.A. (2000) The effects of pasture management practices. In: Follett, R.F., Kimble, J.M. and Lal, R. (eds) *The Potential of USA Grazing Lands to Sequester Carbon and Mitigate the Greenhouse Effect*. Lewis, Boca Raton, Florida, pp. 291–322.

Schomberg, H.H., Stuedemann, J.A., Franzluebbers, A.J. and Wilkinson, S.R. (2000) Spatial distribution of extractable phosphorus, potassium, and magnesium as influenced by fertilizer and tall fescue endophyte status. *Agronomy Journal* 92, 981–986.

Scott, C.A., Walter, M.F., Brooks, E.S., Boll, J., Hes, M.B. and Merrill, M.D. (1998) Impacts of historical changes in land use and dairy herds in water quality in the Catskills Mountains. *Journal of Environmental Quality* 27, 1410–1417.

Self-Davis, M.L., Moore, Jr. P.A., Nichols, D.J., Sauer, T.J., West, C.P., Aiken, G.E. and Edwards, D.R. (2003) Forage species and canopy cover effects on runoff from small plots. *Journal of Soil and Water Conservation* 58, 349–359.

Shah, S.B., Miller, J.L. and Basden, T.J. (2004) Mechanical aeration and liquid dairy manure application impacts on grassland runoff water quality and yield. *Transactions of the American Society of Agricultural Engineers* 47, 777–788.

Sharpley, A.N. and Syers, J.K. (1976) Phosphorus transport in surface runoff as influenced by fertiliser and grazing cattle. *New Zealand Journal of Science* 19, 277–282.

Sharpley, A.N. and Syers, J.K. (1979) Loss of nitrogen and phosphorus in tile drainage as influenced by urea application and grazing animals. *New Zealand Journal of Agricultural Research* 22, 127–131.

Sharpley, A.N., Meisinger, J.J., Breeuwsma, A., Sims, T., Daniel, T.C. and Schepers, J.S. (1998) Impacts of animal manure management on ground and surface water quality. In: Hatfield, J. (ed.) *Effective Management of Animal Waste as a Soil Resource*. Lewis Publishers, Boca Raton, Florida, pp.173–242.

Sharpley, A.N., Withers, P.J.A., Abdalla, C. and Dodd, A. (2005) Strategies for the sustainable management of phosphorus. In: Sims, J.T. and Sharpley, A.N. (eds) *Phosphorus; Agriculture and the Environment*. American Society of Agronomy Monograph. American Society of Agronomy, Madison, Wisconsin, pp. 1069–1101.

Shere, J.A., Bartlett, K.J. and Kaspar, C.W. (1998) Longitudinal study of *Escherichia coli* O157:H7 dissemination on four dairy farms in Wisconsin. *Applied Environmental Microbiology* 64, 1390–1399.

Sollenberger, L.E. and Newman, Y.C. (2007) Grazing management. In: Barnes, R.F. *et al.* (eds) *Forages: The Science of Grassland Agriculture*. Vol. II. 6th edn. Blackwell Publishing, Ames, Iowa, pp. 651–659.

Steenvoorden, J., Fonck, H. and Oosterom, H.P. (1986) Losses of nitrogen from intensive grassland systems by leaching and surface runoff. In: van der Meer, H.G., Ryden, J.C. and Ennik, G.C. (eds) *Nitrogen Fluxes in Intensive Grassland Systems*. Martinus Nijoff Publishers, Dordrecht, the Netherlands, pp. 85–97.

Stout, W.L., Fales, S.L., Muller, L.D., Schnabel, R.R., Priddy, W.E. and Elwinger, G.R. (1997) Nitrate leaching from cattle urine and feces in north-east USA. *Soil Science Society of America Journal* 61, 1787–1794.

Stout, W.L., Fales, S.L., Muller, L.D., Schnabel, R.R., Elwinger, G.F. and Weaver, S.R. (2000) Assessing the effect of management intensive grazing on water quality in the north-east U.S. *Journal of Soil and Water Conservation* 55, 238–243.

Stuedemann, J.A. and Seman, D.H. (2005) Integrating genetics, environment, and management to minimize animal toxicosis. In: Roberts, C.A. *et al.* (eds) *Neotyphodium in Cool-season Grasses*. Blackwell Publishing, Ames, Iowa, pp. 305–324.

US Department of Agriculture and US Environmental Protection Agency. (1999) *Unified National Strategy for Animal Feeding Operations*. 9 March, 1999. Available at: *http://www.epa.gov/owm/finafost.htm*

US Department of Agriculture – Natural Resources Conservation Service. (2003) *Cost Associated with Development and Implementation of Comprehensive Nutrient Management Plans. Part 1 – Nutrient Management, Land Treatment, Manure and Wastewater Handling and Storage, and Recordkeeping*. US Government Printing Office, Washington, DC, 220 p. Available at: http://www.nrcs.usda.gov/technical/land/pubs/cnmp1.html

US Environmental Protection Agency. (1998) *Water Quality Criteria and Standards Plan: Priorities for the Future*. EPA-823-F-98-011. USEPA, Office of Water (4304F), US Government Printing Office, Washington, DC.

US Environmental Protection Agency. (2002) *Protocol for Developing Nutrient TMDLs*. EPA-841-B-99-007. US Environmental Protection Agency, Office of Water (4503F), Washington, DC. Available at: http://www.epa.gov/owow/tmdl/nutrient/pdf/nutrient.pdf

US Environmental Protection Agency. (2004) *Managing Manure Nutrients at Concentrated Animal Feeding Operations*. EPA-821-B-04-006.

US Environmental Protection Agency, Office of Water (4303T), Washington, DC. Available at: http://www.epa.gov/guide/cafo/

Watson, R.H., McCann, M.A., Parish, J.A., Hoveland, C.S., Thompson, F.N. and Bouton, J.H. (2004) Productivity of cow-calf pairs grazing tall fescue pastures infected with either the wild-type endophyte or a non-ergot alkaloid-producing endophyte strain. AR542. *Journal of Animal Science* 82, 3388–3393.

Welsh, D.J. (1992) *Riparian Forest Buffers*. USDA, Forest Service, Forest Resources Management, North-eastern Area, Radnor, Pennsylvania. NA-PR-07-91.

Wesley, I.V., Wells, S.J., Harmon, K.M., Green, A., Schroeder-Tucker, L., Glover, M. and Sudduque, I. (2000) Fecal shedding of *Campylobacter* and *Arcobacter* spp in dairy cattle. *Applied Environmental Microbiology* 66(5), 1994–2000.

West, C.P. and Mallarino, A.P. (1996) Nitrogen transfer from legumes to grasses. In: Joost, R.E. and Roberts, C.A. (eds) *Nutrient Cycling in Forage Systems*. Potash and Phosphate Institute, Norcross, Georgia, pp. 167–175.

West, C.P. and Waller, J.C. (2007) Forage systems for humid transition areas. In: Barnes, R.F. *et al.* (eds) *Forages: The Science of Grassland Agriculture*. Vol. II. 6th edn. Blackwell Publishing, Ames, Iowa, pp. 313–321.

West, C.P., Mallarino, A.P., Wedin, W.F. and Marx, D.B. (1989) Spatial variability of soil chemical properties in grazed pastures. *Soil Science Society of America Journal* 53, 784–789.

Whitehead, D.C. (1995) *Grassland Nitrogen*. CAB International, Wallingford, UK.

Wilkinson, S.R., Stuedemann, J.A. and Belesky, D.P. (1989) Soil potassium distribution in grazed K-31 tall fescue pastures as affected by fertilization and endophytic fungus infection level. *Agronomy Journal* 81, 508–512.

8 The Environmental Impacts of Non-irrigated, Pasture-based Dairy Farming

R.M. Monaghan

AgResearch Invermay, Private Bag 50034, Mosgiel, New Zealand

Introduction

Most modern dairy farming systems continually seek to improve farm business productivity through increasing outputs of saleable product or through decreasing input costs, or both. An inevitable outcome of the former strategy is increased land-use intensity and farm inputs of feed, fertilizer and energy. Inefficient nutrient cycling and increased losses of nutrients to the environment are usually an unfortunate consequence of this pathway to economic productivity. Greater attention is now being focused on some of the off-site impacts of farming activities around the world, particularly the consequences of nutrient enrichment of groundwater and surface waters, and the contribution of greenhouse gases (GHG) such as methane (CH_4) and nitrous oxide (N_2O) to global warming. Intensive agriculture is known to emit significant amounts of nutrients, particularly nitrogen (N) and phosphorus (P), faecal bacteria and sediment (Gillingham and Thorrold, 2000; Watson and Foy, 2001; Oliver *et al.*, 2005a; Monaghan *et al.*, 2007a). While these emissions are typically not large by agronomic standards, transferring pollutants from land to water can significantly impair water quality. These transfers have been shown to increase as farm inputs increase and systems intensify (e.g. Scholefield *et al.*, 1993; Ledgard *et al.*, 1999a; Watson *et al.*, 2000; McDowell *et al.*, 2003b; Monaghan *et al.*, 2005). The political and economic ramifications of the Kyoto protocol have now also placed GHG emissions from agriculture under the spotlight. Coupled with issues of energy-use efficiency in a future where scarcity and cost are likely to become acute, these concerns mean that today's dairy farms are coming under much greater scrutiny, and for a broader range of environmental issues than ever before.

Considerable research over the past four to five decades has helped us understand how nutrients cycle in grazed pastures, and optimize the efficiency of fertilizer nutrients for agronomic productivity. Earlier research on the recycling of nutrients in grazed pasture systems, reviewed by Haynes and Williams (1993), reinforced the concept that nutrients in pasture ingested by the grazing animal are inefficiently utilized in growth, or milk, meat and wool production. The majority of nutrients are excreted in dung or urine. Much of the earlier research also defined relationships between soil test measures of plant available nutrients and pasture production. This led to the development of a suite of soil sampling, testing and interpretation procedures that now underpin many farm fertilizer recommendations (e.g. see reviews by Edmeades *et al.*, 2005, 2006). Much of this information was captured in decision-support models that have been extended to econometric and nutrient budgeting modelling frameworks (Ledgard *et al.*, 1999b). The development of these decision-support tools has shifted nutrient management decision making from simple fertilizer requirement recommendations to more comprehensive farm nutrient management planning on both a block (e.g. group of fields) and whole-farm basis, taking account of nutrient transfers on to and within the farm. These systems have been augmented with plant and animal testing (particularly for trace nutrients), and are now being increasingly adopted into reporting systems by fertilizer companies and soil-testing laboratories.

These decision-support tools have also been progressively modified to address the potential environmental effects of combinations of soil–climate–land management scenarios (e.g. Brown *et al.*, 2005; Ketterings *et al.*, 2006). Evaluations of nutrient-use efficiencies such as nutrient recovery in saleable product, farm nutrient surplus, nutrient loss (kg/ha) and management risks associated with nutrient losses to the environment, are examples of agri-environmental indicators which can assess potential environmental effects. Langeveld *et al.* (2007) and Hanegraaf and Boer (2003) document how some of these types of indicators have been used to guide farm decision making in the Netherlands, although caution that they need to be used with care given that indicators are simplifications of complex and variable processes. To be effective, these tools must accurately encompass the key drivers of nutrient flows on farms, and incorporate the main management practices determining nutrient losses. This chapter reviews our current understandings of some of these key drivers of nutrient flows and losses from non-irrigated dairy farming systems. Potential mitigation measures to decrease these losses are described, including a cost–benefit analysis. We also focus on the impacts of some less-obvious land management practices contributing to farms' discharges of non-nutrient environmental pollutants, such as faecal microorganisms (FMOs) and sediment, and discuss the current state of knowledge of these issues. For brevity, we confine the review to non-irrigated dairy agriculture in temperate regions where pasture is the main source of feed.

Key Sources of Pollutant Losses from Dairy Farms

Much research has focused on the role of dairy farming contributing to nutrient enrichment of groundwater and/or surface waters (Ledgard *et al.*, 1999a; Aarts *et al.*, 2000a,b; Eckard *et al.*, 2004; Humphreys *et al.*, 2004; Wachendorf and Golinski, 2006). While dairy cows are not usually the sole contributor to water quality impairment, this research has shown that inappropriate management of a dairy farm has the potential to cause significant groundwater and stream pollution. It is probably unrealistic to expect clean water to flow from fertile and highly productive land, but there is clearly a desire to minimize losses of nutrients, FMOs and GHGs from dairy farms. Examples of N and P losses from cattle-grazed farmland are given in Table 8.1.

Table 8.1. Losses of N and P to water from grazed pasture systems.

Location and/or stocking rate	N applied (kg N/ha/ year)	Nitrate-N loss (kg N/ha/year)	Total P loss (kg/ ha/year)	Reference
Field/farm scale losses				
Dairy cows, New Zealand				
Otago	88	25	0.08[a]	Monaghan and Smith (2004) and unpublished results
Southland[b]	0	31	0.37	Monaghan *et al.* (2005); Smith and Monaghan (2003)
	200	48	0.43[a]	
	400	58	0.30	
Waikato	0	40		Ledgard *et al.* (1999a)
	200	79		
	400	150		
Taranaki	0	19		Roach *et al.* (2001)
	200	20		
	400	42		
Beef cattle, UK				
Devon: Drained	200	59	2.15	Tyson *et al.* (1997); Scholefield *et al.* (1993); Cuttle and Scholefield (1995); Haygarth *et al.* (1998)
	400	194		
Undrained	200	18	3.15	
	400	74		
Hillsborough, Northern Ireland	100	18		Watson *et al.* (2000)
	200	27		
	400	65		
Antrim, Northern Ireland	300	20	1.1	Jordan and Smith (1985)
Catchment scale losses				
Southland, New Zealand, 0.9 cows/ha[c]		8	0.43	Wilcock *et al.* (2007)
Canterbury, New Zealand, 1.6 cows/ha		9	0.89	Wilcock *et al.* (2007)
Waikato, New Zealand, 1.9 cows/ha		13	0.67	Wilcock *et al.* (2006)
Westland, New Zealand, 1.8 cows/ha		23	5.02	Wilcock *et al.* (2007)
Taranaki, New Zealand, 3.2 cows/ha		26	0.72	Wilcock *et al.* (2007)
Pennsylvannia, USA[d]		56	2.80	Galeone (1999)

[a]Subsurface drainage losses.
[b]Mixed dairy and beef.
[c]Whole catchment equivalent.
[d]Mixed land use.

Pasture-based dairy production systems have historically been driven by soil, plant and climate factors that have been tempered by management inputs, but generally without the constraint of nutrient efficiency. In today's environment nutrient-use efficiency has become one key indicator of sustainable agriculture. Because this indicator is heavily influenced by soil, plant, weather and management variables, manipulation of these factors to achieve an optimal production system is a very complex task. Modern dairy farms receive large inputs of nutrients and cycle most within the farm. As an example, P losses within non-irrigated dairy farms range from 0.3 to 5.0 kg P/ha/year. This typically represents 1–10% of farm P inputs via fertilizer or feed, or much less than 1% of the total mass of P resident within the soil profile. For N, typical leaching losses of between 20 and 40 kg/ha/year represent equivalent losses of less than 10% and 1% of fertilizer inputs and soil N reserves, respectively. The sources of these losses vary, depending on the pollutant and the dairy farm's management systems (Fig. 8.1). Some key aspects of these drivers are outlined below.

Nitrogen

Considerable research into N flows within grazed dairy pastures since the late 1970s clearly shows that the amount of N excreted by animals, and in particular urine N, is the most important determinant of N losses (including leaching

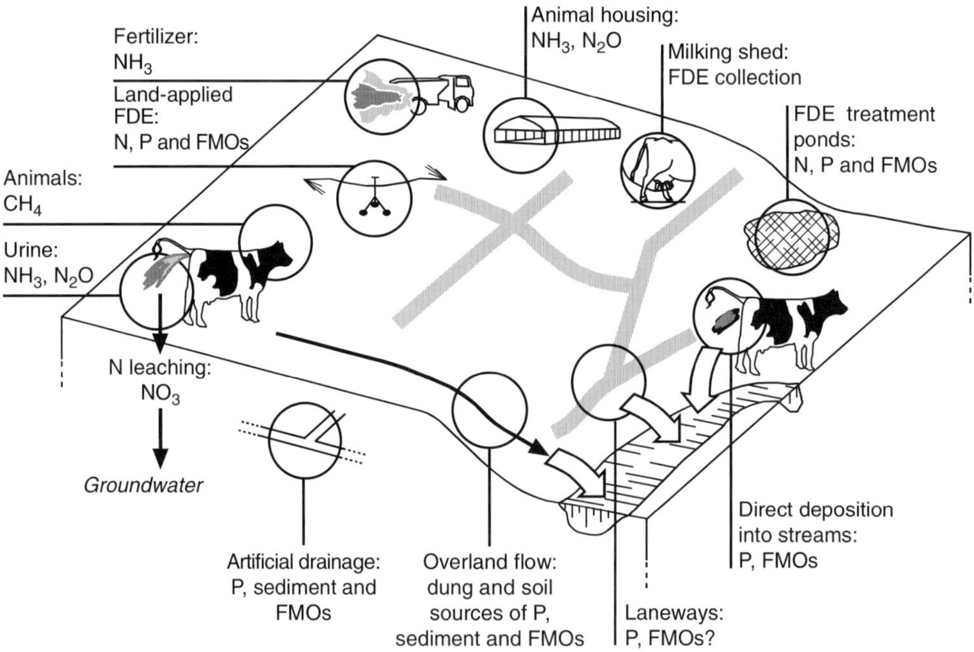

Fig. 8.1. Schematic representation of the main sources of contaminant loss from grazed dairy farms.

to deep drainage or runoff to stream and gaseous losses; e.g. Ledgard, 2001; Di and Cameron, 2002a). Consequently, the amount of N excreted by animals is the primary driving factor of N losses rather than inefficiencies related to N fertilizer usage (Fig. 8.1). The main effect of fertilizer N use on N cycling efficiency in grazed pastures is therefore indirect, with N fertilizer inputs, which allow for an increase in pasture production and animal stocking rate, also increasing urine N excretion. A variety of studies have quantified the potential for N in urine to leach through pasture soils. In an experiment using ^{15}N-labelled urine, Fraser *et al.* (1994) estimated that 8% of urine N leached below 1.2 m depth after 1 year. The timing of urine deposition strongly influences the potential for urine N to leach through pasture soils, with large losses typically observed for urine deposited shortly prior to the onset of drainage. For clover-based dairy pastures, N fertilizer is generally not a major direct source of N loss, as it is applied in relatively low rates and used strategically to supplement N supply from biological N fixation. Direct leaching of fertilizer N is usually low if the rates and timing of applications are correct (Ledgard *et al.*, 1999a; Di and Cameron, 2002b; Monaghan *et al.*, 2005). Soil type exerts a considerable influence on the amount of nitrate-N leached from the soil profile, with greater losses observed for light free-draining soils compared to heavier-textured and/or poorly drained soils where a proportionally greater amount of soil nitrate is removed via denitrification processes.

The amount of N excreted by grazing cows is tied to the amount of N consumed, which in turn is broadly related to animal stocking rate. The relationship between N losses and stocking rate is closer for sheep and beef farms because of the relatively small variation between farms in external N inputs. However, there is a wide variation in external N inputs and per-cow production and intake of feed-N (e.g. approximately threefold) on dairy farms. Stocking rates (animals/ha) are therefore a crude proxy for the magnitude of N loss for a dairy farm. Large losses of nitrate-N in drainage may occur from areas used for forage crop grazing during winter when pasture growth rates are usually low. The greater losses from the wintering part of the system arise due to: (i) relatively large amounts of mineral N remaining in the soil in late autumn following pasture cultivation and forage crop establishment the preceding spring; and (ii) the deposition of much excretal N on to the grazed forage crop during winter when plant uptake is correspondingly low. Dairy grazing systems in cool-temperate regions have accordingly evolved which utilize housing and high inputs of supplemental feed to manage dairy herds through winter months when pasture growth rates are low. Soil poaching is often a concern and nutrient losses in drainage are greatest. These systems contrast with those typically found in warm-temperate regions such as central and northern New Zealand where year-round grazing is commonly practised because of a better match between herd feed demand and pasture growth rates. The type of system employed strongly influences the pathway of nutrient loss. For systems that utilize animal housing, manure management has a major influence on how much N can be lost via ammonia volatilization and denitrification during storage and application to pastures in spring. In contrast, drainage is an important pathway of N loss from excreta voided to pastures during winter under a year-round grazing management system. Although strongly dependent on economic considerations, experience has shown that dairy systems which have utilized housing and feeding systems have also shown a greater propensity for farm intensification, presumably because much of the cost associated with infrastructure required to house and feed animals has already been incurred, and the marginal benefits of additional production outweigh the marginal costs of additional farm inputs.

Ammonia volatilization from recently applied urea and ammonium-based fertilizers, slurry/manure applications or urine deposition can be an important N loss pathway, particularly where soil pH is high and warm conditions favour the removal of ammonia gas from the soil surface (Theobald and Ball, 1984). Urea, derived either from fertilizer or urine applications, has the greatest potential for ammonia loss due to the pH increase associated with urea hydrolysis. Volatilization losses can account for 0–28% of the urine N excreted from dairy livestock, depending upon N application rate, stocking density and duration, climatic conditions and soil properties (Bussink, 1994; Whitehead, 1995; Ledgard *et al.*, 1999a). Losses are also greater for systems where cows are housed for some part of the year and manure storage is required; the magnitude of these losses being strongly influenced by how the manure is stored (e.g. covered tanks versus open lagoons) and environmental conditions (Webb *et al.*, 2005). Ammonia volatilization from pastoral land can contribute to the unwanted nutrient enrichment and acidification of both neighbouring and distant natural ecosystems.

Soil denitrification processes yield both dinitrogen and N_2O gases, the latter being a potent GHG that contributes to global warming (discussed in greater detail in Chapter 1, this volume). Although CH_4 usually represents the major portion of a dairy farm's total GHG emissions inventory, N_2O can account for up to 30% of total losses. Given the limited scope currently for reducing CH_4 emissions from ruminants, there is much ongoing research that seeks to manipulate the N cycle in pastures to reduce N_2O losses. Dairy farm systems can contribute to emissions of N_2O via both on-farm denitrification processes, typically under wet soil conditions, and off-farm and indirect denitrification processes in wetlands and streams. The nitrification of deposited excretal or fertilizer N can also yield significant quantities of this gaseous emission.

Phosphorus and sediment

Sources of P losses from dairy farms tend to vary more than for N. P losses depend heavily on spatial factors and the type of management practices employed on farm, such as how effluent or manures are handled, and the degree of protection of streams banks and beds from erosion and animal treading. P losses from intensively grazed pastures arise from dissolution and loss of particulate material from the soil, washing off of P from recently grazed pasture plants, dung deposits and fertilizer additions. All except from fertilizer additions are influenced by the action of grazing, whether it is the ripping of pasture plants or the influence of treading on soil erosion and surface runoff potential. Clover-based pasture dairy systems typically have relatively large P fertilizer usage to maintain adequate soil P fertility for optimum clover growth. Of the P recycled via the grazing cow, most is excreted in dung and in a soluble form (Kleinman *et al.*, 2005). Dung therefore represents a concentrated form of readily available P that can have a large impact on surface water quality if voided directly into water. Stock access to streams, effluent pond treatment systems and effluent/manure applications to land are therefore key land management practices that can potentially contribute substantially to farm P losses (Hickey *et al.*, 1989; Byers *et al.*, 2005).

Allowing cattle access to streams has historically been one means of providing pastured cattle with drinking water and comfort during hot weather, but is now recognized as a poor management practice from the standpoint of nutrient loss as well as cattle health. Practices such as fencing out riparian areas, providing alternative sources of water and shade, and selection of feeding sites can have a profound effect on the environmental fate of nutrients from the excreta of pastured cattle. McDowell and Wilcock (2007) monitored a 2100 ha catchment in New Zealand containing dairy farms with seasonal milking. They observed elevated concentrations of total P in stream flow that were strongly correlated with stream sediment concentrations, attributing the sediment to trampling and destabilization of the stream bank by stock, as well as to other riparian management factors such as removal of riparian trees that stabilize banks. Elsewhere, James *et al.* (2007) estimated that 2800 kg of P was defaecated directly into pasture streams by dairy cattle every year in a 1200 km^2 catchment in the north-eastern USA with predominantly farms of low-intensity grazing. An additional 5600 kg P was deposited within a 10 m riparian area. Across the catchment, direct deposition of dung P into streams was equivalent to roughly 10% of the annual P loadings attributed to all agricultural sources.

Overland flow processes can also make a large contribution to the total P lost from dairy farms, unlike N. Although overland flow volumes are usually small relative to the volumes of water discharged in subsurface drainage, the entrainment of soil and dung P in this flow makes it a concentrated source of P and other potential stream contaminants such as ammonium-N and FMOs. Despite much research on P loss from agricultural soils, the contributions from overland flow sources are still difficult to define because of problems associated with spatial and temporal variability, making sampling and measurement of flows under field conditions very difficult. Current understandings suggest that near-stream areas are important sources of overland flow, as are areas of land underlain by artificial drainage systems which act as direct conduits between soil and stream. These artificial subsurface drainage systems have been shown to act as important sources of P and sediment, presumably due to the entrainment of particulate and dissolved P as water moves through the macropores and fissures to tile or pipe drains (Sharpley and Syers, 1979; Haygarth *et al.*, 1998; Hooda *et al.*, 1999; Monaghan *et al.*, 2005).

The issue of elevated levels of soil P is widely recognized as a significant source and unnecessary risk factor in P loss from farmland (e.g. Heckrath *et al.*, 1995; McDowell *et al.*, 2003a). The restoration of high-P soils to levels that more closely match agronomic requirements, particularly within high-risk areas, is an important measure that can reduce potential transfers of P from soil to water (Haygarth and Jarvis, 1999). Direct losses of applied fertilizer P are another potentially important source of farm P losses. The greatest risks are when soluble forms of P are applied shortly before overland flow events, or if P fertilizer is inadvertently directly spread on to streams or wetlands. However, improved spreading technology and practices on most dairy farms now mean that accidental P applications to streams are usually small. McDowell *et al.* (2003b) showed how the potential for P losses in either overland flow or drainage from soils fertilized with superphosphate decreased exponentially with time, so that after 30–60 days the concentration of P lost in runoff from superphosphate-treated plots equalled that of non-treated plots.

Faecal microorganisms (FMOs)

The transfer of FMOs from land to water is an area of growing importance in the context of diffuse agricultural pollution. Contamination of water with enteric pathogens such as *Escherichia coli* (*E. coli*) 0157, *Salmonella* spp., *Campylobacter* spp. and *Cryptosporidium parvum* has come under the spotlight in recent times as linkages between agricultural practices and pathogen dissemination in the wider community have been considered (Oliver *et al.*, 2005a). Knowledge of the sources and pathways of transfers of FMOs from dairy farmland to water is poor relative to the understandings for nutrients, particularly N. Dung can be a concentrated source of these organisms, and many of the land management practices which decrease P losses may therefore also decrease transfers of FMOs to waterways. Preliminary studies have identified that, like P, surface and subsurface flow pathways are important sources of FMOs (Oliver *et al.*, 2005b). Unfortunately, we know little about the survival rates of FMOs on pasture and in soil, and of their mobility in overland and subsurface flows. Consequently, it is difficult to identify additional opportunities for management interventions that can decrease land–water transfers. Muirhead *et al.* (2006) demonstrated that *E. coli* bacteria are transported in overland flow as single cells rather than as flocs or attached to sediments, behaving similarly to solutes and negating opportunities for removal via settling or filtration by vegetation. While it is difficult to envisage technological solutions that can reduce the survival or mobility of FMOs deposited in the field, opportunities do exist for treating and disinfecting manures and effluents collected in animal housing units or in the dairy yard (e.g. Craggs *et al.*, 2004).

Mitigating Nutrient Losses

Balancing nutrient inputs and outputs at field and farm levels

The importance of balanced pasture nutrition has been increasingly promoted in recent years, with increasing awareness of some of the less desirable consequences of nutrient surpluses in agriculture. Large nutrient imbalances are particularly apparent for feedlot animal production systems (see Chapter 10, this volume) where feed and feed nutrients are produced in one region but consumed in another, generating large nutrient surpluses and potential environmental problems in the process (FAO, 2006). Nutrient imbalances can occur even within confined dairy operations if large amounts of feed and feed nutrients are transferred between farm blocks, such as from areas used to grow supplements or forages to areas where farm dairy effluent (FDE) or housing/feed-pad manure is applied.

A lack of awareness of the nutrient content of effluent and manures, particularly their N, P and K contents, often leads to widespread nutrient enrichment of soil irrigated with these materials, especially on small land areas. Nutrient budgeting is a valuable tool to account for all nutrient input sources to a farm and their redistribution within a farm. These flows can be used to determine the efficiency of nutrient management on farms, and to examine the potential environmental impacts. To be effective, nutrient budgets must encompass the key drivers of nutrient flows on farms, and incorporate the main management practices which determine nutrient losses. The magnitude, timing and form of nutrient inputs are some of the main drivers of nutrient-use efficiency and loss for a farming system. Nutrient budgeting tools such as the OVERSEER nutrient budgeting model (Wheeler *et al.*, 2003) can be used to guide on-farm decision making and suggest remedial actions for nutrient enriched areas, such as increasing the land area receiving effluent or manure, or adjusting the timing of applications to avoid high-risk periods. They are most effective when coupled with an ongoing soil-testing strategy to confirm if nutrients are accumulating or declining under a nutrient budget surplus or deficit.

Management interventions

Improved fertilization practices

Current N and P fertilization practices aim to ensure that sufficient nutrient is applied to meet pasture growth requirements, and is applied at times when the risk of nutrient loss to water or the atmosphere is lowest. For N fertilization, this means that small tactical applications are made according to anticipated animal feed demands and pasture growth rates in the following 4- to 6-week period. Conversely, fertilization is not recommended during periods of low pasture growth rate (i.e. winter), or at times that precede periods of significant surplus rainfall (such as late autumn) when drainage/overland flow is likely or soils are likely to remain close to saturation. Decision-making guidelines use soil temperature and expected rainfall as criteria for scheduling N fertilization, with applications being discouraged if temperatures are less than 5°C in spring or less than 7°C in autumn. Research in UK pastures has also shown how fertilizer N inputs can be further adjusted to account for soil N supply (Titchen and Scholefield, 1992). In certain situations, it is recommended that fertilizer applications are timed to coincide with anticipated rainfall to ensure that as much of the applied fertilizer is washed into the soil if ammonia volatilization losses from applied N fertilizer are likely to be large, such as in summer-dry regions where urea-based fertilizers are used, reducing the potential for volatilization (Black *et al.*, 1987).

Particular consideration of P fertilization practices is required for dairy systems that yield relatively large volumes of overland flow and potentially substantial P losses from soluble P fertilizers, if flow events coincide with recent fertilization (McDowell *et al.*, 2003b). In these situations, the use of low-solubility P fertilizers can help to decrease P losses from grazed pastures. Scheduling applications of soluble P fertilizers to months when the risk of overland flow is much less is another obvious strategy to minimize the chances of direct losses of soluble P fertilizer.

Nitrogen process inhibitors

Much recent research has examined the role of N process inhibitors for improving N-use efficiencies in pastoral agriculture. One mitigation technology that has been developed to decrease nitrate (NO_3^-) leaching and N_2O emissions from grazed pastoral soils is the use of the nitrification inhibitor dicyandiamide (DCD) to slow the conversion of NH_4^+ to NO_3^-. DCD inhibits the first stage of nitrification, i.e. the oxidation of NH_4^+ to NO_2^-, by rendering the bacteria's enzymes ineffective (Amberger, 1989). Applications of this inhibitor have been observed to substantially decrease N (and cation) leaching losses and N_2O emissions from soil lysimeters treated with dairy cow urine (Di and Cameron, 2002c, 2004; Fig. 8.2). Different formulations of DCD products have been evaluated for their effectiveness in nitrification inhibition (Smith *et al.*, 2005). The longevity of the effect is believed to vary according to temperature and rainfall. In practice, most nitrate leaching occurs from urine patches deposited over many grazing events during autumn, winter and early spring. Therefore, the benefit of DCD use will depend on factors including the timing of DCD applications relative to grazing and the onset of drainage. DCD is therefore currently used in some regions to strategically target this critical period from late autumn to early spring. Because DCD: (i) can help retain nutrients in the soil, making them available for pasture uptake for a longer period; and (ii) has a N content of 69% and thus a small fertilization effect of its own; the use of DCD can also result in increased pasture production (Di and Cameron, 2004, 2005).

Urease inhibitors have also been used to decrease N losses to the environment, particularly ammonia volatilization to the atmosphere (Watson *et al.*, 1998). Urease inhibitors inactivate urease enzymes in the soil, slowing down the rate of hydrolysis of urea to ammonium. Singh *et al.* (2006) described how the dual application of the urease inhibitor Agrotain (N-(n-butyl) thiophosphoric triamide)

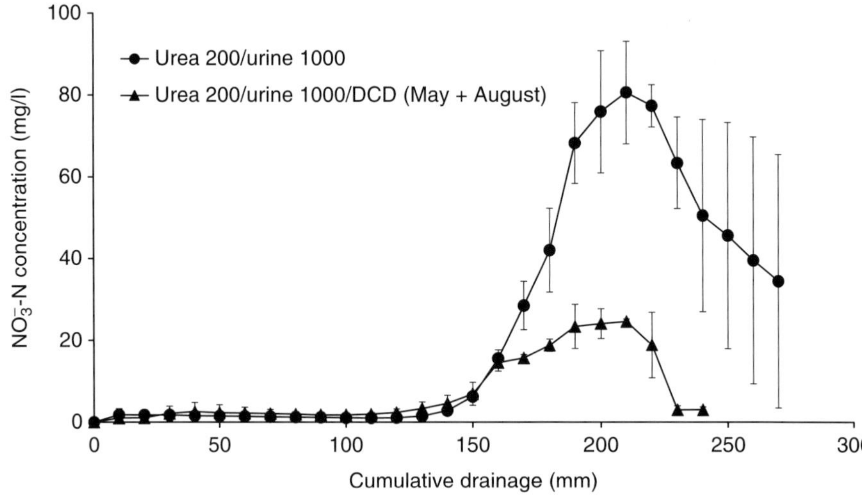

Fig. 8.2. Nitrate-N concentrations in drainage water from soil lysimeters which received dairy cow urine with and without DCD. (From Di and Cameron, 2004.)

and the nitrification inhibitor DCD decreased both N_2O and NH_3 losses from plots treated with urine. As observed for DCD, there are several clear potential benefits of using soil N process inhibitors, and ongoing field research is required to more fully evaluate their practical use and cost-effectiveness in grazed pasture systems. The retention of N within the pastoral system from the use of soil N process inhibitors has implications for subsequent N cycling and losses, and long-term studies are required to define the long-term benefit of this mitigation technology.

Improved effluent and stock management practices

Because effluents and manures are potent sources of N, P and FMOs, mismanagement of these materials during storage and application to land, or during on-farm treatment processing prior to discharge to waterways, can result in large transfers of these pollutants from land to streams. Fortunately, effluents and manures are increasingly recognized as valuable sources of pasture nutrients and management practices have been developed and implemented on farms to ensure that such transfers are minimized. These include recommendations on the appropriate design of effluent storage facilities, maximum annual rates of effluent N applications to land (and the farm area required to receive effluent loadings), split applications, improved application methods and exclusion times for grazing animals after application (e.g. see reviews by Houlbrooke *et al.*, 2004a; Webb *et al.*, 2005). Reviews indicate that effluent application to land is a relatively effective way of recycling nutrients contained in effluent or manure, with only 2–20% of the nutrients applied being lost to water bodies if best practice recommendations are adhered to (Houlbrooke *et al.*, 2004a).

However, it was also recognized that some soils present a greater difficulty for safely applying effluent to land. Soils that have low infiltration rates or artificial subsurface drainage systems have an increased risk of overland or preferential flow of effluent directly to streams, resulting in higher risk of contaminant losses (Monaghan and Smith, 2004), particularly faecal bacteria, dissolved P and ammonium-N. Although ammonium-N losses generally represent 5% or less of total farm N losses to waterways, in some situations the pulsed outputs of ammonium-N in FDE arising from effluent flow through mole-pipe drainage systems can potentially cause ammonia toxicity to aquatic life. Accordingly, more recent research efforts have focused on improved methods of scheduling and applying effluent to these sensitive soil types, such as using deferred and low application rate irrigation strategies (Houlbrooke *et al.*, 2004b; Monaghan and Smith, 2004). Intermittent and low rate sprinkler application systems have been observed to be particularly effective at removing *E. coli*, P and ammonium-N from drainage induced by the irrigation of FDE to artificially drained soils that were close to field capacity (Houlbrooke *et al.*, 2006). The advanced pond system (APS) is another technology developed as an improved effluent treatment system for areas where land application of effluent is unsuitable (Craggs *et al.*, 2004). This four-pond treatment system is particularly efficient at removing *E. coli* from FDE.

Excluding stock from stream margins is a relatively simple way of avoiding the direct deposition of dung P into streams and the erosion of sediment associated-P due to treading damage of stream banks. Belsky *et al.* (1999) document many of the detrimental effects of livestock grazing of streams and stream channels on water quality, channel morphology, vegetation and wildlife. Observations of cattle grazing behaviour suggest that, if allowed, they spend on average 4% of the day in the riparian zone and void about 4% of the expected number of faeces during this time (Bagshaw, 2002). Assuming an animal stocking rate of three cows per hectare, a constant rate of 12 defaecations per day and a dung P content of 6 mg/kg, this direct deposition of faecal P into streams could represent a loading of 0.5–1.0 kg P/ha/year. Stream fencing costs range from US$1/m to US$4/m depending on the type and permanence of the fencing. James *et al.* (2007) reported that conservation initiatives to exclude pastured cattle from streams were estimated to have contributed to a 32% decrease in in-stream deposition of faecal P.

Stock laneways represent another important yet poorly defined potential source of P loss from deposited dung. Calculations based on the length of time animals spend on stock laneways indicate that much excretal P is deposited on these hard surfaces. It would be expected that a high potential for P loss exists where these surfaces drain to adjacent channels or ditches, or directly to a stream. Hively *et al.* (2005) conducted rainfall simulations at various locations on a dairy farm including several pastures and a cow path. They found that surface runoff from the cow path transported large P loads. The path was characterized by a relatively impervious surface as well as a high concentration of cattle dung. As a result, Hively *et al.* (2005) concluded that loads from limited areas of a dairy farm, such as cow paths, contribute significantly to off-farm P loads, particularly during the summer when stream flow is relatively low. Bunding or contouring paths and lanes to ensure that flow is directed away from drainage conduits are practices that are likely to avoid potentially large transfers of faecal P and microorganisms to waterways. Alternatively, siting P-sorbing materials just before entry to a stream may mitigate P losses (McDowell *et al.*, 2006a).

System interventions

IMPROVED UTILIZATION OF DIETARY N AND P Several strategies have been evaluated to improve the poor utilization of dietary N by ruminants, decreasing the large amounts of N excreted (Tamminga, 1996; Davidson *et al.*, 2003). Improved feeding, breeding and animal health have significantly improved the conversion efficiency of farm inputs into saleable products (FAO, 2006). In addition to playing a role in reducing the surpluses of N and P found on a dairy farm, regulating the diet of a cow influences the potential for environmental emissions of nutrients. Regulating the amount of protein fed to dairy cattle can significantly affect urine N content, hence the potential for NH_3 volatilization. Broderick (2003) found that increasing crude protein in the diet of lactating dairy cows from 15–18% increased urinary N from 23% to 35% of dietary N. High-quality pasture contains greater concentrations of N than is needed by the ruminant, and despite efficient synthesis of microbial protein from diets containing high-quality pasture, up to 30% of the ingested N is not metabolized (Beever *et al.*, 1986; Kolver *et al.*, 1998). This loss is due to a high concentration of NH_3 in the rumen that results from rapid and extensive rumen degradation of pasture N that, if not recaptured into microbial protein, is converted to urea and excreted in milk and urine. For pasture-based dairy farming, inclusion of high-energy feeds such as maize, maize silage or grain in the diet has been identified as a viable option for uncoupling farm productivity and N losses. Jarvis *et al.* (1996) and Ledgard *et al.* (2003) document how the inclusion of such high-energy feeds can maintain milk production while decreasing N losses to the environment (Fig. 8.3). An important aspect of these evaluations has been including N losses from areas used for housing animals and support land used for growing feed supplements.

The chemical characteristics of pasture plants can influence the potential for N leaching losses in grazed pasture systems. Desirable N-efficient chemical characteristics in plants include high sugar concentration, decreased N concentration and presence of tannins. Research at sites with cool-temperate winter conditions indicates that high-sugar grasses may potentially increase efficiency of N cycling, through greater N recovery by animals and relatively less N excreted in urine than in dung (e.g. Miller *et al.*, 2001). Research with tannin-containing plants has shown the same potential benefits for N cycling (Barry *et al.*, 1986). However, none of these plant species or characteristics has been evaluated in grazed pasture systems to measure their effects on N losses to the environment.

Supplemental P fed in excess of that required by the cow is not absorbed in the cow's gastrointestinal tract and is excreted in the faeces (Dou *et al.*, 2002). Under controlled experiments, strong relationships have been reported between dietary and faecal P concentrations. In a 308-day lactation trial with 26 multiparous Holstein cows, Wu *et al.* (2000) reported a strong linear relationship between faecal P excretion and P intake (faecal P = 0.643 dietary P − 6.2; units = g/day; $r^2 = 0.81$). However, in the survey of Dou *et al.* (2003), where factors such as breed, life stage and diet were not controlled, faecal P concentrations were only weakly related to P concentrations in the diet (faecal P = 1.89 dietary P + 1.03; units = g/kg; $r^2 = 0.43$). As such, a variety of factors undoubtedly contribute to

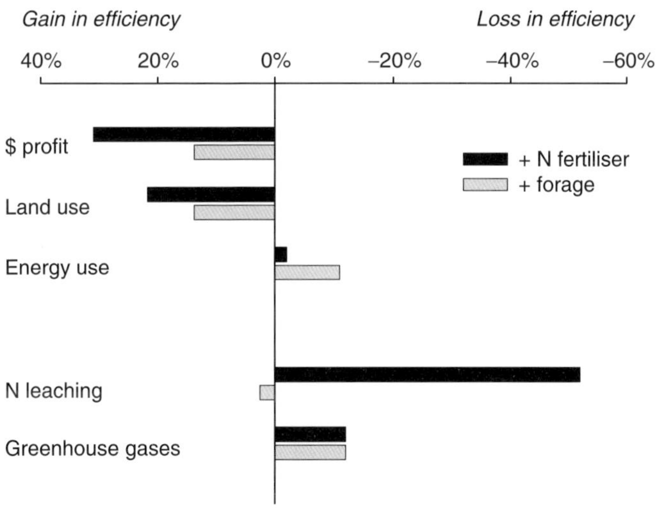

Fig. 8.3. Effect of intensification from 850 (base) to 1020 kg milk solids/ha/year, using extra N fertilizer (+200 kg N/ha/year) or forage (+2 t DM/ha/year as maize + oats silage), on efficiency of resource use and environmental emissions. Data refer to the whole dairy system (dairy farm + grazing + forage land). (From Ledgard *et al.*, 2003.)

faecal P concentrations on working farms, with dietary P an important, but by no means unique, factor. In addition to affecting the total P concentration in the faeces of dairy cattle, excess dietary P increases the water extractable P (WEP) fraction in particular. This fraction is often strongly related to P in surface runoff when manure or dung is deposited on the soil surface (Kleinman *et al.*, 2002). Dou *et al.* (2002), summarizing results of three feeding trials, found that increasing dietary P concentrations with mineral supplements increased the concentration of total P in faeces, largely by increasing WEP concentrations. In that study, diets with 3.4, 5.1 or 6.7 g P/kg dry matter resulted in mean faecal WEP concentrations of 2.9, 7.1 and 10.5 g P/kg (dry weight basis), respectively, with WEP accounting for 56%, 77% and 83% of total P.

MANIPULATING THE TIMING OF EXCRETAL DEPOSITION As most N leaching comes from animal urine patches, particularly those deposited at grazing events closest to the onset of drainage, management systems which target the amount of urinary N deposited or modify the timing of deposition can greatly influence these losses. Management practices for mitigating N losses to waterways therefore focus on avoiding the deposition of urine patches during autumn and winter when the risk of N loss is highest. Dairy farm system grazing studies (de Klein *et al.*, 2001; Chadwick *et al.*, 2002; de Klein *et al.*, 2006) have shown that restricted grazing management strategies where animals are removed from pasture on to a feed-pad from autumn until calving (4 months), with collection of effluent and reapplication during spring/summer, can decrease N leaching losses by up to 60%. McDowell *et al.* (2005) reported that this grazing management strategy can also decrease P losses in overland flow from cropland used for winter forage grazing.

Restricted autumn and/or winter grazing of pastures all have a cost. Desktop analyses indicate that the costs of construction and use of winter feed-pads may be compensated for by a small increase in production, due to more efficient use of the nutrients in effluent returned to pasture, or to savings in cow wintering costs associated with off-farm grazing during winter months (de Klein *et al.*, 2001; Monaghan *et al.*, 2007a). In practice, the use of feed-pads or strategic stand-off pads has increased on many dairy farms, although this has occurred through the desire to decrease pugging damage of pasture, improve animal welfare or in conjunction with increased use of supplement feeding during the milking season. The effectiveness and cost of mitigation practices, such as the use of stand-off/feed-pads, differ and the preferred option, or options, will vary between farms depending on economics and practicality.

INTERCEPTING P, SEDIMENT AND FMOS IN FARM RUNOFF In-stream or near-stream technologies that decrease largely particulate losses of potential stream pollutants include sedimentation ponds (McDowell *et al.*, 2006b) and wetlands (Sukias *et al.*, 2006). Wetlands can also be a significant sink for N contained in land drainage. Their efficiency depends on flow rate and generally decreases as flow rate increases. Buffer strips or riparian areas are also promoted as management practices that can intercept particulate material in surface flows, although their trapping capacity can be negated by flow that converges enough to overwhelm any interaction with the strip, or if the strip is bypassed altogether via subsurface drainage systems

(Verstraeten *et al.*, 2006). Careful consideration is therefore needed on placement and configuration within the landscape to maximize opportunities for intercepting flow. A range of other technologies have been trialled to improve P capture from drainage, overland flow or stream flow, such as the use of absorbent materials like alum or ash and steel smelter slag residues (e.g. McDowell, 2005). Logistical considerations, potential toxicity problems with some of these materials and low cost-effectiveness mean that most of these technologies are unlikely to be widely implemented for field use as yet, however.

Prioritizing Mitigation Expenditure

Implementing mitigation measures to decrease the environmental footprint of dairy farms is complex due to the variety of potential impacts, the varied potential sources of pollutants on farm and the range of management options that could mitigate these sources. Given that farms are operational businesses with a finite budget available for environmental measures, it is imperative that policy and farmer decision making is guided to ensure the best return on investment. Unfortunately, the processes and tools for guiding these decisions are fragmented. There is therefore an urgent need for tools that can ascertain the most appropriate course of action, depending on considerations that include:

- Understanding the *sensitivity* of the surrounding environment, for example: are farm emissions of most concern to air quality or catchment water quality, and if the latter, is nutrient enrichment or faecal pollution of water of most concern? Is groundwater or surface water of most concern, and if the latter, is N or P the most limiting nutrient for eutrophication?
- Knowing where the *likely key sources* of farm pollutants are derived from, and how these sources can be mitigated.
- Information about the *cost-effectiveness* of potential mitigation options to help prioritize expenditure.
- An awareness of the importance of farm *context* in the decision-making process, recognizing how factors such as soils, topography, existing farm infrastructure and lifestyle combine to influence farmer decision making.

Proof of the effectiveness of mitigation measures is also generally required before farmers will voluntarily adopt 'improved' management practices.

As an example, Monaghan *et al.* (2008) have put together existing information on nutrient flows and mitigation options into a 'toolbox' of alternative technologies as an aid to the decision-making process. This approach has been developed and used as a decision-support tool to guide farm planning initiatives on some New Zealand dairy farms located in four contrasting catchments (Monaghan *et al.*, 2008). Farm systems simulation tools were used to compute the cost-effectiveness of several targeted mitigation options to ensure that farm expenditure was prioritized and used most efficiently. Examples of the projected cost-effectiveness, expressed as a dollar cost per kilogram of nutrient conserved, are shown in Table 8.2 for a range of mitigation measures. This metric clearly identifies and ranks the most cost-effective technology available to decrease N and

Table 8.2. The cost-effectiveness ($ cost per kg of nutrient conserved) of some mitigation measures for decreasing losses of N and P from dairy farms in four catchments used for intensive dairy farming in New Zealand. Figures are expressed relative to the control farm system in each catchment (a positive value means a cost and a negative value means an economic benefit). (Adapted from Monaghan *et al.*, 2008.)

	Toenepi	Waiokura	Waikakahi	Bog Burn
Phosphorus				
Optimal soil P fertility	–113	–221	–421	–490
Deferred effluent irrigation to land	22	0	n/a	44
Low rate effluent irrigation to land	8	n/a	n/a	21
Small effluent application depth	n/r	n/a	n/a	24
Irrigation bunding	n/r	n/r	15	n/r
Low-solubility P fertilizer	n/a	n/a	0	0
APS[a]	108	n/a[b]	n/a[b]	n/a[b]
Nitrogen				
Nitrification inhibitors	–33	–30	–51	–45
Restricted autumn/winter grazing	5	5	6	1
Nil N fertilizer input	16	4	1	16
Low N feed	12	13	0	41
Wintering pads	24	36	9	–2
APS	20	n/a[b]	n/a[b]	52

n/a = not applicable; n/r = not relevant.
[a]Advanced pond system for the treatment of farm dairy effluent.
[b]Projected to increase P or N losses from these model farms due to change from land application of FDE to treatment via an APS.

P losses. It is readily apparent that the cost-effectiveness of technologies varies between catchments, due to the contrasting soil types, farm infrastructure and management systems present. The above points demonstrate: (i) there are usually a number of management options available to reduce nutrient leakages from dairy farms; (ii) the relevance of any mitigation measure depends on a farm's soil types and management systems; and (iii) some mitigation options are more cost-effective than others.

Conclusions and Future Issues

Pressures to improve the economic and environmental performances of dairy farms are unlikely to abate, and the tension between these two objectives will dictate much future dairy systems research. Research must strive to ensure that inevitable improvements in milk and pasture yields do not compromise system profitability or the environment (Clark *et al.*, 2007). Future perspectives on farming practices are likely to extend to encompass a broader consideration of whole-system resource (energy, water, nutrients) use efficiencies and a wider range of environmental indices, including GHG emissions and biodiversity. Various studies have shown the importance of considering the environmental costs of all inputs to a farming system when evaluating future options for mitigation. These studies highlight how measures for improving resource-use efficiency in one part of the system can sometimes decrease overall system efficiency, if all components of the production system (e.g. support land used for supplement provision) are considered (e.g. van der Nagel *et al.*, 2003), or can identify if the mitigation measures designed to improve one environmental index may lead to a decline in another. The latter is evident in Fig. 8.3 where a life cycle assessment (LCA) methodology was applied to account for all contributors to projected nutrient use and emissions for a New Zealand dairy farm that intensified via either additional N fertilizer or importing more feed supplement. This shows differences in whole-system efficiencies for different indicators, where estimated profit and N leaching losses were much higher for the +N system, demonstrating that there can be conflicts between economic and environmental efficiency.

Applying this LCA methodology to other mitigation strategies can, however, identify some management systems that do not incur additional financial or environmental costs. An example is the use of covered pads to house animals during winter. This approach is used by some farmers as an alternative to wintering animals outdoors on forage crops and incurs benefits by decreasing soil treading damage and improving animal welfare. It has also been identified as a system that, providing that effluent deposited on the pad is captured, stored and returned to land during spring, can decrease nitrate leaching losses (Monaghan *et al.*, 2008). Applying the LCA methodology to these contrasting wintering strategies indicates that the total system energy requirements per unit of milk produced are actually quite similar, despite the requirement for feeding animals during the time they are on the covered pad. A breakdown of the energy requirements shows that the additional energy required for this feeding operation is more than off-set by savings in energy required for cultivating land for crop establishment (Table 8.3).

Table 8.3. A life cycle assessment of the land and energy requirements and global warming potential of milk production systems using either a forage crop or a covered pad for wintering cows in southern New Zealand. Data modelled for a 630-cow herd. The computer program SimgaPro (version 6.04) was used to manage the inventory database and to model farm production from each system into a series of unit processes which were then connected to form a structure culminating in the production of 1 l of milk. Each unit process contained the input flows of materials and energy and output flows of emissions.

	Forage crop[a]	Covered winter pad[b]
Land requirement (m^2/l milk)		
Pasture	0.83	0.83
Forage crop	0.16	0
Off-farm supplement areas	0.02	0.16
Total	1.01	0.99
Energy requirement (kJ/l milk)		
Milk harvesting	293	293
Diesel use (non-crop areas)	586	586
Nitrogen (urea) fertilizer	595	543
Other fertilizer	319	284

continued

Table 8.3. Continued

	Forage crop[a]	Covered winter pad[b]
Supplement production	9	95
Forage crop production	164	0
Animal transport to forage crop	26	0
Winter pad operations and capital	0	112
Total	1992	1913
Global warming potential (g CO_2-equivalents/l milk)		
Feed digestion (CH_4)	547	538
Feed production (N_2O mainly)	245	236
Diesel use (CO_2)	61	56
Nitrogen (urea) fertilizer (CO_2)	24	22
Electricity generation (CO_2)	16	16
Other processes (including pad construction)	4	6
Total	897	873

[a]Assuming a forage kale crop yielding 11 t DM/ha which is located 35 km from the home farm.
[b]Covered sawdust loafing and concrete feeding-pad areas totalling 6 m^2/cow.

Land requirements and global warming potential estimates for the pad system are also slightly lower than estimated for the forage crop wintering system.

Although environmental targets are typically set at a catchment scale or greater, sustainable and competitive pasture-based dairy farming will only be achieved by addressing many of the issues at a farm and field scale using decision-support systems that take account of the unique soil, weather and management conditions that prevail on individual farms. Voluntary adoption of improved practices will rely on farmers placing a value on external impacts and seeking tools and systems that reduce these impacts while meeting their other goals. Experience with adoption of environmental management shows that a coordinated approach of regulation and enforcement, industry direction, market mechanisms, education and communication is the most effective means to bring about change (Clark *et al.*, 2007).

References

Aarts, H.F.M., Habekotte, B. and Keulen, H. (2000a) Nitrogen (N) management in the 'De Marke' dairy farming system. *Nutrient Cycling in Agroecosystems* 56, 231–240.

Aarts, H.F.M., Habekotte, B. and Keulen, H. (2000b) Phosphorus (P) management in the 'De Marke' dairy farming system. *Nutrient Cycling in Agroecosystems* 56, 219–229.

Amberger, A. (1989) Research on dicyandiamide as a nitrification inhibitor and future outlook. *Communications in Soil Science and Plant Analysis* 20, 1933–1955.

Bagshaw, C.S. (2002) Factors influencing direct deposition of cattle faecal material in riparian zones. *MAF (New Zealand Ministry of Agriculture and Forestry) Technical Paper No. 2002/19*. Available at: http://www.maf.govt.nz/mafnet/rural-nz/sustainable-resource-use/resource-management/cattle-faecal-material/index.htm

Barry, T.N., Manley, T.R. and Duncan, S.J. (1986) The role of condensed tannins in the nutritional value of *Lotus pedunculatus* for sheep. 4. Sites of carbohydrate and protein digestion as influenced by dietary reactive tannin concentration. *British Journal of Nutrition* 55, 123–137.

Beever, D.E., Losada, H.R., Cammell, S.B., Evans, R.T. and Haines, M.J. (1986) Effect of forage species and season on nutrient digestion and supply in grazing cattle. *British Journal of Nutrition* 56, 209–225.

Belsky, A.J., Matzke, A. and Uselman, S. (1999) Survey of livestock influences on stream and riparian ecosystems in the western United States. *Journal of Soil and Water Conservation* 54, 419–431.

Black, A.S., Sherlock, R.R. and Smith, N.P. (1987) Effect of timing of simulated rainfall on ammonia volatilization from urea, applied to soil of varying moisture content. *Journal of Soil Science* 38, 679–687.

Broderick, G.A. (2003) Effects of varying dietary protein and energy levels on the production of lactating dairy cows. *Journal of Dairy Science* 86, 1370–1381.

Brown, L., Scholefield, D., Jewkes, E.C., Lockyer, D.R. and Prado, A.d. (2005) NGAUGE: a decision support system to optimise N fertilisation of British grassland for economic and environmental goals. *Agriculture, Ecosystems and Environment* 109, 20–39.

Bussink, D.W. (1994) Relationships between ammonia volatilization and nitrogen fertilizer application rate, intake and excretion of herbage nitrogen by cattle on grazed swards. *Fertilizer Research* 38, 111–121.

Byers, H.L., Cabrera, M.L., Matthews, M.K., Franklin, D.H., Andrae, J.G., Radcliffe, D.E., McCann, M.A., Kuykendall, H.A., Hoveland, C.S. and Calvert, V.H. (2005) Phosphorus, sediment, and Escherichia coli loads in unfenced streams of the Georgia Piedmont, USA. *Journal of Environmental Quality* 34, 2293–2300.

Chadwick, D.R., Ledgard, S.F. and Brown, L. (2002) Nitrogen flows and losses in dairy farms in New Zealand and the UK: effects of grazing management. In: Currie, L.D. and Loganathan, P. (eds) *Dairy Farm Soil Management.* Fertilizer and Lime Research Centre, Massey University, Palmerston North, New Zealand, pp. 319–332.

Clark, D.A., Caradus, J.C., Monaghan, R.M., Sharp, P. and Thorrold, B.S. (2007) Issues and options for future dairy farming in New Zealand. *New Zealand Journal of Agricultural Research* 50, 203–221.

Craggs, R.J., Sukias, J.P., Tanner, C.T. and Davies-Colley, R.J. (2004) Advanced pond system for dairy-farm effluent treatment. *New Zealand Journal of Agricultural Research* 47, 449–460.

Cuttle, S.P. and Scholefield, D. (1995) Management options to limit nitrate leaching from grassland. *Journal of Contaminant Hydrology* 20, 299–312.

Davidson, S., Hopkins, B.A., Diaz, D.E., Bolt, S.M., Brownie, C., Fellner, V. and Whitlow, L.W. (2003) Effects of amounts and degradability of dietary protein on lactation, nitrogen utilization, and excretion in early lactation Holstein cows. *Journal of Dairy Science* 86, 1681–1689.

de Klein, C.A.M., Paton, J. and Ledgard, S.F. (2001) An evaluation of a strategic de-stocking regime for dairying to improve nitrogen efficiency and reduce nitrogen losses from dairy farms in nitrate-sensitive areas. *Proceedings of the New Zealand Grassland Association* 62, 105–110.

de Klein, C.A.M., Smith, L.C. and Monaghan, R.M. (2006) Restricted autumn grazing to reduce nitrous oxide emissions from dairy pastures in Southland, New Zealand. *Agriculture, Ecosystems and Environment* 112, 192–199.

Di, H.J. and Cameron, K.C. (2002a) Nitrate leaching in temperate agroecosystems: sources, factors and mitigating strategies. *Nutrient Cycling in Agroecosystems* 64, 237–256.

Di, H.J. and Cameron, K.C. (2002b) Nitrate leaching and pasture production from different nitrogen sources on a shallow stony soil under flood irrigated dairy pasture. *Australian Journal of Soil Research* 40, 317–334.

Di, H.J. and Cameron, K.C. (2002c) The use of a nitrification inhibitor, dicyandiamide (DCD), to reduce nitrate leaching from cow urine patches in a grazed dairy pasture under irrigation. *Soil Use and Management* 18, 395–403.

Di, H.J. and Cameron, K.C. (2004) Treating grazed pasture soil with a nitrification inhibitor, 'eco-nTM, to decrease nitrate leaching in a deep sandy soil under spray irrigation – a lysimeter study. *New Zealand Journal of Agricultural Research* 47, 351–361.

Di, H.J. and Cameron, K.C. (2005) Reducing environmental impacts of agriculture by using a fine particle suspension nitrification inhibitor to decrease nitrate leaching from grazed pastures. *Agriculture, Ecosystems and Environment* 109, 202–212.

Dou, Z., Knowlton, K.F., Kohn, R.A., Wu, Z., Satter, L.D., Zhang, G., Toth, J.D. and Ferguson, J.D. (2002) Phosphorus characteristics of dairy faeces affected by diets. *Journal of Environmental Quality* 31, 2058–2065.

Dou, Z., Ferguson, J.D., Fiorini, J., Toth, J.D., Alexander, S.M., Peterson, A.B., Sims, J.T. and Wu, Z. (2003) Phosphorus feeding levels and critical control points on dairy farms. *Journal of Dairy Science* 86, 3787–3795.

Eckard, R.J., White, R.E., Edis, R., Smith, A. and Chapman, D.F. (2004) Nitrate leaching from temperate perennial pastures grazed by dairy cows in south-eastern Australia. *Australian Journal of Agricultural Research* 55, 911.

Edmeades, D.C., Thorrold, B.S. and Roberts, A.H.C. (2005) The diagnosis and correction of sulphur deficiency and the management of sulphur requirements in New Zealand pastures: a review. *Australian Journal of Experimental Agriculture* 45, 1205–1223.

Edmeades, D.C., Metherell, A.K., Waller, J.E., Roberts, A.H.C. and Morton, J.D. (2006) Defining the relationships between pasture production and soil P and the development of a dynamic P model for New Zealand pastures: a review of recent developments. *New Zealand Journal Agricultural Research* 49, 207–222.

FAO (2006) Livestock's long shadow: environmental issues and options. Available at: http://www.fao.org/ag/magazine/0612sp1.htm

Fraser, M., Cameron, K.C. and Sherlock, R.R. (1994) Lysimeter study of the fate of nitrogen in animal urine returns to irrigated pasture. *European Journal of Soil Science* 45, 439–447.

Galeone, D.G. (1999) Surface water quality: calibration of paired basins prior to streambank fencing of pasture land. *Journal of Environmental Quality* 28, 1853–1863.

Gillingham, A.G. and Thorrold, B.S. (2000) A review of New Zealand research measuring phosphorus in runoff from pasture. *Journal of Environmental Quality* 29, 88–96.

Hanegraaf, M.C. and Boer, D.J.d. (2003) Perspectives and limitations of the Dutch minerals accounting system (MINAS). *European Journal of Agronomy* 20, 25–31.

Haygarth; P.M. and Jarvis, S.C. (1999) Transfer of phosphorus from agricultural soils. *Advances in Agronomy* 66, 195–249.

Haygarth, P.M., Hepworth, L. and Jarvis, S.C. (1998) Forms of phosphorus transfer in hydrological pathways from soil under grazed grassland. *European Journal of Soil Science* 49, 65–72.

Haynes, R.J. and Williams, P.H. (1993) Nutrient cycling and soil fertility in the grazed pasture ecosystem. *Advances in Agronomy* 49, 119–199.

Heckrath, G., Brookes, P.C., Poulton, P.R. and Goulding, K.W.T. (1995) Phosphorus leaching from soils containing different phosphorus concentrations in the Broadbalk Experiment. *Journal of Environmental Quality* 24, 904–910.

Hickey, C.W., Quinn, J.M. and Davies-Colley, R.J. (1989) Effluent characteristics of dairy shed oxidation ponds and their potential impacts on rivers. *New Zealand Journal of Marine and Freshwater Research* 23, 569–584.

Hively, W.D., Bryant, R.B. and Fahey, T.J. (2005) Phosphorus concentrations in overland flow from diverse locations on a New York dairy farm. *Journal of Environmental Quality* 34, 1224–1233.

Hooda, P.S., Moynagh, M., Svoboda, I.F., Edwards, A.C., Anderson, H.A. and Sym, G. (1999) Phosphorus loss in drainflow from intensively managed grassland soils. *Journal of Environmental Quality* 28, 1235–1242.

Houlbrooke, D.J., Horne, D.J., Hedley, M.J., Hanly, J.A. and Snow, V.O. (2004a) A review of literature on the land treatment of farm-dairy effluent in New Zealand and its impact on water quality. *New Zealand Journal of Agricultural Research* 47, 499–511.

Houlbrooke, D.J., Horne, D.J., Hedley, M.J., Hanly, J.A., Scotter, D.R. and Snow, V.O. (2004b) Minimising surface water pollution resulting from farm-dairy effluent application to mole-pipe drained soils. I. An evaluation of the deferred irrigation system for sustainable land treatment in the Manawatu. *New Zealand Journal of Agricultural Research* 47, 405–415.

Houlbrooke, D.J., Monaghan, R.M., Smith, L.C. and Nicolson, C. (2006) Reducing contaminant losses from land applied farm dairy effluent using K-line irrigation systems. In: Currie, L.D. and Hanly, J.A. (eds) *Implementing Sustainable Nutrient Management Strategies in Agriculture.* Fertiliser and Lime Research Centre, Massey University, Palmerston North, New Zealand, pp. 290–300.

Humphreys, D., Darmody, P., O'Connell, K., Casey, I.A. and Watson, C.J. (2004) Lowering nitrogen inputs to grassland-based dairy production. In: Luscher, A., Jeangros, B., Kessler, W., Huguenin, O., Lobsiger, M., Millar, N. and Suter, D. (eds) *Land Use Systems in Grassland Dominated Regions.* Proceedings of the 20th General Meeting of the European Grassland Federation, Luzern, Switzerland, 21–24 June 2004.

James, E., Kleinman, P., Veith, T., Stedman, R. and Sharpley, A. (2007) Phosphorus contributions from pastured dairy cattle to streams of the Cannonsville Watershed, New York. *Journal of Soil and Water Conservation* 62, 40–47.

Jarvis, S.C., Wilkins, R.J. and Pain, B.F. (1996) Opportunities for reducing the environmental impact of dairy farming managements: a systems approach. *Grass and Forage Science* 51, 21–31.

Jordan, C. and Smith, R.V. (1985) Factors affecting leaching of nutrients from an intensively managed grassland in County Antrim, Northern Ireland. *Journal of Environmental Management* 20, 1–15.

Ketterings, Q.M., Albrecht, G.L., Rasmussen, C.N. and Czymmek, K.J. (2006) Cornell Cropware: decision support tool for fertilizer and manure nutrient management planning. *Journal of Natural Resources and Life Sciences Education* 35, 140–151.

Kleinman, P.J.A., Sharpley, A.N., Moyer, B.G. and Elwinger, G.F. (2002) Effect of mineral and manure phosphorus sources on runoff phosphorus. *Journal of Environmental Quality* 31, 2026–2033.

Kleinman, P.J.A., Wolf, A.M., Sharpley, A.N., Beegle, D.B. and Saporito, L.S. (2005) Survey of water-extractable phosphorus in livestock manures. *Soil Science Society of America Journal* 69, 701–708.

Kolver, E.S., Muller, L.D., Varga, G.A. and Cassidy, T.W. (1998) Synchronization of ruminal degradation of supplemental carbohydrate with pasture nitrogen in lactating dairy cows. *Journal of Dairy Science* 81, 2017–2028.

Langeveld, J.W.A., Verhagen, A., Neeteson, J.J., Keulen, H.v., Conijn, J.G., Schils, R.L.M. and Oenema, J. (2007) Evaluating farm performance using agri-environmental indicators: recent experiences for nitrogen management in the Netherlands. *Journal of Environmental Management* 82, 363–376.

Ledgard, S.F. (2001) Nitrogen cycling in low input legume-based agriculture, with emphasis on legume/grass pastures. *Plant and Soil* 228, 43–59.

Ledgard, S.F., Penno, J.W. and Sprosen, M.S. (1999a) Nitrogen inputs and losses from clover/grass pastures grazed by dairy cows, as affected by nitrogen fertilizer application. *Journal of Agricultural Science, Cambridge* 132, 215–225.

Ledgard, S.F., Williams, P.H., Broom, F.D., Thorrold, B.S., Wheeler, D.M. and Willis, V.J. (1999b) OVERSEER – a nutrient budgeting model for pastoral farming, wheat, potatoes, apples and kiwifruit. In: Currie, L.D., Hedley, M.J., Horne, D.J. and Loganathan, P. (eds) *Best Soil Management Practices for Production.* Fertilizer and Lime Research Centre, Massey University, Palmerston North, New Zealand, pp. 143–152.

Ledgard, S.F., Finlayson, J.D., Gavin, J., Blackwell, M.D., Carran, R.A., Wedderburn, M.E. and Jollands, N.A. (2003) Resource use efficiency and environmental emissions from an average Waikato dairy farm, and impacts of intensification using nitrogen fertiliser or maize silage. *Proceedings of the New Zealand Grassland Association* 65, 185–189.

McDowell, R.W. (2005) The effectiveness of coal fly-ash to decrease phosphorus loss from grassland soils. *Australian Journal of Soil Research* 43, 853–860.

McDowell, R.W. and Wilcock, R.J. (2007) Sources of sediment and phosphorus in stream flow of a highly productive dairy farmed catchment. *Journal of Environmental Quality* 36, 540–548.

McDowell, R.W., Monaghan, R.M. and Morton, J. (2003a) Soil phosphorus concentrations to minimise P loss to surface waters in Southland. *New Zealand Journal of Agricultural Research* 46, 239–253.

McDowell, R.W., Monaghan, R.M. and Carey, P.L. (2003b) Potential phosphorus losses in overland flow from pastoral soils receiving long-term applications of either superphosphate or reactive phosphate rock. *New Zealand Journal of Agricultural Research* 46, 329–337.

McDowell, R.W., Drewry, J.J., Muirhead, R.W. and Paton, R.J. (2005) Restricting the grazing time of cattle to decrease phosphorus, sediment and *E. coli* losses in overland flow from cropland. *Australian Journal of Soil Research* 43, 61–66.

McDowell, R.W., Hawke, M.F. and McIntosh, J.J. (2006a) Assessment of a novel technique to remove phosphorus from streamflow. In: *Soils and Society*. Proceedings of the New Zealand Society of Soil Science Conference, Rotorua, New Zealand, p. 29.

McDowell, R.W., McGrouther, N., Morgan, G., Srinivasan, M.S., Stevens, D., Johnson, M. and Copland, R. (2006b) Monitoring of the impact of farm practices on water quality in the Otago and Southland deer focus farms. *Proceedings of the New Zealand Grasslands Association* 68, 183–188.

Miller, L.A., Moorby, J.M., Davies, D.R., Humphreys, M.O., Scollan, N.D., Macrae, J.C. and Theodorou, M.K. (2001) Increased concentration of water-soluble carbohydrate in perennial ryegrass (*Lolium perenne* L.): milk production from late-lactation dairy cows. *Grass and Forage Science* 56, 383–394.

Monaghan, R.M. and Smith, L.C. (2004) Minimising surface water pollution resulting from farm-dairy effluent application to mole-pipe drained soils. II. The contribution of preferential flow of effluent to whole-farm pollutant losses in subsurface drainage from a West Otago dairy farm. *New Zealand Journal of Agricultural Research* 47, 417–429.

Monaghan, R.M., Paton, R.J., Smith, L.C., Drewry, J.J. and Littlejohn, R.P. (2005) The impacts of nitrogen fertilisation and increased stocking rate on pasture yield, soil physical condition and nutrient losses in drainage from a cattle-grazed pasture. *New Zealand Journal of Agricultural Research* 48, 227–240.

Monaghan, R.M., Wilcock, R.J., Smith, L.C., Tikkisetty, B., Thorrold, B.S. and Costall, D. (2007a) Linkages between land management activities and water quality in an intensively farmed catchment in southern New Zealand. *Agriculture, Ecosystems and Environment* 118, 211–222.

Monaghan, R.M., de Klein, C.A.M. and Muirhead, R.W. (2008) Prioritisation of farm scale remediation efforts for reducing contaminant losses to waterways: a case study of New Zealand dairy farming. *Journal of Environmental Management* 87, 609–622.

Muirhead, R.W., Collins, R.P. and Bremer, P.J. (2006) The association of *E. coli* and soil particles in overland flow. *Water Science and Technology* 54, 153–159.

Oliver, D.M., Clegg, C.D., Heathwaite, A.L. and Haygarth, P.M. (2005a) Assessing the potential for pathogen transfer from grassland soils to surface waters. *Advances in Agronomy* 85, 125–180.

Oliver, D.M., Heathwaite, A.L., Haygarth, P.M. and Clegg, C.D. (2005b) Transfer of *Escherichia coli* to water from drained and undrained grassland after grazing. *Journal of Environmental Quality* 34, 918–924.

Roach, C.G., Longhurst, R.D. and Ledgard, S.F. (2001) Land application of farm dairy effluent for sustainable dairy farming. *Proceedings of the New Zealand Grassland Association* 63, 53–57.

Scholefield, D., Tyson, K.C., Garwood, E.A., Armstrong, A.C., Hawkins, J. and Stone, A.C. (1993) Nitrate leaching from grazed grassland lysimeters: effects of fertilizer input, field drainage, age of sward and patterns of weather. *Journal of Soil Science* 44, 601–613.

Sharpley, A.N. and Syers, J.K. (1979) Loss of nitrogen and phosphorus in tile drainage as influenced by urea and grazing animals. *New Zealand Journal of Agricultural Research* 22, 127–131

Singh, J., Saggar, S., Bolan, N.S. and Zaman, M. (2006) Influence of urease and nitrification inhibitors on ammonia and nitrous oxide emissions under field conditions. In: Currie, L.D. and Hanly, J.A. (eds) *Implementing Sustainable Nutrient Management Strategies in Agriculture*. Fertiliser and Lime Research Centre, Massey University, Palmerston North, New Zealand, pp. 162–170.

Smith, L.C. and Monaghan, R.M. (2003) Nitrogen and phosphorus losses in overland flow from a cattle-grazed pasture in Southland. *New Zealand Journal of Agricultural Research* 46, 225–238.

Smith, L.C., Monaghan, R.M., Ledgard, S.F. and Catto, W.D. (2005) The effectiveness of different nitrification inhibitor formulations in limiting nitrate accumulation in a Southland pastoral soil. *New Zealand Journal of Agricultural Research* 48, 517–530.

Sukias, J.P.S., Tanner, C.C. and McKergow, L. (2006) Dairy farm drainage nitrate attenuation wetlands and filters. In: Currie, L.D. and Hanly, J.A. (eds) *Implementing Sustainable Nutrient Management Strategies in Agriculture.* Fertiliser and Lime Research Centre, Massey University, Palmerston North, New Zealand, pp. 301–307.

Tamminga, S. (1996) A review on environmental impacts of nutritional strategies in ruminants. *Journal of Animal Science* 74, 3112–3124.

Theobald, P.W. and Ball, P.R. (1984) Nitrogen lost by ammonia volatilisation, and the effectiveness of urea and ammonium sulphate fertilisers. *Proceedings of the New Zealand Grassland Association* 45, 236–238.

Titchen, N.M. and Scholefield, D. (1992) The potential of a rapid test for soil mineral nitrogen to determine tactical applications of fertilizer nitrogen to grassland. *Aspects of Applied Biology* 30, 223–229.

Tyson, K.C., Scholefield, D., Jarvis, S.C. and Stone, A.C. (1997) A comparison of animal output and nitrogen leaching losses recorded from drained fertilized grass and grass/cloves pasture. *Journal of Agricultural Science, Cambridge*, 129, 315–323.

van der Nagel, L.S., Waghorn, G.C. and Forgie, V.E. (2003) Methane and carbon emissions from conventional pasture and grain-based total mixed rations for dairying. *Proceedings of the New Zealand Society of Animal Production* 63, 128–132.

Verstraeten, V., Poesen, J., Gillijns, K. and Govers, G. (2006) The use of riparian vegetated filter strips to reduce river sediment loads: an overestimated control measure? *Hydrological Processes* 20, 4259–4267.

Wachendorf, M. and Golinski, P. (2006) Towards sustainable intensive dairy farming in Europe. (Grassland Science in Europe, Volume 11). In: Lloveras, J., González-Rodriguez, A., Vazquez-Yanez, O., Pineiro, J., Santamaria, O., Olea, L. and Poblaciones, M. J. (eds) *Sustainable Grassland Productivity.* Proceedings of the 21st General Meeting of the European Grassland Federation, Badajoz, Spain, 3–6 April 2006, pp. 624–634. Sociedad Espanola para el Estudio de los Pastos (SEEP), Madrid, Spain.

Watson, C.J. and Foy, R.H. (2001) Environmental impacts of nitrogen and phosphorus cycling in grassland systems. *Outlook on Agriculture* 30, 117–127.

Watson, C.J., Poland, P. and Allen, M.B.D. (1998) The efficacy of repeated applications of the Urease inhibitor N-(n-butyl) thiophosphoric triamide for improving the efficiency of urea fertilizer utilization on temperate grassland. *Grass and Forage Science* 53, 137–145.

Watson, C.J., Jordan, C., Lennox, S.D., Smith, R.V. and Steen, R.W.J. (2000) Organic nitrogen in drainage water from grassland in Northern Ireland. *Journal of Environmental Quality* 29, 1233–1238.

Webb, J., Menzi, H., Pain, B.F., Misselbrook, T.H., Dammgen, U., Hendriks, H. and Dohler, H. (2005) Managing ammonia emissions from livestock production in Europe. *Environmental Pollution* 135, 399–406.

Wilcock, R.J., Monaghan, R.M., Quinn, J.M., Campbell, A.M., Duncan, M.J., Thorrold, B.S., McGowan, A.W. and Betteridge, K. (2006) Land use impacts and water quality targets in the intensive dairying catchment of the Toenepi Stream, New Zealand. *New Zealand Journal of Marine and Freshwater Research* 40, 123–140.

Wilcock, R.J., Monaghan, R.M., Thorrold, B.S., Meredith, A.S., Betteridge, K. and Duncan, M.J. (2007) Land–water interactions in five contrasting dairy catchments: issues and solutions. *Land Use and Water Resources Research* 7, 2.1–2.10.

Wheeler, D.M., Ledgard, S.F., deKlein, C.A.M., Monaghan, R.M., Carey, P.L., McDowell, R.W. and Johns, K.L. (2003) OVERSEER nutrient budgets – moving towards on-farm resource accounting. *Proceedings of the New Zealand Grassland Association* 65, 191–194.

Whitehead, D.C. (1995) *Grassland Nitrogen.* CAB International, Wallingford, UK.

Wu, Z., Satter, L.D. and Sojo, R. (2000) Milk production, reproductive performance, and faecal excretion of phosphorus by dairy cows fed three amounts of phosphorus. *Journal of Dairy Science* 83, 1028–1041.

9 Impacts of Irrigated Dairying on the Environment

D. NASH[1,2] AND K. BARLOW[3]

[1]Victorian Department of Primary Industries, Ellinbank, Victoria 3821, Australia; [2]eWater CRC, University of Canberra, GPO Canberra 2601, Australia; [3]Victorian Department of Primary Industries, Rutherglen, Victoria 3685, Australia

Introduction

Pasture-based dairying is a major land use of the agricultural industry in both Australia and New Zealand. As of 2006, Australia had *c*.2 million cows on 9250 dairy farms producing over 10 billion l of milk with a value of US$2.5 billion at the farm gate (Dairy Australia, 2006). In the same year in New Zealand, *c*.4 million cows in 12,000 herds produced over 1.2 million t milk solids with a value of US$3.8 billion at the farm gate (Australian Bureau of Agricultural and Resource Economics and Ministry of Agriculture and Forestry, 2006; Ministry of Agriculture and Forestry, 2006).

A major impediment to the expansion of the dairy industry in these countries is the lack of water, especially in summer and early autumn. The lack of water, particularly soil moisture, limits pasture productivity and pasture-based grazing, and increases production costs because of the need to source alternative feed supplies (Dillon *et al.*, 2005). Border-check (also called border-dyke or flood) irrigation and spray irrigation can be used to offset water deficiencies and increase production.

While 23% of Australia's dairy farms are classed as 'irrigated' (Dairy Australia, 2006), 52% of dairy farmers supplement natural rainfall with irrigation, either from major irrigation schemes where water is delivered to the farm gate or from other sources including catchments within the farm and groundwater bores (Dairy Australia, 2005). Dairy farms using irrigation water as the basis for fodder production are concentrated in south-eastern Australia including the lower reaches of the Murray River in south-eastern South Australia (Lower Murray), the Murray River plains in northern Victoria and Southern New South Wales and the Macalister Irrigation District in Gippsland, Victoria (Fig. 9.1). In other regions, irrigation is often used to supplement grass production in primarily rain-fed dairy systems. In New Zealand, irrigated pasture production is increasingly prevalent especially around Canterbury and Otago (Parliamentary Commissioner for the Environment, 2002).

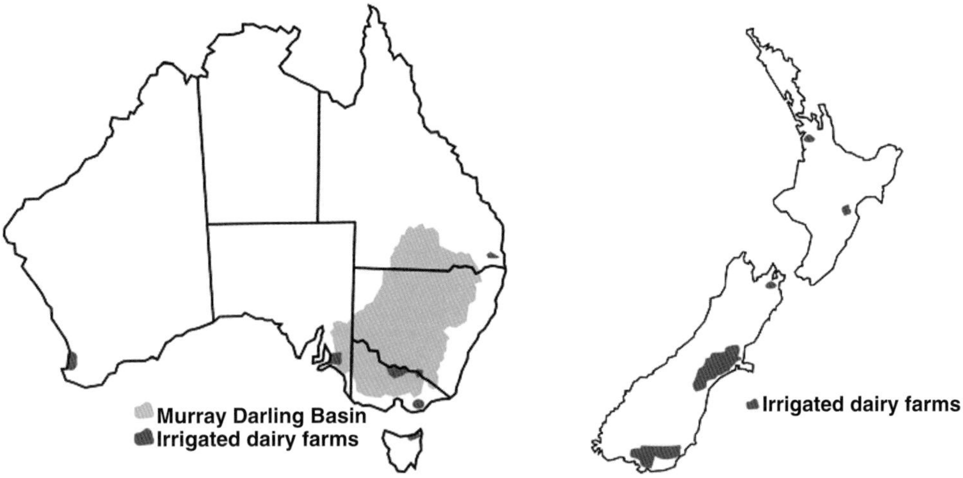

Fig. 9.1. Irrigated dairying areas of Australia and New Zealand. Figures not to scale.

The sustainability of irrigation is threatened by processes that degrade the environment and the economic constraints imposed on any farming system. In this chapter, we examine the properties of border-check and spray irrigation systems, particularly the hydrology of the different systems, the management challenges and the application of alternative irrigation technologies. The chapter then investigates the sustainability of irrigated pastures for dairy production in terms of both environmental constraints (e.g. deep drainage and pollutant export) and productivity constraints, both of which will ultimately affect their economic viability.

Irrigation Systems and Their Hydrology

Irrigation is practised on a range of soils in both Australia and New Zealand. For example, in southern New South Wales and northern Victoria, soils deposited as a result of prior stream activity are irrigated. Coarser soils with higher infiltration rates tend to occur on the levees while finer-textured clay soils, with low permeability, occur on the flood plain (Skene and Poutsma, 1962; Lyle *et al.*, 1986). The properties of these soils, their infiltration rate and hydraulic conductivity, as well as the source of the water (i.e. groundwater or gravity-fed channels) and the existing infrastructure, all affect the methods of irrigation that are used.

Understanding the sustainability (economic and environmental) of the various irrigation systems requires an understanding of the hydrology of the system, the associated risks of adverse impacts on and off the farm and the infrastructure required to support the systems. This section investigates the hydrology of the main irrigation systems used for irrigated pasture production: border-check and spray irrigation. In Australia, border-check is the most common irrigation method used for fodder production (Wood and Finger, 2006), while spray irrigation is the more common irrigation method used on pastures in New Zealand.

Border-check irrigation

The aim of border-check irrigation is to restore the root zone to field capacity (Finger, 2005), with water applied when the soil water deficit, generally measured as evaporation – rainfall (estimated using Class A pan evaporation and normalized for the region), is 40–50 mm. Water is applied to the top of a bay (commonly *c.*350 × 40 m but can be more than 1000 m in length and 30–50 m wide) in excess of the soil infiltration rate and moves down the bay as infiltration excess surface runoff. The water is confined on the bay by check banks (i.e. raised earthen ridges) which run down the sides of the bay. Water is usually applied 10–20 times from late spring to early autumn, with annual applications of between 5 and 10 ML/ha.

Irrigation water moves into the soil through a combination of 'bypass' flow through channels and cracks, and saturated and unsaturated matrix flow. The highest infiltration rates occur at the wetting front where the water initially passes into dry soil, and decline behind the wetting front (Austin, 1998; Fig. 9.2). A major deficiency of border-check irrigation is that water needs to traverse the

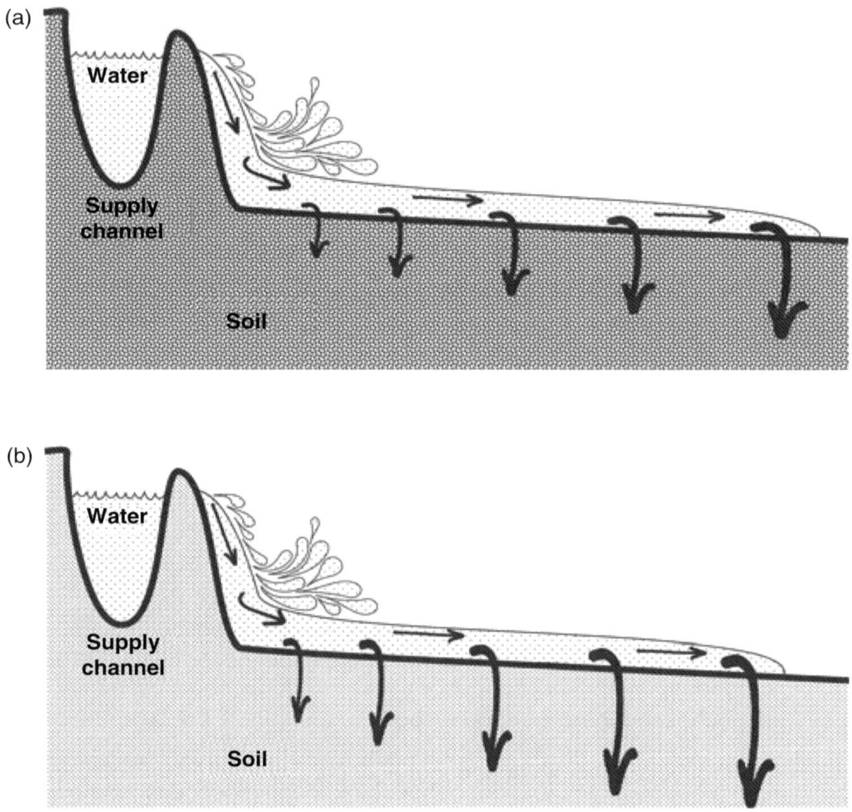

Fig. 9.2. A diagrammatic representation of water movement in border-check irrigation systems on (a) heavy, low-infiltration soils and (b) light, high-infiltration soils.

soil surface for the full length of the bay to ensure that the entire bay is irrigated. This results in *c*.20% more water than required being applied (Nexhip *et al.*, 1997) and passing from the foot of the bay into drainage channels. This surface water is commonly referred to as surface runoff and can include re-emergent interflow (Nash *et al.*, 2002).

In practice, border-check irrigation rarely distributes water as evenly down a bay as might be hoped due to soil variability, particularly infiltration rates, and the variable time that water is ponded on the surface. To assist in the even distribution of water in a bay, most border irrigation systems are graded to a slope of 1:400 to 1:1000, depending on the region and site. While laser grading can manufacture constant slopes on the bays, preferential flow paths and variable infiltration characteristics still occur, affecting the distribution of water and the efficiency of the irrigation system. For example, in some regions, such as the Macalister Irrigation District, bays may traverse two or more soil types with differing infiltration characteristics. Animal traffic also affects drainage via localized soil structural decline, especially when wet, and areas of low infiltration due to animal tracking. This is a particular problem at the foot of bays where surface flow can accumulate and waterlogging can suppress pasture production. Spinner drains (i.e. shallow, <10 cm, scalloped-shaped drains extending longitudinally down bays) are used in some areas to improve surface drainage and fortunately, on many farms, the adverse effects of cattle traffic during the irrigation season are minimized by only grazing when the soil surface is dry.

The water application efficiency of border-check irrigation systems has been enhanced through the use of laser grading to improve water distribution on the paddock, whole-farm planning, the installation of reuse systems that collect outwash (surface runoff) and the availability of higher flow rates that decrease the time available for infiltration (Ewers, 1988; Water Force Victoria, 1990; Malano and Patto, 1992; Douglass and Poulton, 2000). The use of high flow rates during application causing pulses rather than a continual stream has been shown on some soil types to result in more uniform water application and less infiltration below the root zone as water passes over pre-wetted soil (with a lower infiltration rate; Turral, 1993). In addition to soil type, the effectiveness of surge flow irrigation appears to be affected by factors such as water salinity, sodicity and sediment (Heydari *et al.*, 2001; Wang *et al.*, 2005).

The hydrology of border-check irrigation suggests that on many soil types drainage below the root zone and water draining from the foot is to be expected if the whole bay is to be irrigated. It follows that border-check irrigation is most efficient (measured as production per unit water applied) on heavier soils. Where infiltration rates are higher, the probability of water moving below the root zone increases along with the consequences of deep drainage such a rising water tables.

Sprinkler irrigation

Sprinkler systems distribute water much like rainfall. Sprinkler systems and the related infrastructure vary dramatically between regions depending on the size and shape of the area to be irrigated, the topography, physical obstructions such as trees and buildings, the availability of labour, the necessary application rate and the source of the water. The types of spray irrigation used in pastoral industries include fixed, operator shift low-pressure (bike-shift) sprinkler gun, centre-pivot and lateral move sprinkler systems (side-roll; Wood and Martin, 2000).

The basic hydrology of sprinkler irrigation is similar to rainfall. In general, water is added to soil at a rate below the soil infiltration rate and penetrates to a depth determined by the flow characteristics of the soil and the irrigation management. The notable exception is at the circumference of larger centre-pivot irrigators where due to the higher ground speed, greater water application rates are required and some temporary ponding may occur in the immediate vicinity of the sprays. As water moves predominantly in a vertical direction (Fig. 9.3) and there is no requirement for lateral flow, the loss of water and associated pollutants in irrigation surface runoff (re-emergent interflow and overland flow) should be minimal (Ebbert and Kim, 1998).

Theoretically sprinkler irrigation provides more control over water distribution than systems such as border-check due to the ability to match application rates and infiltration characteristics in a sprinkler system (Burt *et al.*, 2000). This maximizes irrigation efficiency and presumably, for example, in the Shepparton Irrigation Region of northern Victoria, centre-pivot irrigators would be expected to operate with 75–90% efficiency while the equivalent border-check irrigation systems are 55–90% efficient (Department of Primary Industries, 2004). While environmental factors including wind affect sprinklers, the efficiency of sprinkler systems largely depends on management particularly in systems that require significant operator intervention, such as low-pressure sprinklers that require the operator to shift them often. Even where sprinkler irrigation is uniform, undulating soil and varying soil infiltration rates can lessen overall irrigation efficiency. This is most noticeable in small depressions where water may temporarily collect. Grazing cattle tend to preferentially compact soil in these areas decreasing the infiltration rate (i.e. poaching) and exacerbating the problem. As a result, a mosaic of small wet areas may develop where the vegetation is different to other sections of the paddock.

An important issue, particularly where centre-pivot and travelling irrigators are used on sloping ground is the lateral flow of subsurface water down slope. Unless irrigation rates are modified in different areas, water moving past the root

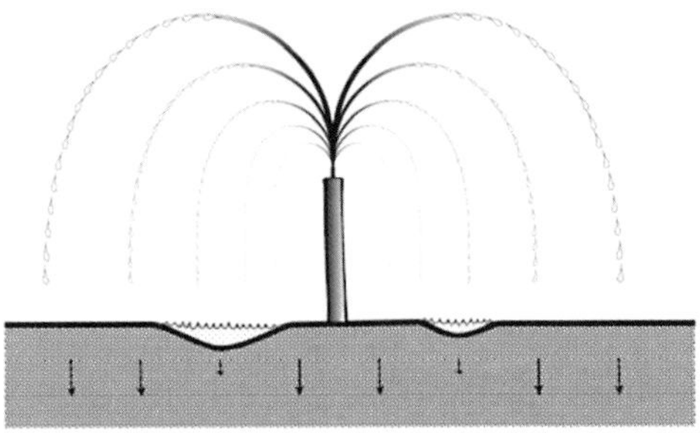

Fig. 9.3. A diagrammatic representation of water movement under spray irrigation.

zone through preferential pathways accumulates down slope (as interflow) leading to either saturated soil (or soaks) and/or subsurface drainage that can export contaminants off site. Unfortunately, in many of these situations there are often few alternatives to this form of irrigation.

Alternative irrigation technologies

With increasing pressure on limited water supplies, there is increasing interest in micro-irrigation technologies such as subsurface drip. Subsurface drip irrigation has been successfully applied across a range of industries, with increased yields and decreases in water use compared to other systems (Ayars *et al.*, 1999; Alam *et al.*, 2002; Lamm and Trooien, 2003). In one trial, subsurface drip irrigation used 200 mm/year less irrigation water than border-check and produced approximately 1.0 t dry matter (DM)/ha/year more pasture (Finger and Wood, 2006). Compared to sprinkler irrigation, subsurface drip can decrease evaporation (Alam *et al.*, 2002), decrease erosion (Bosch *et al.*, 1992) and lessen scalding by entrained salts (Cetin and Bilgel, 2002).

Adverse Impact of Irrigated Dairy Pastures

Rising water tables, salinity and nutrient exports (particularly nitrogen (N) and phosphorus (P)) are major problems that threaten the environmental sustainability of irrigated pasture production. All of these environmental impacts are linked to the hydrology of irrigation systems. Deep drainage contributes to rising water tables and associated soil salinity, and N leaching, while surface runoff can transport N, P and other contaminants into aquatic ecosystems.

An issue of increasing prominence, especially under border-check irrigation where nitrogenous fertilizers are used, is the production of nitrous oxides. Nitrous oxides are powerful greenhouse gases and whether they are produced in significant quantities under irrigation is the subject of continuing research (de Klein *et al.*, 2001; Dalal *et al.*, 2003; Phillips *et al.*, 2007).

This chapter focuses the adverse impacts from deep drainage and surface runoff, while Chapter 1 (this volume) investigates greenhouse gas emissions from pasture systems.

Deep drainage

Deep drainage from irrigation is a major environmental challenge in many parts of Australia and New Zealand. Deep drainage, the movement of water below the root zone of plants and into the vadose zone en route to groundwater, contributes to rising water tables and the associated problem of salinization, as well as the leaching of N and other potential pollutants, including pesticides, to groundwater systems.

Rising water tables and salinity is a significant challenge for irrigated regions in Australia (Lyle *et al.*, 1986). The impacts of deep drainage are well documented in the Murray Darling Basin in Australia, which in the mid-1980s had approximately 96,000 ha of irrigated land showing visible signs of salinization as a result of high water tables. It was anticipated prior to the recent drought that the area affected by high water tables in the Murray Darling Basin would have increased to 869,000 ha of land by 2015 (Blackmore *et al.*, 1999).

The challenge in minimizing the risks associated with deep drainage is that in all irrigation systems some drainage is essential to remove salts from the root zone that have been concentrated via the evapotranspiration of irrigation water (Richards, 1954). For optimum irrigation efficiency, such leaching would result from natural rainfall in the non-irrigation season. However, irrigation management for zero drainage is difficult to achieve, particularly as all but the most unstructured soils have preferential flow paths (macropores) that transmit water below the root zone ensuring some deep drainage occurs (Nash *et al.*, 2002). While small amounts of deep drainage may seem unimportant, it is worth noting that 10 mm of drainage may result in water tables rising 200 mm as the water only occupies pore space in the soil.

The proportion of water draining below the root zone is primarily a function of the soil type, irrigation method and water application rate (i.e. irrigation management) and is generally the result of unsaturated flow processes, which are difficult to measure in the field or to describe quantitatively (Hillel, 2004). Deep drainage is often estimated on the basis of a soil water balance, fluctuations in soil water content, mathematical models of unsaturated flow in soils or through the use of lysimeters (e.g. Bethune and Wang, 2004).

Estimates of deep drainage on heavier soils used for border-check irrigation in the Murray Darling Basin are commonly in the range of 0–100 mm with most around 50 mm. Several studies have estimated that <10 mm of deep drainage occurred annually on the heavier soils in the basin (Holmes and Watson, 1967; Gilfedder *et al.*, 2000; Bethune and Wang, 2004). Shallow water tables may have confounded the results in some of these cases. The estimates for deep drainage on more permeable levee soils using similar irrigation methods range between 100–500 mm. Similar rates of drainage (100–600 mm) have been estimated in trials in New Zealand, where the majority of leaching occurred over winter (Ledgard *et al.*, 1996; Di *et al.*, 1998).

The amount of deep drainage is affected by the irrigation system used, its management as well as soil properties, with a number of studies comparing deep drainage under different irrigation systems. For example, border-check, sprinkler, subsurface drip and surge (i.e. high flow) irrigation have been compared in Northern Victoria on a flood plain soil (Bethune *et al.*, 2003; Wood and Finger, 2006). During the trial, irrigation of the border-check irrigation bays was scheduled when Class A pan evaporation exceeded rainfall by 50 mm, while a 30 mm deficit was used to schedule sprinkler irrigation. The temporal pattern of root zone water depletion for these two systems is shown in Fig. 9.4, with the border-check irrigated bays replenished to field capacity following irrigation, while under sprinklers, soil water depletion was managed within a narrower range and never reached field capacity. Deep drainage was estimated to be 12 and 7 mm for the sprinkler irrigated bays and 248 and 64 mm for the border-check irrigated bays. Other findings from the study were that:

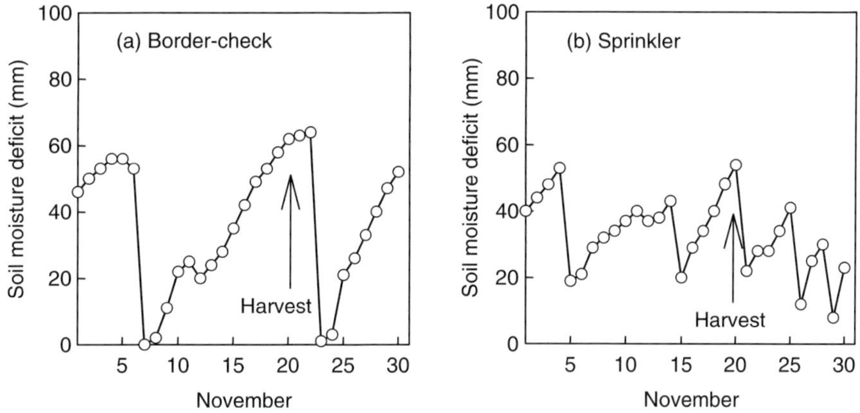

Fig. 9.4. Variation in root zone soil water depletion for (a) border-check-irrigated and (b) sprinkler-irrigated bays. (Adapted from Bethune *et al.*, 2003.)

- Subsurface drip and sprinkler irrigation used on average 20% less irrigation water than border-check irrigation (assuming surface drainage was reused).
- Subsurface drip and sprinkler irrigation had *c.*15% greater water-use efficiency (defined as the quantity of DM produced per megalitre of infiltrated depth plus rainfall) than border-check irrigation.

N leaching associated with deep drainage can affect groundwater quality and can also be discharged into neighbouring water bodies. This problem is particularly acute in parts of New Zealand (Parliamentary Commissioner for the Environment, 2005), where groundwater nitrate concentrations have been found to exceed the World Health Organization drinking water limit of 10 mg N/l in many regions. Intensive agricultural activities, such as dairy farming, are considered to be major non-point sources of nitrate to groundwater systems.

Minimizing deep drainage is important; however, it is possible to recover deep drainage by pumping groundwater. This can lessen the negative effects of irrigation on the environment by lowering water tables and removing salts from the soil. However, the groundwater is often saltier than the original irrigation water as the concentration of salts through evapotranspiration is unavoidable. In some hydrogeologic settings, such as the Murray Darling Basin, salts from other areas may also contribute to salt in groundwater (Greg Hoxley, March 2007, personal communication). The use of saline water for irrigation of pastures either alone or mixed (i.e. shandied) with better quality water (termed conjunctive water use) is one way of lessening the impact of deep drainage from pasture-based grazing systems (Bethune *et al.*, 2004).

Saline drainage can also be used to irrigate field crops (Tanji and Karajeh, 1993). One rather innovative system for managing deep drainage and associated saline irrigation water is Serial Biological Concentration (SBC; Heath and Heuperman, 1996; Heuperman, 1999). These systems concentrate drainage water by reuse on successively more salt tolerant crops and ultimately dispose of the salt in evaporation basins.

Surface runoff

The export of pollutants, especially N, P and microbial pathogens, in surface runoff is a major environmental issue for irrigated dairy systems. There are limited data on the microbial composition of drainage. Loss of the faecal indicator bacteria, *Escherichia coli*, in outwash from border-check irrigated pastures is estimated at about 1.5×10^7 coliform forming units/m^2 per irrigation event (McDowell *et al.*, 2008). However, nutrient exports, both nutrient loads and nutrient concentrations, have been extensively studied (Drewry *et al.*, 2006) and a number of remedial programmes have been implemented (Department of Natural Resources and Environment, 1998). While nutrients loads are the product of the nutrient concentration and flow volume, flow is the major determinant of nutrient loads as it is highly variable (i.e. can vary from almost zero to several orders of magnitude greater than the average), whereas nutrient concentrations are generally within a comparatively narrow range (i.e. vary by an order of magnitude or less; Nexhip *et al.*, 1997; Haygarth *et al.*, 2004). The exception is where a readily available nutrient source such as fertilizer has recently been applied to the pasture (Nash *et al.*, 2005).

At the bay/field scale, there is no question that the nutrient loads exported from most irrigation systems will exceed those from a rain-fed system in the same area (i.e. a non-irrigated farm with the same rainfall). This is especially true for border-check irrigation. Any form of irrigation will increase soil moisture, decrease the soil infiltration rates compared to dry, unirrigated soil and increase surface runoff. Equally important, plants take up nutrients from soil water. To grow higher-yielding plants under irrigation, fertilizers are generally applied to increase nutrient concentrations in soil water. These nutrients are mobilized when runoff occurs. Not surprisingly, at the bay or field scale in the same area, irrigated pastures generally have greater fertility and productivity than rain-fed systems, but also produce a greater volume of surface drainage with a higher nutrient concentration.

Nutrient concentrations in surface runoff are affected by nutrient sources, mobilization processes (including demobilization) and the hydrology of the system. For well-managed irrigated pastures, physical detachment and transport of soil particles (i.e. erosion) and associated nutrients should not be a major contributor to nutrient exports. Consequently, it is the export of dissolved nutrients ($<0.45\ \mu m$) that is the major concern for well-managed farms.

Nutrient concentrations in surface runoff from irrigation systems depend on the scale at which they are measured. While nutrient loads have often been measured at the bay/field scale and under different management regimes, the associated mobilization and transport processes have received less detailed study. At the bay/field scale, the concentrations of nutrients in irrigation-induced surface runoff increase as water moves down the bay, especially at the wetting front (Fig. 9.5). A simple explanation for the increasing nutrient concentrations in the wetting front would be that labile nutrient stores at the soil surface are being mobilized and that more nutrient is being mobilized than is infiltrated. Consequently, the further the water moves the greater its concentration becomes. This hypothesis is consistent with studies where surface-applied, labile P has been shown to rapidly infiltrate soil (Bush and Austin, 2001). A similar hypothesis would explain why nutrient

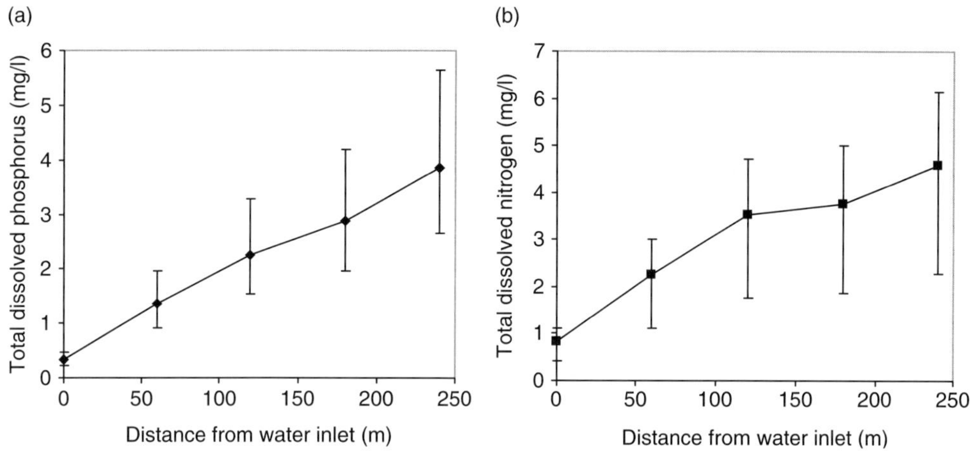

Fig. 9.5. Wetting front (a) total dissolved phosphorus and (b) total dissolved nitrogen concentrations with distance down a border-check irrigation bay from the water inlet.

concentrations also increase behind the wetting front where infiltration rates decline but infiltration does not cease altogether. But is it reasonable that only the nutrients at the surface are being mobilized, especially behind the wetting front?

An alternative explanation for nutrient concentrations increasing with path length is that behind the wetting front the flow of dissolved nutrients into the soil is opposed by intermittent turbulence near the surface (*c.* 5 mm) and the quasi-diffusion of nutrients from within topsoil layers into surface runoff (Fig. 9.6). This implies that nutrient concentrations in surface runoff are a function of the soil hydrology, the rate of nutrient release from its primary source, its location relative to the soil surface (i.e. vertical path length and tortuosity), and factors affecting diffusion such as source solubility and demobilization (i.e. fixation) reactions, rather than simply the size of the nutrient source and its solubility. For example, soluble nutrients contained in organic matter may well avoid infiltration at the wetting front. Their subsequent diffusion into surface runoff may result in greater overall nutrient concentrations than would otherwise be the case. Such an explanation may help explain the large between-storm variability often encountered in field studies.

There is circumstantial evidence that processes similar to the one proposed in Fig. 9.6 operate in border-check irrigation systems. In field experiments using within bay sampling to compare two fertilizers with different dissolution rates, single-superphosphate and di-ammonium phosphate were shown to affect dissolved P concentrations at, and possibly behind, the wetting front (Nash *et al.*, 2003b, 2004). In model studies where vertical fluxes had been largely eliminated, dissolved nutrient concentrations in surface flow have been shown to initially increase and then decrease to a concentration well above zero for the remainder of the experiment (Doody *et al.*, 2006). It is difficult to explain such results based on variable solubilities of nutrient sources alone and such studies suggest that the kinetics of nutrient diffusion from the source to flowing water is also having an effect. Similarly, increasing flow path length has been shown to increase

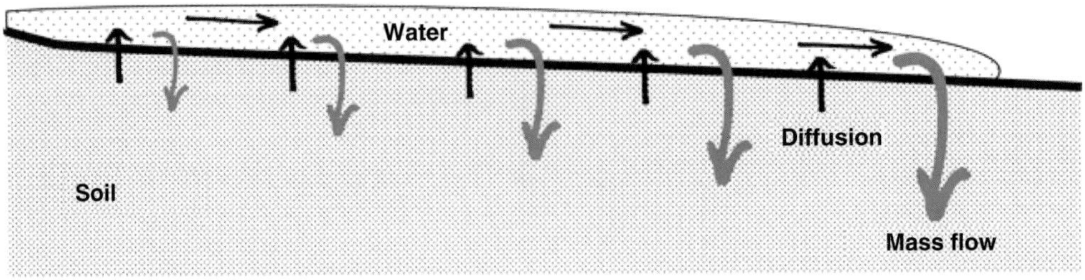

Fig. 9.6. A schematic representation of the roles of mass flow and diffusion in contaminant mobilization in surface runoff.

dissolved nutrient concentrations, while increasing flow rates generally decrease dissolved nutrient concentrations, probably due to dilution. Such a mechanism would also help explain the insensitivity of P concentrations in surface runoff to runoff volume in some field-scale (*c.*2 ha) studies (Nash *et al.*, 2005) and why in some larger-scale and model studies dissolved nutrient concentrations paralleled rain-induced surface runoff rates when, due to dilution, they would be expected to decline (Pote *et al.*, 1999; Heathwaite and Dils, 2000; Lazzarotto *et al.*, 2005). In addition to the physical processes described in Fig. 9.6, increased flow rates are likely to be associated with increasingly larger source areas leading to longer horizontal path lengths, longer residence times, access to additional nutrient sources and therefore, higher concentrations. In some cases re-emergent interflow may also have contributed to nutrient concentrations.

The processes responsible for nutrient mobilization are important because they may provide a guide to opportunities for lessening nutrient exports from irrigated dairying. For example, if the main factor is the rate at which nutrients physically intersect surface runoff, then this will be affected by path length. This implies that location of nutrient sources (i.e. depth) relative to the flow will be as important as the solubility of the chemical species in water in determining nutrient concentrations in surface runoff. Detailed testing of this hypothesis using a conventional rainfall simulator may be difficult as the physical impact of the water droplets would increase surface turbulence and the effective depth of interaction (Ahuja and Lehman, 1983).

Nutrients accumulate at the surface of pasture soils and model studies have suggested that de-stratification (i.e. mixing surface and subsurface soil) can lessen nutrient concentrations in surface runoff (Dougherty *et al.*, 2006; Sharpley, 2003). Recent studies of border-check irrigation in the Macalister Irrigation District of south-eastern Australia have confirmed that hypothesis (Nash *et al.*, 2007). Changes in soil P (0–20 mm), soil water P and N, and P and N concentrations in surface runoff were measured in four recently laser-graded (<1 year) and four established (>10 years) irrigated pastures in south-eastern Australia after 4 years of irrigated dairy production. Laser grading, which involves cultivation and mixing of surface soil, initially lowered soil surface (0–20 mm) total P, Olsen P, Colwell P, water extractable P, calcium chloride extractable P, organic P, P sorption saturation and total C and increased P sorption compared to established pastures but did not affect Olsen P and Colwell P concentrations in the root zone (0–100 mm).

Over the 3 years of the study on the lasered bays, Olsen P, Colwell P and P sorption decreased and water extractable P and P sorption saturation increased while on the untreated bays only Olsen P and Colwell P decreased. These results presumably reflect the inputs and outputs being in approximate balance, incorporation of subsoil into the surface layer and a general decline in P availability.

Three years after laser grading, soil water total dissolved P (TDP) concentrations were greater on the established bays while dissolved reactive P (DRP) concentrations were unaffected. Soil water organic P (estimated as TDP-DRP and also called dissolved unreactive P) comprised 70% and 32% of TDP for the established and lasered bays, respectively. These soil water data were reflected in the surface runoff where after 3 years, compared to established bays, laser grading decreased TDP, total dissolved N, total P and total N exports in wetting front drainage by 40%, 29%, 41% and 36%, respectively. This is an important result for management of dairy systems as it suggests that the regular cultivation used to renovate pasture on more intensive (>2cows/ha) dairy farms probably decreases the short-term exports of P and N compared to an otherwise similar, non-cultivated alternative. But would the results have been the same if this were sprinkler irrigation?

At the farm scale, compared to many rain-fed grazing systems, irrigation farms are often in the unique position of being able to control both irrigation and rain-induced drainage. To prevent waterlogging of the bays, border-check irrigation farms generally have a well-developed drainage network that can be used to collect and recycle surface runoff (also termed outwash). The effectiveness of these systems depends on their management, but in some studies, nutrient exports have been virtually eliminated through drainage reuse (Barlow *et al.*, 2005). It follows that, depending on the farm infrastructure and management, farm-scale nutrient exports from irrigated dairying need not be any worse than from other land uses.

When comparing the overall environmental performance of different irrigation systems, the ability to recycle outwash is a major point of distinction between border-check and other forms of irrigation. At the bay scale, border-check irrigation outwash is almost always greater than from sprinkler irrigation of the same land. However, the volume of rainfall-induced runoff from irrigated areas depends on soil moisture and therefore is a function of both the annual rainfall pattern and, during the irrigation season, irrigation management. In border-check irrigation, soil is intermittently saturated. In spray irrigation, soil is maintained below field capacity, but well above the minimum moisture content used as a trigger before border-check irrigation occurs. It follows that where rain falls immediately after irrigation, runoff will be greater from border-check irrigation bays. However, where rain occurs at the end of an irrigation cycle (i.e. immediately before the next irrigation), runoff will be greater from spray irrigation areas (Nash *et al.*, 2003a).

At the farm scale, the impact of a grazing system on water quality in the surrounding catchment depends primarily on drainage from the farm rather than the bays and there is no reason to believe that any water application system will always generate less farm-scale runoff. It is the irrigation management system, including the reuse system, rather than the water application system that determines the volume of drainage and nutrient loads discharged from irrigated grazing farms. This is increasingly recognized by irrigation agencies who are upgrading infrastructure to support more flexible irrigation management.

Conclusions and the Future of Irrigated Dairying in Australasia

The major impacts of irrigated dairying on the environment are the result of inefficient resource utilization. Water is a limited resource and the more water lost from pasture systems as deep drainage and surface runoff, the less water is available for pasture production. Improved water-use efficiency could therefore yield both environmental and productivity improvements. For the purpose of this discussion, production water-use efficiency (PWUE) will be defined as the amount of milk produced per volume of water (i.e. milk fat + protein per megalitre of rainfall and irrigation). Economic water-use efficiency (EWUE) will be defined as the margin between income generated from pasture and variable costs of producing pasture per megalitre of water. Improving water-use efficiency has been the focus of considerable research (e.g. (Wood and Martin, 2000) with many studies investigating better management of existing infrastructure, often border-check irrigation, or conversion of farms from border-check to spray irrigation.

A survey of water-use efficiency on 170 randomly selected dairy farms in northern Victoria and southern New South Wales (Armstrong *et al.*, 2000) provides some important insights into the improvements possible using existing border-check irrigation infrastructure. High-PWUE farms produced the same amount of milk for approximately two-thirds the water, half the land and grazed a similar number of cows to low-PWUE farms. There was a strong ($r = 0.97$) correlation between PWUE and EWUE: income from pasture of the top 10% of farms had two-and-half times greater EWUE than the bottom 10%. Similar results were obtained in a benchmarking study of the Macalister Irrigation District in south-eastern Victoria (McAinch, 2003). Clearly there are both economic and environmental benefits to improved management (Armstrong *et al.*, 1998). Subsequent analyses of these and other data suggest that decreasing water availability by one-third (i.e. from >150% of the water for which delivery is contracted to 100–120%) had little impact on water-use efficiency (Linehan *et al.*, 2004). In Australia, a water right is defined as a formally established or legal authority to take water from a water body and to retain the benefits of its use. Rights may be attenuated in a number of ways and are referred to in different jurisdictions as licences, concessions, permits, access entitlements or allocations (Productivity Comission, 2003). The impact of the recent drought in these areas on water-use efficiency remains to be seen given the structural changes that may occur in the industry.

Changing infrastructure has been considered an important policy option for enhancing environmental performance in a number of areas and, in Australia in particular, significant resources have been committed to assisting farmers change from border-check to spray irrigation. The potential for sprinkler systems to decrease deep drainage runoff is widely recognized (Cockroft and Mason, 1987; Collis-George, 1991). Improved farm management, productivity, lifestyle and marketability of farms are perceived by farmers as the key benefits of converting border-check irrigation to centre-pivot or sprinkler systems in northern Victoria. Capital cost, operation and maintenance costs, farm layout and unreliability of systems were perceived as the key barriers to adopting sprinkler technology (Maskey *et al.*, 2006). A detailed economic analysis of conversion from border-check to centre-pivot irrigation in the same area suggests that if conversion resulted in 20% less water use and 10% more pasture growth, conversion was profitable. However, returns on investment depended heavily on the land that could be irrigated using centre-pivots, the actual changes in PWUE and EWUE and energy and milk prices (Wood *et al.*, 2007).

It is questionable whether irrigated dairying can survive in some of the areas in which it is currently located. Increasing pressure on the use of water resources and the effects irrigated dairying may be having on them may well result in the fundamental changes over and above those that have been discussed here. Moving from pasture-based grazing to cut-and-carry systems where forage crops are grown to feed animals elsewhere has the potential to increase PWUE (Greenwood *et al.*, 2007). The real question will be how the EWUE of those systems compare to pasture-based grazing and the value of the water used for alternative purposes in the context of other changes that may be occurring in the industry (Garcia and Fulkerson, 2007).

References

Ahuja, L.R. and Lehman, O.R. (1983) The extent and nature of rainfall soil interaction in the release of soluble chemicals to runoff. *Journal of Environmental Quality* 12, 34–40.

Alam, M., Trooien, T.P., Dumlier, T.J. and Rogers, D.H. (2002) Using subsurface drip irrigation for alfalfa. *Journal of the American Water Resources Association* 38, 1715–1721.

Armstrong, D., Knee, J., Doyle, P., Pritchard, K. and Gyles, O. (1998) *A Survey of Water-Use Efficiency on Irrigated Dairy Farms in Northern Victoria and Southern New South Wales.* Department of Natural Resources and Environment, Melbourne, Australia.

Armstrong, D.P., Knee, J.K., Doyle, P.T., Pritchard, K.E. and Gyles, O.A. (2000) Water-use efficiency on irrigated dairy farms in northern Victoria and southern New South Wales. *Australian Journal of Experimental Agriculture* 40, 643–653.

Austin, N.R. (1998) Sustainable use of water and fertiliser in the irrigated dairy industry of Australia. PhD thesis, University of Melbourne, Melbourne, Australia.

Australian Bureau of Agricultural and Resource Economics and Ministry of Agriculture and Forestry. (2006) *Agricultural Economies of Australia and New Zealand: Past, Present Future.* ABARE, Canberra, Australia, August.

Ayars, J.E., Phene, C.J., Hutmacher, R.B., Davis, K.R., Schoneman, R.A., Vail, S.S. and Mead, R.M. (1999) Sub-surface drip irrigation for row crops: a review of fifteen years research at the Water Management Research Laboratory. *Agricultural Water Management* 42, 1–27.

Barlow, K., Nash, D. and Grayson, R. (2005) Phosphorus export at the paddock, farm-section and whole farm scale on an irrigated dairy farm in south-eastern Australia. *Australian Journal of Agricultural Research* 56, 1–9.

Bethune, M.G. and Wang, Q.J. (2004) A lysimeter study of the water balance of border-check irrigated perennial pasture. *Australian Journal of Experimental Agriculture* 44, 151–162.

Bethune, M., Wood, M., Finger, L. and Wang, Q.J. (2003) *Sprinkler, Subsurface Drip and Surge Experiment – Final Report of DAV11163.* Final Report. Department of Primary Industries, Tatura, Victoria, Australia.

Bethune, M.G., Gyles, O.A. and Wang, Q.J. (2004) Options for management of saline groundwater in an irrigated farming system. *Australian Journal of Experimental Agriculture* 44, 181–188.

Blackmore, D.J., Goss, K.F., Newman, R.J. and Powell, J. (1999) *Australian Academy of Sciences and Technological Engineering Symposium: Spirit of the Snowy – Fifty years on, November 1999.* Australian Academy of Sciences and Technological Engineering, Parkville, Victoria, Australia.

Bosch, D.J., Powell, N.L. and Wright, F.S. (1992) An economic comparison of subsurface microirrigation with centre pivot irrigation. *Journal of Production Agriculture* 5, 431–437.

Burt, C.M., Clemmens, A.J., CBleiesner, R., Merriam, J.L. and Hardy, L. (2000) *Selection of Irrigation Methods for Agriculture.* American Society of Civil Engineers On-Farm Irrigation Committee, Reston, Virginia.

Bush, B.J. and Austin, N.R. (2001) Timing of phosphorus fertilizer application within an irrigation cycle for perennial pasture. *Journal of Environmental Quality* 30, 939–946.

Cetin, O. and Bilgel, L. (2002) Effects of different irrigation methods on shedding and yield of cotton. *Agricultural Water Management* 54, 1–15.

Cockroft, B. and Mason, W. (1987) Irrigated agriculture. In: Connor, D. and Smith, D.F. (eds) *Agriculture in Victoria. Australian Institute of Agricultural Science*, Melbourne, Australia.

Collis-George, N. (1991) Drainage and soil structure: a review. *Australian Journal of Soil Research* 29, 923–933.

Dairy Australia. (2005) Australian dairy industry in focus 2005. Available at: www.dairyaustralia.com.au

Dairy Australia. (2006) *Dairy 2006: Situation and Outlook.* Report to the Australian Dairy Industry.

Dalal, R.C., Wang, W., Robertson, G.P. and Parton, W.J. (2003) Nitrous oxide emission from Australian agricultural lands and mitigation options: a review. *Australian Journal of Soil Research* 41, 165–195.

de Klein, C.A.M., Sherlock, R.R., Cameron, K.C. and van der Weerden, T.J. (2001) Nitrous oxide emissions from agricultural soils in New Zealand – a review of current knowledge and directions for future research. *Journal of The Royal Society of New Zealand* 31, 543–574.

Department of Natural Resources and Environment. (1998) *Macalister Irrigation District Nutrient Reduction Plan.* Department of Natural Resources and Environment, Maffra, Victoria, Australia.

Department of Primary Industries. (2004) From 'Border-check' to Sprinkler? [Online]. Available by Department of Primary Industries (posted April 2004; verified 01/03/07).

Di, H.J., Cameron, K.C., Moore, S. and Smith, N.P. (1998) Nitrate leaching from dairy shed effluent and ammonium fertilser applied to a free-draining soil under spray or flood irrigation. *New Zealand Journal of Agricultural Research* 41, 263–270.

Dillon, P., Roche, J.R., Shalloo, L. and Horan, B. (2005) Proceedings of a satellite workshop of the XXth International Grassland Congress. In: Murphy, J.J. (ed.) Cork, Ireland. Wageningen Academic Publishers, Wageningen, the Netherlands.

Doody, D., Moles, R., Tunney, H., Kurz, I., Bourke, D., Daly, K. and O'Reagan, B. (2006) Impact of flow path length and flow rate on phosphorus loss in simulated overland flow from a humic gleysol grassland soil. *Science of the Total Environment* 372, 247–255.

Dougherty, W., Davies, P., Chittleborough, D.J., Cox, J. and Nash, D.M. (2006) *Soil Phosphorus De-stratification to Reduce Export of Phosphorus in Surface Runoff a Sub-catchment Used for Intensive Grazing.* 18th World Congress of Soil Science, Convention Centre, Philadelphia, Pennsylvania.

Douglass, W. and Poulton, D.C. (2000) Living with limited water – towards efficient water use on irrigated dairy pastures. *ICID Journal* 49, 29–40.

Drewry, J.J., Newham, L.T.H., Greene, R.S.B., Jakeman, A.J. and Croke, B.F.W. (2006) A review of nitrogen and phosphorus export to waterways: context for catchment modelling. *Marine and Freshwater Research* 57, 757–774.

Ebbert, J.C. and Kim, M.H. (1998) Relation between irrigation methods, sediment yields and losses of pesticides and nitrogen. *Journal of Environmental Quality* 27, 372–380.

Ewers, C. (1988) *Laser Grading in the Goulburn-Murray Irrigation District: An Innovation Diffusion Study.* Monash University, Melbourne, Victoria, Australia.

Finger, L. (2005) *Integrating Irrigation and Plant Systems for Victoria's Dairy Industry: Literature review ISBN: 1 741146 394 7.* Victorian Department of Sustainability and Environment Melbourne, Victoria, Australia.

Finger, L. and Wood, M. (2006) 7th International Micro Irrigation Congress, PWTC, 10–16 September, Kuala Lumpur, Malaysia.

Garcia, S.C. and Fulkerson, W.J. (2007) Opportunities for future Australian dairy systems: a review. *Australian Journal of Experimental Agriculture* 45, 1041–1055.

Gilfedder, M., Connell, L.D. and Mein, R.G. (2000) Border irrigation field experiment. I: water balance. *Journal of Irrigation and Drainage Engineering* 126, 85–91.

Greenwood, K.L., Mundy, G.N. and Kelly, K.B. (2007) On-farm measurement of the water and productivity of maize. *Australian Journal of Experimental Agriculture* 48, 274–284.

Haygarth, P.M., Turner, B.L., Fraser, A.I., Jarvis, S.C., Harrod, T.R., Nash, D.M. and Halliwell, D.J. (2004) Temporal variability in phosphorus transfers: classifying concentration-discharge event dynamics. *Hydrology and Earth System Sciences* 8, 88–97.

Heath, J. and Heuperman, A. (1996) Serial biological concentration of salts – a pilot project. *Australian Journal of Soil and Water Conservation* 9, 27–31.

Heathwaite, A.L. and Dils, R.M. (2000) Characterising phosphorus loss in surface and subsurface hydrological pathways. *The Science of the Total Environment* 251/252, 523–538.

Heuperman, A. (1999) Serial biological concentration: a new management system for saline drainage water. *Irrigation Australia* 14, 21–22, 27.

Heydari, N., Das Gupta, A. and Loof, R. (2001) Salinity and sodicity influences on infiltration during surge flow irrigation. *Irrigation Science* 20, 165–173.

Hillel, D. (2004) *Introduction to Environmental Soil Physics*. Elsevier Academic Press, Oxford.

Holmes, J.W. and Watson, C.L. (1967) The water budget of irrigated pasture land near Murray Bridge, South Australia. *Agricultural Meteorology* 4, 177–188.

Lamm, F.R. and Trooien, T.P. (2003) Subsurface drip irrigation for corn production: a review of ten years research in Kansas. *Irrigation Science* 22, 195–200.

Lazzarotto, P., Prasuhn, V., Butscher, E., Crespi, C., Fluhler, H. and Stamm, C. (2005) Phosphorus export dynamics from two Swiss grassland catchments. *Journal of Hydrology* 304, 139–150.

Ledgard, S.F., Sprosen, M.S., Brier, G.J., Nemaia, E.K.K. and Clark, D.A. (1996) Nitrogen inputs and losses from New Zealand dairy farmlets, as affected by nitrogen fertiliser application: year one. *Plant and Soil* 181, 65–69.

Linehan, C.J., Armstrong, D.P., Doyle, P.T. and Johnson, F. (2004) A survey of water use efficiency on irrigated dairy farms in northern Victoria. *Australian Journal of Experimental Agriculture* 44, 131–136.

Lyle, C.W., Chalmers, D.J., Wildes, R.A. and Heuperman, A.F. (1986) Causes of shallow watertables in the irrigation regions of the Shepparton Irrigation Region of Northern Victoria. *Journal of the Australian Institute of Agricultural Science* 52, 23–27.

Malano, H.M. and Patto, M. (1992) Automation of border irrigation in south-east Australia: an overview. *Irrigation and drainage systems* 6, 9–26.

Maskey, R., Bethune, M. and Lawler, D. (2006) Landowners' perceptions of the benefits and barriers to adopting centre pivot and lateral move irrigators on dairy farms. *Australian Journal of Experimental Agriculture* 46, 1433–1437.

McAinch, J. (2003) *Benchmarking Production Water-use Effectiveness on Irrigated Dairy Farms in the Macalister Irrigation District*. Project Report. Department of Primary Industries, Melbourne, Australia.

McDowell, R., Houlbrooke, D.J., Muller, K., Muirhead, R.W., Shepherd, M. and Cuttle, S.P. (2008) *Grazed Pastures and Surface Water Quality*, 1st edn. Nova Science Publishers, Hauppauge, New York.

Ministry of Agriculture and Forestry. (2006) *Situation and Outlook for New Zealand Agriculture and Forestry*. Ministry of Agriculture and Forestry, Wellington, New Zealand.

Nash, D., Clemow, L., Hannah, M., Barlow, K. and Gangaiya, P. (2005) Modelling phosphorus exports from rain-fed and irrigated pastures in southern Australia. *Australian Journal of Soil Research* 43, 745–755.

Nash, D.M., Halliwell, D. and Cox, J. (2002) Hydrological mobilisation of pollutants at the slope/field scale. In: Haygarth, P.M. and Jarvis, S.C. (eds) *Agriculture, Hydrology and Water Quality*. CAB International, Wallingford, UK, pp. 225–242.

Nash, D.M., Savage, G., Clemow, L.M. and Webb, B.J. (2003a) *Macalister Research Farm Water*

Balance. Department of Primary Industries, Ellinbank, Victoria.

Nash, D.M., Hannah, M., Clemow, L., Halliwell, D., Webb, B.J. and Chapman, D. (2003b) A laboratory study of phosphorus mobilisation from commercial fertilisers. *Australian Journal of Soil Research* 41, 1201–1212.

Nash, D.M., Hannah, M., Clemow, L., Halliwell, D., Webb, B.J. and Chapman, D. (2004) A field study of phosphorus mobilisation from commercial fertilisers. *Australian Journal of Soil Research* 42, 313–320.

Nash, D.M., Webb, B., Hannah, M., Adeloju, S., Toifl, M., Barlow, K., Robertson, F., Roddick, F. and Porter, N. (2007) Changes in nitrogen and phosphorus concentrations in soil, soil water and surface run-off following grading of irrigation bays used for intensive grazing. *Soil Use and Management* 23, 374–383.

Nexhip, K.J., Mundy, G.N., Collins, M.D. and Austin, N.R. (1997) *Development of Nutrient Water Quality Targets for Irrigated Pasture Subcatchments*. Final Report Project No.15037. Department of Natural Resources and Environment, Institute of Sustainable Irrigated Agriculture, Tatura, Australia.

Parliamentary Commissioner for the Environment. (2002) Creating Our Future: Sustainable Development for New Zealand [Online]. Available by Parliamentary Commissioner for the Environment (verified 01-03-07).

Parliamentary Commissioner for the Environment. (2005) Growing for Good: Intensive Farming, Sustainability and New Zealand's Environment [Online]. Available by Parliamentary Commissioner for the Environment (verified 01/03/07).

Phillips, F.A., Leuning, R., Baigent, R., Kelly, K.B. and Denmead, O.T. (2007) Nitrous oxide flux measurements from an intensively managed irrigated pasture using micrometeorological techniques. *Agricultural and Forest Meteorology* 143, 92–105.

Pote, D.H., Daniel, T.C., Nichols, D.J., Sharpley, A.N., Moore, P.A., Miller, D.M. and Edwards, D.R. (1999) Relationship between phosphorus levels in three ultisols and phosphorus concentrations in runoff. *Journal of Environmental Quality* 28, 170–175.

Productivity Comission. (2003) *Water Rights Arrangements in Australia and Overseas*. Comission Research Paper, Productivity Comission, Melborune, Australia.

Richards, L.A. (1954) *Diagnosis and Improvements of Saline and Alkaline Soils*. Agriculture Handbook 60 USDA, Washington, DC.

Sharpley, A.N. (2003) Soil mixing to decrease surface stratification of phosphorus in manured soils. *Journal of Environmental Quality* 32, 1375.

Skene, J.K.M. and Poutsma, T.J. (1962) *Soils and Land Use in Part of the Goulburn Valley, Victoria*. Department Of Agriculture, Melbourne, Victoria, Australia.

Tanji, K.K. and Karajeh, F.F. (1993) Saline drain water reuse in agroforestry systems. *Journal of irrigation and drainage engineering* 119, 170–180.

Turral, H.N. (1993) The evaluation of surge flow in border irrigation (with special reference to cracking soils). PhD thesis, University of Melbourne, Melbourne, Australia.

Wang, W., Luo, W. and Wang, Z. (2005) Surge flow irrigation with sediment-laden water in northeastern China. *Agricultural Water Management* 75, 1–9.

Water Force Victoria. (1990) *Irrigation for Profit Irrigation Association of Australia*. Numurkah, Victoria, Australia.

Wood, M., Wang, Q.J. and Bethune, M. (2007) An economic analysis of conversion from border-check to centre pivot irrigation on dairy farms in the Murray Dairy Region, Australia. *Irrigation Science* (in press) 26, 9–20.

Wood, M.L. and Finger, L. (2006) Influence of irrigation method on water use and production of perennial pastures in northern Victoria. *Australian Journal of Experimental Agriculture* 46, 1605–1614.

Wood, M.L. and Martin, M. (2000) *Improved Irrigation Practices for Forage Production: Module 2. Alternative Irrigation Technology Desktop Analysis*. Department of Natural Resources and Environment, Melbourne, Victoria, Australia.

10 The Impact of Hybrid Dairy Systems on Air, Soil and Water Quality: Focus on Nitrogen and Phosphorus Cycling

P.J.A. Kleinman and K. Soder

US Department of Agriculture, Agricultural Research Service, University Park, Pennsylvania, USA

Introduction

In the latter half of the 20th century, the use of confined animal feeding operations (CAFOs) for intensive dairying increased. This increase was a result of trends towards expansion of dairy farm operations, supported by advantages of economies of scale, development of new milking methods, changing distribution systems, intensification of production and changing agricultural policies, all of which were major forces in the declining role of grazing in dairy farming in some countries. Increasingly, however, climbing input costs combined with stagnant or decreasing milk prices have prompted farmers to consider alternatives to intensification, reversing grazing trends in the dairy industry (Rauniyar and Parker, 1999), and turn to grazing, with the potential of lower input costs outweighing marginal declines in milk yields associated with less-intensive production (Fales *et al.*, 1993).

In the 1990s, new management-intensive grazing systems were exported from New Zealand to other temperate dairy producing regions (Schmit *et al.*, 2001; Gloy *et al.*, 2002; Kriegl and McNair, 2005). Attractive to small family farms, management-intensive grazing systems improve the productivity and use of pastures and therefore lower feed costs. This increased use of pastures had potential to increase farm profitability despite the potential for decreased milk production per cow (Emmick and Toomer, 1991; Parker *et al.*, 1992). The adoption of management-intensive grazing has been attributed to a variety of other factors including: (i) providing a more 'natural' environment for the cows; (ii) improved quality of life for farmers (less time spent on hauling manure, etc.); (iii) human health benefits associated with consuming dairy products produced by grazing cows; and (iv) environmental benefits (Clancy, 2006; Soder and Muller, 2007).

A litany of environmental benefits is cited by advocates of grazing, largely along the lines that properly managed grazing systems mimic natural grassland ecosystems and are better at recycling resources than confined operations (Rook *et al.*, 2005; Tallowin *et al.*, 2005). Grazing can profoundly affect how nutrients cycle in dairy farming systems, with the fate of nutrients tied to a variety of management variables such as diet, storage and handling of manure, and pasture management. However, inefficiencies in converting nutrient resources into milk remain an ongoing target of grazing managers who must juggle nutrient management priorities with other farm priorities including herd health, profitability, quality of life, farm management objectives and socio-economic pressures. Thus, a growing body of work sheds light on the potential for grazing systems to adversely impact soil and water quality (e.g. Stout *et al.*, 2000; McDowell, 2006; James *et al.*, 2007).

This chapter examines aspects of nutrient management related to the role of grazing in dairy systems not covered in Chapters 8 and 9 (this volume). We contrast *grazing* and *total confinement* components of hybrid (i.e. mixed confined/grazing) dairy farms, recognizing that such farms cover a broad management spectrum.

Grazing Dairy Systems

As they were developed in New Zealand, modern management-intensive grazing systems originally consisted of a seasonally calved herd (where all animals calved in a short time frame) to take advantage of lush spring pastures, with little to no feed supplementation for the herd. While such systems worked well in the physiographic conditions found in New Zealand, conditions elsewhere were sufficiently different to hamper their unilateral adoption. Specifically, differences in climate and forages restricted the establishment of pastures with similar quality as the ryegrass-clover pastures in New Zealand (Parker *et al.*, 1991). In addition, animal genetics in other areas often differed from New Zealand dairy genetics. Large-framed Holsteins genetically selected for high milk production in North America often lose body condition rapidly on an all-pasture diet, potentially impairing health, milk production and reproduction, and perhaps raising animal welfare issues (Kolver *et al.*, 2000; Harris and Kolver, 2001; Kolver, 2003).

While some grazing dairies have adopted a ‘no concentrate’ policy, the majority of grazing dairies supplement the diets of their pastured herds to improve milk production. Additional benefits of supplementation include improved body condition, which may in turn improve reproductive efficiency and animal health. The type of supplement can vary widely, but include: concentrates, non-forage fibre sources and total mixed ration (TMR). There is considerable flexibility in formulating rations around pasture-based diets, but forage testing of pastures is required to efficiently meet the nutrient requirements of the grazing dairy cow.

The amount of total dry matter intake obtained from pasture is quite variable (Soder and Muller, 2007). Throughout the year, supplementation is often required, such as during winter or periods of drought when pasture productivity or quality is low. Supplement is typically fed twice a day (either before or after milking), but can occur on pasture as well.

Milk yield is often (although not always) lower in grazing herds than in confined herds. This decrease (if any) can be attributed to many variables, including: (i) variable pasture quality and quantity; (ii) less control over total dietary intake; (iii) use of animal genetics with lower milk production potential or with poor grazing adaptation; (iv) increased energy expenditure by the cow for grazing and walking; and (v) management ability of the producer.

Pasture quality and pasture management varies widely across farms. Many farms utilize a forage mixture that contains grasses and legumes, and some utilize warm-season annuals (such as Sorghum-Sudan grass) to feed cows during the heat of summer when most cool-season grasses decrease in production. Fertilization practices also vary widely; some farms use commercial fertilizers while others use only manure from the barn or dung deposited on the pasture by the cows.

Total Confinement Dairy Systems

Modern confinement operations represent the evolution of dairies that traditionally housed and milked cows in stanchion barns. Many of these confinement operations have expanded the herd from 1000 to 10,000+ cows to take advantage of economies of scale. The advent of the TMR in confined herds improved the efficiency of feeding dairy cows (Coppock *et al.*, 1981; Muller, 1992). Prior to feeding TMRs, 'component feeding' was normal where concentrates and forages were fed separately. While the component feeding system worked fairly well, feeding large amounts of grain at once (commonly called 'slug feeding') sometimes resulted in digestive upsets and metabolic problems, as cows were able to sort feeds (Muller, 1990). With TMR, feed ingredients are thoroughly blended, providing the cow with a nutritionally balanced diet in every bite (Muller, 1990, 1992).

Confinement of cattle to barns and lots creates challenges with regard to managing manure and manure nutrients. Modern confinement operations have large manure storage facilities and require large areas of land for manure application to maintain compliance with nutrient management regulations (Fulhage, 1997). Increasingly, new technologies are emerging to take advantage of energy (e.g. methane digesters), nutrients (composting) and other resources (recycling of bedding) in manure (Day and Funk, 1998). Adoption of these technologies has been primarily by larger operations with sufficient waste volumes to justify the significant investment in capital of newer technologies (Lazarus and Rudstrom, 2007).

Hybrid Dairy Systems

Many dairy farming operations combine grazing, like that mentioned in the previous section, with confinement of cattle. These hybrids can take different forms, most commonly: (i) pasturing in the growing season and confinement in the winter months; or (ii) supplementing high pasture forage intake with stored feeds (such as grain, hay or silage). In the latter case, there can be considerable range in supplementation depending upon the land base and supply of pasture forages. Not surprisingly, grazing management can vary widely across hybrid operations. Some herds are kept on pasture continuously except when they are milked (20+ h/day), while others are provided pasture for short periods each day. For instance, grazing herds may only be provided access to pasture at night during hot summer months, spending their days in well-ventilated barns with stored forages or a TMR.

Graziers sometimes supplement pasture intake with a partial TMR (pTMR; White, 2000; Soriano *et al.*, 2001; Bargo *et al.*, 2002). The advantage of a pTMR is that producers utilize pasture resources while maintaining a dietary control by feeding a percentage of the forage and all of the concentrate, vitamins and minerals in a mixed diet in the barn.

Utilizing a pTMR can increase farm profitability over using pasture plus concentrate (Soder and Rotz, 2003). One challenge with a pTMR is maintaining a balanced diet with constantly changing pasture quality and quantity throughout the grazing season; graziers must be flexible in providing the right amount and type of pTMR to balance the nutritional qualities/deficiencies of the variable pasture. Another reason for utilizing a pTMR is due to limited land/pasture base. If sufficient pasture cannot be supplied to the herd within a reasonable walking distance of the milking parlour, a pTMR can be used to 'substitute' stored forages thereby increasing the carrying capacity of the limited pasture land.

As with herd management, manure management will vary greatly depending on individual farm management. If cows are housed in the barn for part of the day (i.e. during the heat of the day), manure hauling costs will be greater than for a herd that is pastured for the entire day and night. The trade-off may be in milk yield. Cows housed in a well-designed barn during the heat of the day may not be as heat-stressed as those outdoors, and may produce more milk, or may re-breed more quickly. The economic trade-offs of these management decisions must be weighed to determine the best management practice for any individual farm.

Economic Factors Affecting the Selection of Alternative Dairy Strategies

Numerous studies compare the economics of grazing-based dairies with hybrid and total confinement dairy systems (Parker *et al.*, 1992; Hanson *et al.*, 1998; Dartt *et al.*, 1999; Tucker *et al.*, 2001; Soder and Rotz, 2003; Tozer *et al.*, 2003). Several studies showed that moderately sized dairies (80–100 cows) can remain competitive by decreasing expenses, particularly expenses associated with feed, crop labour and machinery (Emmick and Toomer, 1991; Smith, 1994). Dartt *et al.* (1999) reported that management-intensive grazing operations had greater economic profit than confinement dairies, primarily by being more efficient in asset use, operating practices, and labour use, and suggested that management-intensive systems could provide a sustainable alternative management tool for the dairy industry. Elbehri and Ford (1995) reported that increased profitability with management-intensive grazing was primarily due to decreasing the cost of production of milk (usually between US$0.54/kg and US$0.64/kg for milk produced in confinement). Therefore, in spite of the potential for decreased milk production per cow (typically 3–5%), with management-intensive grazing (Cunningham, 1993; Parker *et al.*, 1993), profitability could still be greater. If milk production decreases >5%, then non-grazing forage systems often become preferred by some producers (Elbehri and Ford, 1995).

In some areas, the ratio of milk price to concentrate price approaches 1.0 or less, meaning that 1 kg of concentrate costs about the same (or more) as 1 kg of milk (Parker *et al.*, 1992). At lower quantities of grain feeding, US research has shown that 1 kg of concentrate results in approximately 1 additional kg of milk at low levels of supplementation, with the law of diminishing returns taking effect at greater levels of supplementation (Kolver *et al.*, 1998; Bargo *et al.*, 2003). At this ratio, concentrate feeding is not profitable because input cost equals output value, ignoring other ramifications such as improved body condition. In the USA, the milk/concentrate feed price ratio has historically been close to, or greater than, 2:1 (meaning that the milk price is usually double the concentrate price on a per kilogram basis). Therefore, it makes economic sense to feed up to 7–9 kg of concentrate daily to pastured high-genetic-merit Holstein cows to improve milk production and profitability while maintaining rumen function. Using a whole farm simulation model, Soder and Rotz (2001) showed that increasing concentrate supplementation (up to 9 kg/head/day) resulted in greater profitability of a grazing dairy farm, and that at the higher levels of supplementation (9 kg/head/day), profitability was greater than that of a confinement farm.

The objective of most dairy producers is to produce sufficient milk to generate an income that can sustain a desired lifestyle. Therefore, the feeding system utilized in any dairy business should be the one that best achieves the business' goals while ensuring long-term economic and environmental sustainability. Management-intensive grazing has proven to be a sustainable alternative management tool for small- to mid-size dairies to remain competitive in times of tighter profit margins.

Nitrogen and Phosphorus Cycling at a Farm Scale

The cycling of nutrients on dairy farms, particularly those with large confinement components, is largely a misnomer, as modern systems entail a myriad of off-farm nutrient transactions in which there is limited cycling between the original nutrient source and the farm itself. Chapter 4 (this volume) gives an outline of methods to account for this life-cycle assessment. Even so, grazing of dairy cattle can play a key role in modifying disconnected flows in confinement systems and can contribute to an improved balance of nutrient imports and exports at the farm gate.

Nitrogen

Nitrogen (N) is a dynamic element that transforms readily between gas, solution and solid phases. N, in the form of crude protein, is a critical component of feed for dairy cows, as well as a critical fertilizer for pasture and cultivated forages. The introduction of grazing to a dairy operation has the potential to shift sources of dietary crude protein to the pasture, translocate N excretion on the farm and transform sources of N applied to field crops. Leaching and volatilization of N represent the major environmental loss pathways of N on grazing and non-grazing dairies alike, with surface runoff a lesser pathway.

In confinement operations, contained N 'cycling' occurs primarily between barn and field, with environmental losses distributed across the farm (Fig. 10.1a). Depending upon the nature of the operation, crude protein may be derived from purchased sources, making feed the primary input of N at the farm gate. Alternatively, it can be derived from forages produced on farm, with fertilizer sources substituting, in part, for purchased feed sources of N. In systems with a lot of confinement cattle, excretion occurs primarily in and around the barn. Hence, volatilization of NH_3 from impervious surfaces is a key pathway of environmental losses from confinement operation. Additional losses of NH_3 derive from the

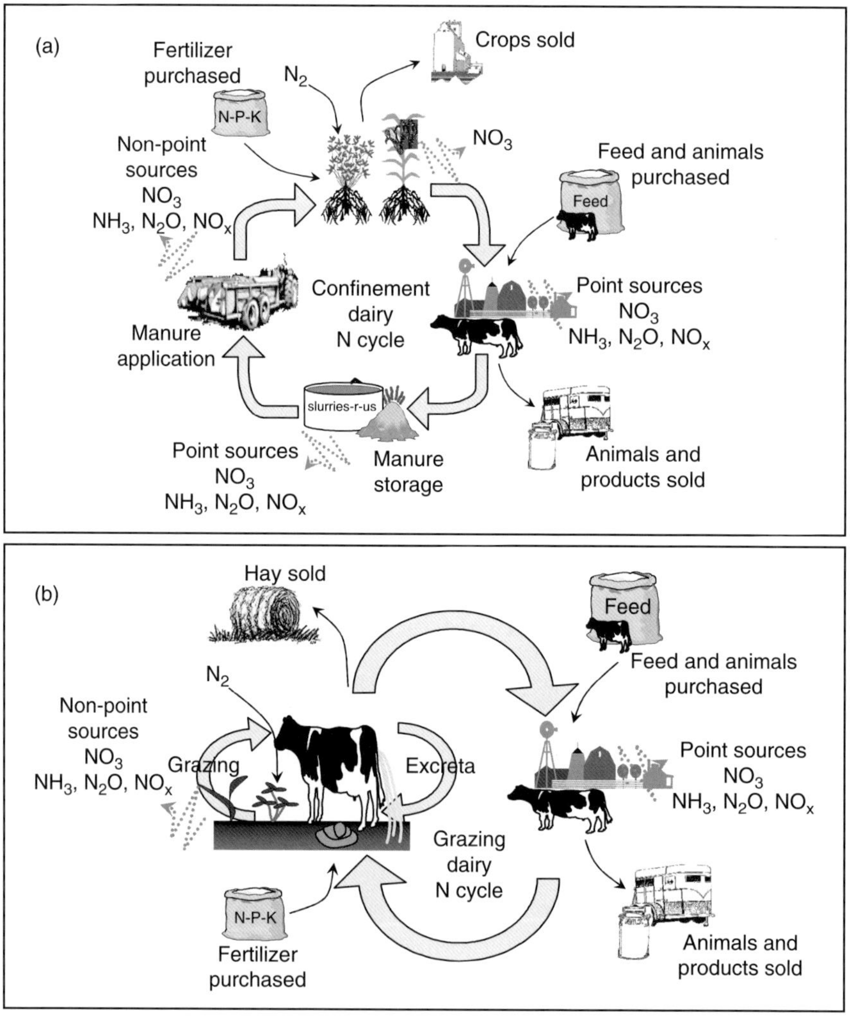

Fig. 10.1. The nitrogen cycle on confinement (a) and grazing (b) dairy farms.

processes of handling, storing and applying manure. Once applied to field soils, both manure and fertilizer sources of N contribute to losses of NO_3 and NH_3, as well as to emissions of N_2 and N_2O.

In grazing operations, pastures can substitute for both purchased and cultivated sources of crude protein, helping to contain N cycling within paddock areas (Fig. 10.1b). In pasture swards with substantial legume components, biological fixation of atmospheric N_2 is a major input of N to pasture soils, which also receive inputs of N from fertilizers and excreta of grazing cattle. Grazing can have a profound effect on NH_3 volatilization potential on a dairy farm by shifting excreta-N from the impervious surfaces of barns and manure storages to permeable pasture soils. Although losses of NO_3-N via denitrification and leaching remain a constant concern of pastures and cultivated fields alike, these losses are of particular concern to grazing operations due to the patchiness of urine deposition. High concentrations of NO_3 have been shown to leach below urine patches exceeding the capacity of pasture species to recover the N (Di and Cameron, 2000; Stout *et al.*, 2000; Oenema *et al.*, 2001). Such patchiness in the deposition of excreta on pastures lowers nutrient-use efficiency when supplemental fertilization is needed to ensure uniform sward production.

Given the dynamic nature of N, quantifying a partial N balance at the farm scale is complicated, requiring a mix of measured and modelled findings. In addition, accounting for internal sources and transfers of N in grazing systems can be quite difficult. Thus, the literature on dairy farm N balances related to grazing is varied, with inconsistencies in the accounting of biologically fixed N. Some authors do not account for biologically fixed N as a farm N input, skewing balances in favour of those operations with extensive legume forage production. When N fixation in pastures is factored into the N account of a grazing operation, as in Table 10.1, consistent differences between grazing and confined dairy operations are difficult to detect. Indeed, confinement operations reported in the literature often possess the smallest on-farm N surpluses, as other management variables such as stocking density may be more important to balancing N at the farm gate (Watson *et al.*, 2002).

Despite difficulty in generalizing farm gate N balances, prudent use of pasture N resources can help to decrease purchased sources of N, and in some cases, decrease on-farm N surpluses. The potential for such improvements, as well as the case-dependent nature of N management, is documented by Rotz *et al.* (2002) who used the Dairy Forage Simulation Model (DAFOSYM) to model nutrient flows on 100- and 800-cow dairy farms representative of various conditions found in the north-eastern USA. For the 100-cow farm, N imports at the farm gate decreased from 193 kg/ha with low-intensity grazing to 175 kg/ha with management-intensive grazing. Although N fertilizer was used to improve pasture productivity with management-intensive grazing, the improved quality of pasture forages and utilization of these forages by grazing cows enabled less crude protein to be imported on to the management-intensive grazing operation. At the same time, N exports from the two types of grazing operations remained nearly constant (49–50 kg/ha), equivalent to 26% and 28% of N imports for the low-intensity and management-intensive grazing herds, respectively.

For the 800-cow farm simulated by Rotz *et al.* (2002), increased pasture productivity through increased N fertilization resulted in a decline in N imports from

Table 10.1. Summary of literature reporting N balances on dairy functions in relationship to role of grazing.

Location	Farm characteristics	Stocking density	N imports			N exports	N surplus[a]	Environmental losses		Source
			Total	Purchased	Bio. fixation			NH_3-N	NO_3-N	
		(animal units/ha)		(kg/ha)		(kg/ha)	(kg/ha)	(kg/ha)		
USA	105 cow hybrid									Rotz *et al.* (2002)
	Low-intensity grazing	1.3	193	NA	NA	50	143	71	45	
	Management-intensive grazing	1.3	175	NA	NA	50	125	76	35	
	800 cow hybrid									
	Low-intensity grazing	0.8	119	NA	NA	24	95	32	31	
	Management-intensive grazing	0.8	86	NA	NA	24	62	33	27	
	70 cow confinement	1.7	NA	NA	NA	NA	NA	57	24	Soder and Rotz (2001)
	125 cow grazing (no feed supplements)	3.0	NA	NA	NA	NA	NA	117	21	
	103 cow grazing (low feed supplements)	2.5	NA	NA	NA	NA	NA	99	20	
	90 cow grazing (medium feed supplements)	2.2	NA	NA	NA	NA	NA	84	19	
	81 cow grazing (high feed supplements)	2.0	NA	NA	NA	NA	NA	73	18	
	80 cow confinement	0.9	37	37	87	36	1	NA	NA	Saporito and Lanyon (1998)

	65 cow confinement	2.8	143	143	79	68	75	NA	NA	Bacon *et al.* (1990)
	52 cow hybrid									Ghebremichael *et al.* (2007)
	No precision P feeding	1.1	136	NA	NA	27	109	NA	25	
	Precision P feeding	1.1	136	NA	NA	27	109	NA	25	
	Precision P feeding, improved forage use	1.1	183	NA	NA	27	156	NA	32	
	102 cow hybrid									
	No precision P feeding	1.8	155	NA	NA	59	96	NA	29	
	Precision P feeding	1.8	155	NA	NA	59	96	NA	29	
	Precision P feeding, improved forage use	1.8	191	NA	NA	60	131	NA	34	
Sweden	55 cow confinement	2.0	270	218	52	106	164	80	9	Granstedt (1997)
Austria	Confinement	NA	329	179	150	17	312	NA	NA	Weiser *et al.* (1996)[b]
Canada	Confinement	NA	160	134	26	86	75	NA	NA	Goss and Goorahoo (1995)
	Hybrid (low-intensity grazing)	NA	42	2	41	42	1	NA	NA	
Norway	Unknown	NA	97	45	52	35	62	NA	NA	Ebbesvick (1998)[b]
		NA	107	61	46	34	73	NA	NA	

NA = not available.
[a]N surplus includes purchased and biologically fixed N.
[b]As summarized in Watson *et al.*, 2002.

119 to 86 kg/ha. Again, N exports remained constant across the two scenarios (24 kg/ha), equivalent to 20% and 28% of imported N from the low- and high-intensity grazing dairies, respectively. Changes in the N balance at the farm gate of the 800-cow operation can be tied to increased pasture productivity, which required less land for grazing, as well as decreased demand for imported crude protein due to improvements in pasture forage quality.

Annual surpluses of N on dairy farms, both grazing and confinement, are largely offset by the magnitude of environmental losses. As summarized in Table 10.1, estimated losses of N from dairy farms by leaching and volatilization range from 21% to 97% of annual on-farm N surpluses. These estimates are modelled, as opposed to measured, and therefore subject to the model's limitations. Despite this, substantial environmental losses of N occur from grazing operations as well as from confinement operations, indicating that inclusion of grazing on a dairy farm does not automatically decrease water and air quality impacts.

Phosphorus

Phosphorus (P) cycling and environmental fate on dairy farms differs from that of N, largely due to the absence of a gaseous phase for P, the relative insolubility of P in water and the tendency for native soil P sources to be in limited supply in many physiographic regions. As with N, the introduction of grazing to a dairy farm serves to shift major components of the P cycle from the barn to the landscape, with diffuse losses of P from pastures (Fig. 10.2b) growing in importance relative to point sources of P and land-applied manures (Fig. 10.2a). Unlike N, surface runoff and erosion are usually dominant pathways of environmental P losses, although significant losses can also occur via sub-surface routes like artificial drainage. Strategies to minimize such losses can target critical source areas where high concentrations of P and high potential for transport coincide. Shifting from confinement to grazing operations does not eliminate the major pathways of P loss. As most P in surface runoff derives from a shallow zone at the soil surface (Sharpley, 1985), dung deposition on pastures can be a significant source of P in runoff water (McDowell, 2006, 2007). Even so, improved utilization of pasture resources has been shown to offer a key means of balancing P at the farm gate (Rotz *et al.*, 2002).

In comparison with other modern livestock production systems (e.g. swine, poultry), dairy farming systems tend to have large land bases, therefore a better ability to assimilate imported P. Furthermore, microbial symbioses in the rumen serve to improve the efficiency of metabolizing forage P compared to N. Historically, P has been fed to dairy cattle in excess of recommended levels. In a survey of 612 dairy farms in the north-eastern USA, of which roughly half reported TMR feeding, Dou *et al.* (2003) observed that dietary P concentrations for lactating cows ranged from 3.6 to 7.0 g/kg of feed dry matter. The mean concentration (4.4 g/kg) was roughly one-third greater than that recommended by the National Research Council (2001) for dairy cows with the same average milk yield as those in the survey (27.9 kg/day). In that survey, dietary P concentrations were not associated with milk yields ($r^2 = 0.06$), and excess dietary P was primarily a result of

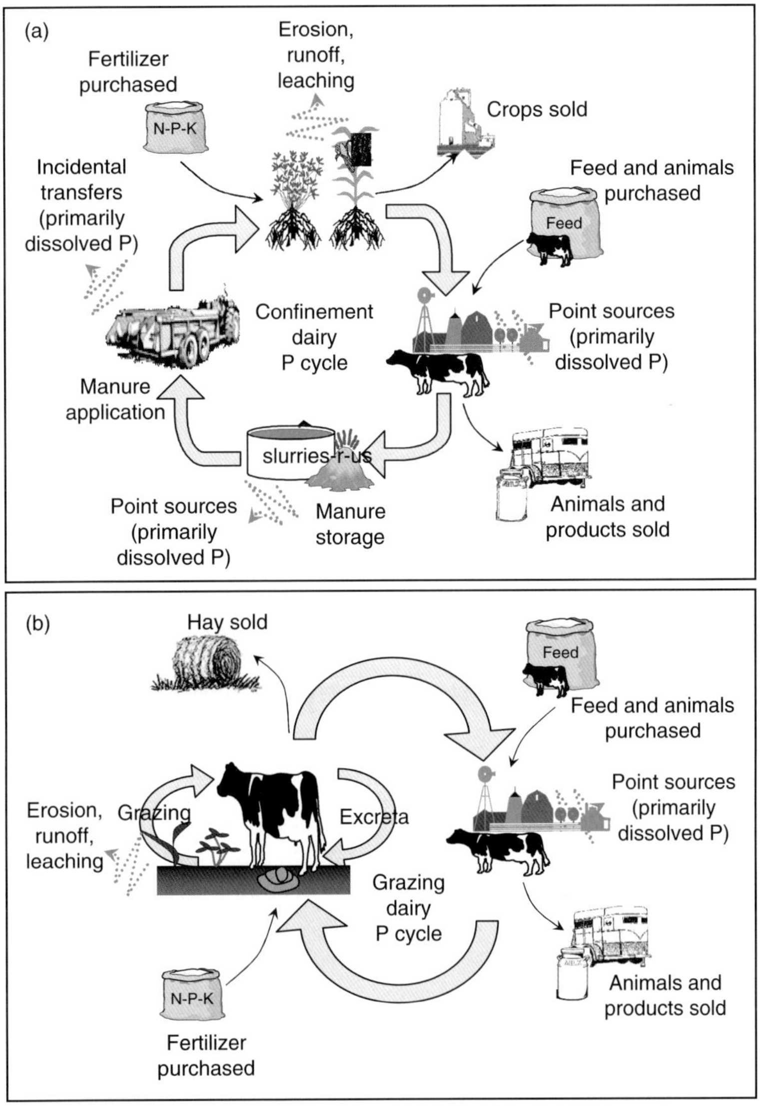

Fig. 10.2. The phosphorus cycle on confinement (a) and grazing (b) dairy farms.

recommendations by consultants. The most frequent reason for overfeeding P to dairy cattle is the perception that high-P rations improved reproductive efficiency (Tallam *et al.*, 2005). While it has been shown that severe P deficiency can impair reproduction (Goff, 1998), there is no evidence that feeding P in excess of recommended levels benefits reproduction (Wu and Satter, 2000; Wu *et al.*, 2000).

A broad body of research has documented P balances between dairy farming systems. Despite historical tendencies of farmers to overfeed P to dairy cattle, dairy farms tend to have much lesser surpluses of P on an annual basis than of N. As summarized in Table 10.2, reported P balances on dairy farms (measured and modelled) range considerably (from −7 to 54 kg/ha). Many variables contribute to these balances, including stocking density, feeding regime and grazing management. Because environmental P losses tend to be low in comparison to other pools of P on the farm, P accumulation on dairy farms represents either a long-term concern or an immediate concern of fields and pastures where manure is applied.

Significant opportunities exist both for confinement and grazing operations to decrease import of feed P and better use P that is on the farm. Ghebremichael *et al.* (2007) modelled 52- and 102-cow hybrid grazing operations, comparing baseline feeding regimes with regimes designed to decrease P import, finding that on-farm P surpluses could be decreased by 43–60%. When precision feeding was combined with substitution of on-farm forages for concentrates (improved forage management), P surpluses could be decreased by >90%. In that study, the additional forages were not derived from pastures, but pasture forages could likely have been substituted for supplements as well. Indeed, simulations by Rotz *et al.* (2002), described in the context of N cycling above, indicate that P surpluses can be lowered from 10% to 44% by intensifying pasture management (including increasing pasture productivity) on hybrid dairy farms while also increasing farm profitability. Overall, however, grazing offers but one set of controls for P management at the farm gate; other variables, some of them site-specific, contribute to on-farm P import.

Managing Nutrient Losses from Hybrid Dairies

Increased reliance upon grazing on dairy farms must be accompanied with improved management not only to optimize milk production, but also to prevent significant environmental losses of nutrients. Sound management of hybrid operations requires considerable attention to dairy cattle, confinement infrastructure and pastures. Proper accounting of nutrients derived from forages and compliance with established dietary recommendations can decrease farm nutrient imports and even increase profitability by lowering the cost of purchased feeds.

Diet Regulation

In addition to playing a role in surpluses of N and P found on a dairy farm, regulating the diet of a cow influences the potential for environmental emissions of nutrients. Regulating the amount of protein fed to dairy cattle can significantly affect urine N content, hence the potential for NH_3 volatilization. Broderick (2003) found that increasing crude protein in the diet of lactating dairy cows from 15% to 18% increased urinary N from 23% to 35% of dietary N. For hybrid operations, ration formulation to minimize excess protein generally lowers feed costs. High-quality pasture contains greater concentrations of N than is needed by the ruminant, and despite efficient synthesis of microbial protein from diets containing high-quality pasture, up to 30% of the ingested N is not metabolized (Beever *et al.*, 1986; Kolver *et al.*, 1998).

Table 10.2. Summary of literature reporting P balances on dairy functions in relationship to role of grazing.

Farm							
Location	characteristics	Stocking density	P imports	P exports	P surplus	Environ-mental losses	Source
		(animal units/ha)	(kg/ha)	(kg/ha)	(kg/ha)	(kg/ha)	
USA	105 cow hybrid						Rotz *et al.* (2002)
	Low-intensity grazing	1.3	14.5	8.4	6.1	1.0	
	Management-intensive grazing	1.3	11.7	8.3	3.4	0.0	
	800 cow hybrid						
	Low-intensity grazing	0.8	7.9	4.3	3	0.6	
	Management-intensive grazing	0.8	7.6	4.3	2.7	0.6	
	70 cow confinement	1.7	NA	NA	1	NA	Soder and Rotz (2003)
	125 cow grazing (no feed supplements)	3.0	NA	NA	3	NA	
	103 cow grazing (low feed supplements)	2.5	NA	NA	–2	NA	
	90 cow grazing (moderate feed supplements)	2.2	NA	NA	–6	NA	
	81 cow grazing high feed supplements	2.0	NA	NA	–7	NA	
	Confinement	NA	NA	NA	3.0	NA	Soder and Rotz (2001)
	Grazing	NA	NA	NA	1.0	NA	
	80 cow confinement	0.9	14	7	7	NA	Saporito and Lanyon (2003)
	65 cow confinement	2.8	30	13	17	NA	Bacon *et al.* (1990)
	52 cow hybrid						Ghebremichael *et al.* (2007)
	No precision P feeding	1.1	9.9	4.6	5.3	1.9	
	Precision P feeding	1.1	7.6	4.6	3	1.8	

continued

Table 10.2. Continued

Farm: Location	Farm: characteristics	Stocking density	P imports	P exports	P surplus	Environ-mental losses	Source
	Precision P feeding, improved forage use	1.1	5.2	4.7	0.5	1.7	
	102 cow hybrid						
	Precision P feeding	1.8	13.8	10	3.8	2.2	
	Precision P feeding, improved forage use	1.8	10	10	0	2.1	
UK	129 cow hybrid	3.2	43.6	17.2	26.4	1.0	Haygarth *et al.* (1998)
	36 cow hybrid						Withers *et al.* (1999)
	High feed supplements	1.9	90.4	36.7	53.7	0.2	
	Moderate feed supplements	1.9	76.6	32.7	43.9	0.4	
	Low feed supplements	1.6	51.3	30.1	23.4	0.3	
Sweden	55 cow confinement	2.0	24	22	2	0	Granstedt (1997)
Austria	Confinement	NA	4.5	3.75	0.75	NA	Weiser *et al.* (1996)[a]
Norway	Hybrid	NA	9	7	2	NA	Ebbesvick (1998)[a]
		NA	19	7	12	NA	

NA = not available.
[a]As summarized in Watson *et al.*, 2002.

This loss is due to a high concentration of NH_3 in the rumen that results from rapid and extensive rumen degradation of pasture N that, if not recaptured into microbial protein, is converted to urea and excreted in milk and urine (Van Vuuren *et al.*, 1991). Adding an energy source to the rumen (such as a concentrate) that is synchronized with intake of N-rich pasture has been shown to increase yields of microbial protein by greater recapture of excess N in the rumen and decreased excretion (Herrera-Saldana *et al.*, 1990; Aldrich *et al.*, 1993; Kolver *et al.*, 1998); however, the effects on milk production have been mixed (Aharoni *et al.*, 1993; Aldrich *et al.*, 1993; Kolver *et al.*, 1998).

Supplemental P fed in excess of that required by the cow is not absorbed in the cow's gastrointestinal tract and is excreted in the faeces (Dou *et al.*, 2002). Knowlton and Herbein (2002) reported that decreasing dietary P from 0.52% to 0.34% (dry matter basis) resulted in 200% less manure P excretion in early lactation cows. Under controlled experiments, strong relationships have been reported between dietary and faecal P concentrations. In a 308-day lactation trial with 26 multiparous Holstein cows, Wu *et al.* (2000) reported a strong linear relationship between faecal P excretion and P intake (faecal P = 0.643 dietary P − 6.2; units = g/day; $r^2 = 0.81$). However, in the survey of Dou *et al.* (2003), where factors such as breed, life stage and diet were not controlled, faecal P concentrations were only weakly related to P concentrations in the diet (faecal P = 1.89 dietary P + 1.03; units = g/kg; $r^2 = 0.43$). As such, a variety of factors undoubtedly contribute to faecal P concentrations on working farms, with dietary P an important, but by no means unique, factor.

Grazing Component of Hybrid Systems

Poorly managed pastures can produce substantial nutrient emissions to soil, air and water. Even under prudent management, environmental losses from grazing dairy cattle occur as excreta containing high concentrations of N and P in mobile forms. Uneven distribution of excreta on pasture soils produces patchiness in nutrient pools that hampers uniform sward management. Thus, Ball *et al.* (1979) concluded that even an intensively grazed pasture system is 'far from a closed system', estimating that more than half of urine N from grazing animals is readily lost to volatilization and leaching. Note that an expanded review of mitigation strategies in grazed dairies is available in Chapter 8 (this volume).

Management-intensive grazing

Management of intensive grazing, or intensive rotational grazing, can lessen the N content of consumed pasture forage, as brief visits at high stocking density decrease the potential for cattle to preferentially select forage. However, during these visits, losses of N and P can be significant (Sharpley and Syers, 1976, 1979; Bussink and Oenema, 1998). Losses of nutrients from intensively grazed pastures can be difficult to check. Rotz (2004) identified a range of practices to decrease N losses related to grazing animals, including: (i) optimize stocking rate; (ii) move watering, feeding and shade devices to improve excreta-N distribution; (iii) avoiding grazing during periods when plant uptake is low; and (iv) synchronizing carbohydrate source with high-N pasture in the rumen to recapture greater amounts of NH_3. Most of these recommended practices would also serve to limit P losses.

The patchy distribution of urine is a key cause of N leaching under pastures. Reported losses of N under urine patches range from 8% to 30% of urine N (Whitehead and Bristow, 1990; Fraser *et al.*, 1994; Stout, 2003). Indeed, Silva *et al.* (1999) observed the annual equivalent of 124 kg N/ha could be leached under a urine patch, corresponding to 33 kg N/ha across an entire paddock when areas not receiving urine were considered. Management factors such as breed size (hence, urine volume) and timing of urine deposition all affect the potential for urine N to leach through pasture soils (Stout, 2003).

Urine patches are also the site of considerable NH_3 volatilization. According to Bussink and Oenema (1998), summarizing literature results, NH_3 losses from grazing cattle range from 0% to 28% of N in urine and dung deposited on pastures, depending upon N application rate, stocking density and duration, climatic conditions and soil properties. They surmised that intensive rotational grazing may decrease NH_3 volatilization potential from pastured cattle relative to less-intensive systems of grazing.

P losses from intensively grazed pastures occur primarily in dissolved form, arising from dissolution of soil, dung and fertilizer P at or above the soil surface, and washing-off of P from recently grazed pasture plants. In degraded systems, erosion may contribute to particulate P losses. All except fertilizer additions are influenced by the action of grazing; whether it is the ripping of pasture plants or the influence of treading on soil erosion and surface runoff potential. McDowell *et al.* (2007) found that in a grazed pasture in New Zealand, approximately 10%, 30%, 20% and 40% of P losses in surface runoff during a year were attributable to fertilizer, dung, pasture plants and soil components, respectively. Hively *et al.* (2005) conducted rainfall simulations at various locations on a dairy farm including several pastures and a cow path. They found that surface runoff from the cow path transported large P loads due to low permeability of the path and dung accumulation. At the same time loads from pastures under intensive rotational grazing were almost negligible. Areas of low permeability due hoof action and high dung concentrations, such as feeding areas and shade, can be widespread in pastures with little to no rotation of cattle (Peterson and Gerrish, 1996). White *et al.* (2001) recorded high densities of excreta around water troughs of intensively grazed pastures.

Riparian exclusions

Allowing cattle access to streams has historically been one means of providing pastured cattle with drinking water and comfort during hot weather, but is now recognized as a poor management practice from the standpoint of nutrient loss as well as cattle health. Practices such as fencing out riparian areas, providing alternative sources of water and shade and selection of feeding sites can have a profound effect on the environmental fate of nutrients from the excreta of pastured cattle. James *et al.* (2007) estimated that 2800 kg of P was defaecated directly into pasture streams by dairy cattle every year in a 1200 km^2 watershed in the north-eastern USA with predominantly hybrid farms (low-intensity grazing). Across the watershed, direct deposition of dung P into streams was equivalent to roughly 10% of the annual P loadings attributed to all agricultural sources. James *et al.* (2007) estimated that installation of stream bank fencing as part of ongoing conservation activities had already decreased in-stream dung P deposition by 32%. In another case study, Meals (2000) observed a 15% decline in P concentrations when dairy cattle were fenced out of streams in another north-eastern US watershed dominated by dairy farms.

Pasture fertility

Sound management of pasture fertility requires regular testing of soil and forages, and prescriptive application of fertilizer nutrients at rates catering to pasture production with practices that maximize the availability of applied nutrients to growing plants while minimizing losses by erosion, surface runoff and leaching (Sharpley and Halvorson, 1994; Rotz, 2004). In addition, timing of fertilizer application to pasture soils should be adjusted to meet the demands of pasture forages, with summertime applications potentially preferable to spring. The purchase of fertilizers can contribute to on-farm nutrient surpluses, especially when nutrient resources from dung and manure are not fully exploited. Accounting for the contributions of leguminous pasture species can substantially decrease the need for N fertilizer.

Because excess fertilizer N can affect forage quality in pastures, frugal application of fertilizers to pastures can indirectly limit NH_3 volatilization from the urine of grazing cattle (Bussink and Oenema, 1998). Bussink (1992) found that increasing fertilizer N application to grazed swards was associated with a concomitant increase in NH_3 volatilization from the swards. The use of nitrification inhibitors such as dicyandiamide has shown considerable potential for decreasing N losses from grazed pastures. Di and Cameron (2002) observed a roughly 60% decrease in NO_3 leaching losses and 75–85% decrease in NOx emissions from simulated urine patches treated with a nitrification inhibitor. The authors later concluded that for New Zealand's climate conditions, nitrification inhibitors are best applied from late autumn to early spring (Di and Cameron, 2004).

Confinement Component of Hybrid Systems

Confinement of dairy cattle and concentration of their excreta by waste-handling systems introduce additional management responsibilities to minimize environmental losses of N and P from dairy farms. Point sources on dairy farms are typically found in association with barns, barnyards, silos and manure storage infrastructure. In most dairy systems, small volumes of dry manure accumulate in barns, and grey water from milking parlours can be highly enriched with dissolved nutrients (e.g. Kim *et al.*, 2006). Management practices that target these areas can be among the most cost-effective pollution-control techniques on a dairy farm (Meals, 1993). More difficult, and time consuming, is management of losses associated with manure. N/P ratios in manures do not match those required of most pasture or row crops, so that use of manures to meet crop N requirements typically results in over application of P (Smith *et al.*, 1998). Although most dairy systems have large land bases enabling adequate dilution of nutrients in applied manures, the accumulation of nutrients (especially P) in the soils of certain fields is well documented, particularly those in close proximity to barns (e.g. Aarons *et al.*, 2004). The long-term accumulation of manure nutrients and short-term availability of those nutrients to surface runoff, leaching and atmospheric pathways has been a major target of improved nutrient management strategies (Bouldin and Klausner, 1998).

Point sources

Barns and barnyards

The combination of impervious surfaces and high concentrations of manure around barns and barnyards results in large potential for nutrient runoff and volatilization. Because urea hydrolysis begins as soon as urine is excreted, the potential for NH_3 emissions from barn and barnyard areas where cattle concentrate is great. In general, closed barns emit less NH_3 than free stall barns with barnyards which, in turn, emit less than open-sided barns (Bussink and Oenema, 1998). Losses of NH_3 from cattle excreta are positively correlated to the feeding of crude protein (Krober *et al.*, 2000), and can range widely on a seasonal basis, particularly for free stall barns, with up to 50% of excreted N volatilized as NH_3 during summer months (Moreira and Satter, 2006).

A variety of management practices affect NH_3 emissions from dairy barns. Use of bedding materials that absorb urine can potentially decrease NH_3 volatilization, although reported results are decidedly mixed (Bussink and Oenema, 1998). Jeppsson (1999) found that using a combination of peat and chopped straw bedding decreased NH_3 emissions by 57% relative to long straw, which was comparable to solid flooring in NH_3 volatilization. As summarized by Rotz (2004), NH_3 losses from solid floors are affected by floor shape and surface characteristics, and although NH_3 losses from solid floors are relatively unaffected by scraping, they can be decreased by spraying the floor with water following scraping.

Strategies to minimize surface runoff losses of nutrients from barn areas generally involve routing clean surface runoff water around facilities, regular cleaning of areas that can serve as possible sources of nutrient emissions and treating surface runoff that exits barnyards and feedlots to decrease its pollution potential (Wright, 1996). Brown *et al.* (1989) estimated that P loss from barnyards found on hybrid farms typical of the north-eastern USA could be lowered by 50–90% using practices that decreased the volume of barnyard surface runoff.

Practices to treat barnyard surface runoff and milk house effluent include filter strips, constructed wetlands and sediment basins. Filter strips serve to promote infiltration, and sedimentation of influent. Schwer and Clausen (1989) reported >90% decrease in total P and total Kjeldahl N of milk house waste water treated with a vegetative filter strip. However, excessive loading of nutrients and channellized flow can readily undermine the effectiveness of a filter strip. Thus, Schellinger and Clausen (1992) found that only <20% of total P and total Kjeldahl N entering a filter strip were retained by the strip. Wetlands and detention basins arrest barn area surface runoff, promote sediment deposition and promote the uptake of nutrients by plants, assuming adequate retention time. In addition, wetlands can be very effective in promoting denitrification, hence removal of NO_3^--N from influent (Xue *et al.*, 1999; Hunt *et al.*, 2003). Regular removal of sediments and plant biomass are necessary to sustain N and P treatment efficiencies of constructed wetlands (Uusi-Kamppa *et al.*, 2000; Braskerud, 2002). Newman (1997) observed a 43% decrease in NO_3^--N and 27% decrease in total Kjeldahl N by a constructed wetland treating milk house effluent, while Shamir *et al.* (2001) reported only a 17% decrease in total N by a constructed wetland treating dairy waste water. Smith *et al.* (2006) found >90% removal of total P by two newly constructed wetlands treating dairy effluent, while Serodes and Normand (1999) recorded a 63% decrease in total P by a newly constructed wetland treating similar dairy effluent.

Manure handling and storage systems

Properly designed and maintained manure storage systems should not serve as a direct source of nutrients to water. Even so, spills and discharges do occur in and around storage structures, particularly those that are uncovered. Manure handling and storage systems are considered major potential sources of NH_3 emissions (Bussink and Oenema, 1998). The loss of NH_3 during manure storage largely depends upon the storage structure, with the lowest losses associated with underground pits and bedded packs and the highest losses associated with anaerobic lagoons (Powell *et al.*, 2004). Petersen *et al.* (1998) monitored cattle manure stacked in a pile under summertime and autumnal conditions, finding an average of 3.2% of manure N was lost by leaching and 4.1% by NH_3 volatilization.

Permeable (e.g. straw, geotextile) and impermeable (e.g. plastic) covers exist for manure storage systems that can decrease NH_3 emissions by >90% (Bicudo *et al.*, 2004). Covered systems are mandated in some countries (e.g. Denmark, the Netherlands). Because dairy slurry readily forms a crust, simple maintenance of the crust by not agitating the surface can be a no-cost means of reducing NH_3 emissions. Losses of NH_3 from open storages can be curtailed by maintaining a small surface area of stored manure that is exposed to the atmosphere, promoting the formation of surface crusts that are relatively impermeable to NH_3, or acidifying the manure to inhibit the conversion of NH_4 to NH_3. Other key management factors affecting NH_3 loss potential from stored manure include residence time, temperature (highest losses in summer) and mixing (Bussink and Oenema, 1998).

Manure management

Manures are a heterogeneous mix of nutrients in various forms and concentrations, making the delivery of manure nutrients to field crops inefficient in comparison with commercial fertilizers. Manures are different to the effluent produced from all-grazing systems in that they often are more enriched with nutrients (especially P) and are drier. Tying manure application to soil with crop demands can be difficult, as nutrient ratios in manures generally do not match those required of crops and transformations of some forms of nutrients in manure are required before they are available for crop uptake. On hybrid operations with row crop fields and pastures, the application of manure to pastures may be dictated by factors other than the nutrition of pasture forages, such as the absence of other fields to which to spread (especially farms that must apply manure daily).

Manure application to pastures

Surface application of manure (broadcasting) results in the maximum potential for NH_3 volatilization and P loss in surface runoff. Surface applying manure to pastures can also temporarily decrease the palatability of forages and substantially increase environmental losses of nutrients. For instance, Bussink and Oenema (1998) summarized NH_3 losses from dairy farming systems, reporting a range of 1–100% of applied N for grasslands receiving surface application of manures and a range of 3–70% of applied N for row crop lands. Due to the high concentration of water-extractable P (WEP) in manure, dissolved P losses in surface runoff are of particular concern following surface application of manure (Dougherty *et al.*, 2004). In addition, low-density organic matter fractions in manure are highly susceptible to erosion when manure is broadcast (McDowell and Sharpley, 2002).

Alternatives to surface application methods exist, which incorporate slurries into pasture soils with minimal disturbance or concentrate surface-applied slurries within the thatch layer to promote contact with the soil. The direct ground injection (DGI) system, invented in Norway, uses high pressure to incorporate slurries containing <12% solids with minimal disturbance of the soil surface and little exposed slurry left on the soil surface after application (Morken and Sakshaug, 1997). Morken and Sakshaug (1998) reported 60% less NH_3 emissions from dairy slurry applied with the DGI system than by surface application. Elsewhere, Wulf *et al.* (2002) evaluated a variety of surface application systems that placed the dairy slurry within the thatch layer providing better contact with the soil surface (trailing hose and trailing shoe). They found NH_3 from the trailing hose and shoe systems to be 20–60% of those from broadcast manure. Still other systems pair manure application with soil aeration to increase storage of the slurry in the surface soil and infiltration. For instance, Bittman *et al.* (2003) observed a 48% decrease in NH_3 with one aeration infiltration system in comparison with broadcasting. Although the literature on manure application methods to pastures is dominated by NH_3 measurements, these systems can also be expected to impact P loss in surface runoff, as has been documented for various injection technologies in decreased tillage row crop systems (e.g. Daverede *et al.*, 2004). To date, no systems exist to incorporate dry manures (e.g. bedded pack) into pasture soils, but the use of aeration equipment has been shown to significantly decrease surface runoff under infiltration excess conditions; hence, this would also suggest decreased P losses following surface application of manure (Moore *et al.*, 2005).

In addition to the method of manure application to a pasture soil, application rate and application timing can fundamentally influence losses of N and P. The application rate of surface-applied manures has been shown to be well correlated with NH_3 volatilization as well as with dissolved P in surface runoff (Webb, 2001; Kleinman and Sharpley, 2003). Therefore, distributing manure across a dairy farm can substantially affect the potential for acute losses of nutrients from individual fields. Furthermore, the availability of nutrients in applied manure to atmospheric and surface runoff pathways tends to decline quickly with time: the maximum availability and potential for loss is soon after application. During periods when rainfall is frequent, environmental losses of recently applied manure nutrients (also termed, 'incidental transfers') can be substantial. Preedy *et al.* (2001) observed that 1.8 kg/ha of P losses in runoff occurred over the 7 days after dairy slurry was surface-applied. With time, slurry infiltration, non-surface runoff-producing rainfall washing manure nutrients into the soil, manure crusting and invertebrate action all serve to protect N and P from volatilization and surface runoff removal (Vadas *et al.*, 2007). Thus, Mueller *et al.* (1984) reported a 76% decline in dissolved P concentrations in surface runoff from soils broadcast with dairy manure during 2 months of a growing season. Declines in NH_3 emissions from land-applied manures can be even more rapid, with most NH_3 volatilization generally occurring within the first few days of application (e.g. Wulf *et al.*, 2002).

Manure export

Increasingly, confinement and hybrid dairy operations are looking to export their manure off-farm to handle excess nutrients. This is usually prodded by regulations, court orders and neighbours (e.g. Texas Water Resources Institute, 2006). From the standpoint of nutrient emissions, exporting manure decreases the quantity of manure that can potentially be released from storage facilities on dairy farms and offers an alternative to repeated, heavy application of manure nutrients to farm soils above levels necessary for crop production. Manure export can also alleviate pressures caused by limited manure storage capacity to apply dairy manure at times of high risk of incidental transfer to the environment (Preedy et al., 2001). However, depending where the manure is sent, it can merely shift the problems of nutrient losses to the recipient.

Due to the high cost of hauling dairy manure, which is generally in liquid or semi-solid form, de-watering is an important first step in processing dairy manure for export. A large variety of dewatering methods exist, most of which require substantial capital investment (Day and Funk, 1998). Composting, vermicomposting and pelletizing have all been used to add commercial value to dairy manure while decreasing the initial moisture content. Notably, these practices result in significant loss of manure N, primarily due to volatilization of NH_3, and therefore a lowered N/P ratio in the final product (Gassman and Bouzaher, 1995; Hamilton and Sims, 1995; Osei *et al.*, 2000). Industrial uses of exported dairy manure include bioenergy generation (methane and fuel pellets for power plants), turfgrass production, landscaping and aquaculture (Edwards, 1980; Day and Funk, 1998; Vietor *et al.*, 2002).

Conclusions

For many livestock production systems, there has been a long-term trend in geographic intensification of animal production. While dairy systems continue to be relatively extensive in their land base, the trend towards intensification of hybrid-dairy production systems has produced local accumulations of nutrients, often in areas that lack the land base to use the manure nutrients in crop production without negatively impacting water quality. The focus of nutrient management, formerly centred on field management, has shifted to broader scales that involve solutions that cannot be resolved by individual farmers. Increasingly, policies are emerging that recognize the need for action at levels greater than the farm scale.

Regulations in the Netherlands require that nutrient balances are used to characterize the risk of nutrient loss from a farm (Neeteson *et al.*, 2001). Using a mineral accounting system (MINAS), farms with more than 2 LU/ha (livestock units per hectare) must participate in a MINAS programme to calculate, estimate and report major inputs and outputs to the farm. Farms exceeding permitted N or P surpluses must pay a fee to the agency responsible for the MINAS programme. With fees as an incentive, it is hoped that farmers will innovate new strategies to achieve a balance of nutrients at the farm gate, and, at the same time decreasing the long-term risk of N and P losses to the environment.

Nutrient retention and recycling within the hybrid dairy system is important to decrease the need of imported nutrients and decrease nutrient losses to the environment. Current and future policies, including restrictive zoning and nutrient management of confinement operations, may make grazing more profitable and preferred, particularly as the urban–rural interface becomes more and more entwined. As described in this chapter, many management options exist that affect nutrient retention and utilization, including altering stocking rates (i.e. higher stocking rates and shorter duration in each paddock to more evenly distribute manure and urine), shifting production time so that maximum livestock density occurs during the growing season, supplemental feeding, bedding shelters with substances that increase absorption of nutrients, and incorporating legumes into existing pasture to increase N_2 fixation. Precision feeding strategies (matching the nutrients provided in the total ration versus animal nutrient requirements) are one of the foremost areas upon which animal managers can improve in order to decrease nutrient excretion and have the added benefit of decreasing feed costs. Ultimately, social and economic factors must be accounted for to enable prudent management of nutrients on grazing dairy operations.

References

Aarons, S.R., Melland, A.R. and Gourley, C.J.P. (2004) Nutrient distribution within a dairy farm. Third Australian New Zealand Soil Conference, SuperSoil 2004. Available at: http://www.regional.org.au/au/asssi/supersoil2004/s16/oral/1622_aaronss.htm

Aharoni, Y., Arieli, A. and Tagari, H. (1993) Lactational response of dairy cows to change of degradability of dietary protein and organic matter. *Journal of Dairy Science* 76, 3514–3522.

Aldrich, J.M., Muller, L.D. and Varga, G.A. (1993) Nonstructural carbohydrates and protein effects on rumen fermentation, nutrient flow, and performance of dairy cows. *Journal of Dairy Science* 76, 1091–1105.

Bacon, S.C., Lanyon, L.E. and Schlauder, R.M. Jr. (1990) Plant nutrient flow in the managed pathways of an intensive dairy farm. *Journal of Agronomy* 82, 755–761.

Ball, R., Keeney, D.R., Theobald, P.W. and Nes, P. (1979) Nitrogen balance in urine-affected areas of a New Zealand pasture. *Agronomy Journal* 71, 309–314.

Bargo, F., Muller, L.D., Delahoy, J. and Cassidy, T. (2002) Performance of high producing dairy cows with three different feeding systems combining pasture and total mixed rations. *Journal of Dairy Science* 85, 2948–2963.

Bargo, F., Muller, L.D., Kolver, E.S. and Delahoy, J.E. (2003) Invited review: production and digestion of supplemented dairy cows on pasture. *Journal of Dairy Science* 86, 1–42.

Beever, D.E., Losada, H.R., Cammell, S.B., Evans, R.T. and Haines, M.J. (1986) Effect of forage species and season on nutrient digestion and supply in grazing cattle. *British Journal of Nutrition* 56, 209–225.

Bicudo, J.R., Schmidt, D.R. and Jacobson, L.D. (2004) Using covers to minimize odors and gas emissions from manure storage. University of Kentucky Cooperative Extension Bulletin AEN 84. Available at: http://www.ca.uky.edu/agc/pubs/aen/aen84/aen84.pdf

Bittman, S., van Vliet, L.J.P., Kowalenko, C.G., McGinn, S., Lau, A.K., Patini, M., Forge, T., McLaughlin, N.B., Hunt, D.E., Bounaix, F. and Friesen, A. (2003) A low disturbance technique for applying slurry on forage land. In: *Controlling N Flows and Losses*. Proceedings of Grassland and Environmental Research (IGER) Nitrogen Workshop. Academic Press, Wageningen, the Netherlands, pp. 490–492.

Bouldin, D.R. and Klausner, S.D. (1998) Managing nutrients in manure: general principles and applications to dairy manure in New York. In: Hatfield, J.L. and Stewart, B.A. (eds) *Animal Waste Utilization: Effective Use of Manure as a Soil Resource*. Ann Arbor Press, Chelsea, Michigan, pp. 65–88.

Braskerud, B. (2002) Factors affecting phosphorus retention in small constructed wetlands treating agricultural non-point source pollution. *Ecological Engineering* 19, 41–61.

Broderick, G.A. (2003) Effects of varying dietary protein and energy levels on the production of lactating dairy cows. *Journal of Dairy Science* 86, 1370–1381.

Brown, M.P., Longabucco, P., Rafferty, M.R., Robillard, P.D. and Walter, M.F. (1989) Effects of animal waste control practices on nonpoint-source phosphorus loading in the West Branch of the Delaware River watershed. *Journal of Soil and Water Conservation* 44, 67–70.

Bussink, D.W. (1992) Ammonia volatilization from grassland receiving nitrogen fertilizer and rotationally grazed by dairy cattle. *Fertilizer Research* 33, 257–265.

Bussink, D.W. and Oenema, O. (1998) Ammonia volatilization from dairy farming systems in temperate areas: a review. *Nutrient Cycling in Agroecosystems* 51, 19–33.

Clancy, K. (2006) *Greener Pastures: How Grass-fed Beef and Milk Contribute to Healthy Eating.* Union of Concerned Scientists, Cambridge, Massachusetts.

Coppock, C.E., Bath, D.L. and Harris, B. Jr. (1981) From feeding to feeding systems. *Journal of Dairy Science* 64, 1230.

Cunningham, L.C. (1993) An economic comparison of grazing intensity for lactating dairy cattle in northeastern Pennsylvania. MS thesis, The Pennsylvania State University, University Park, Pennsylvania.

Dartt, B.A., Lloyd, J.W., Radke, B.R., Black, J.R. and Kaneene, J.B. (1999) A comparison of profitability and economic efficiencies between management intensive grazing and conventionally managed dairies in Michigan. *Journal of Dairy Science* 82, 2412–2420.

Daverede, I.C., Kravchenko, A.N., Hoeft, R.G., Nafziger, E.D., Bullock, D.G., Warren, J.J. and Gonzini, L.C. (2004) Phosphorus runoff from incorporated and surface-applied liquid swine manure and phosphorus fertilizer. *Journal of Environmental Quality* 33, 1535–1544.

Day, D.L. and Funk, T.L. (1998) Processing manure: physical, chemical and biological treatment. In: Hatfield, J.L. and Stewart B.A. (eds) *Animal Waste Utilization: Effective Use of Manure as a Soil Resource.* Ann Arbor Press, Chelsea, Michigan, pp. 243–282.

Di, H.J. and Cameron, K.C. (2000) Nitrate leaching in temperate agroecosystems: sources, factors and mitigating strategies. *Nutrient Cycling in Agroecosystems* 46, 237–256.

Di, H.J. and Cameron, K.C. (2002) The use of a nitrification inhibitor, dicyandiamide (DCD), to reduce nitrate leaching and nitrous oxide emissions in a simulated grazed and irrigated grassland. *Soil Use and Management* 18, 395–403.

Di, H.J. and Cameron, K.C. (2004) Effects of temperature and application rate of a nitrification inhibitor, dicyandiamide (DCD), on nitrification rate and microbial biomass in a grazed pasture soil. *Australian Journal of Soil Research* 42, 927–932.

Dou, Z., Knowlton, K.F., Kohn, R.A., Wu, Z., Satter, L.D., Zhang, G., Toth, J.D. and Ferguson, J.D. (2002) Phosphorus characteristics of dairy feces affected by diets. *Journal of Environmental Quality* 31, 2058–2065.

Dou, Z., Ferguson, J.D., Fiorini, J., Toth, J.D., Alexander, S.M., Peterson, A.B., Sims, J.T. and Wu, Z. (2003) Phosphorus feeding levels and critical control points on dairy farms. *Journal of Dairy Science* 86, 3787–3795.

Dougherty, W.J., Fleming, N.K., Cox, J.W. and Chittleborough, D.J. (2004) Phosphorus transfer in surface runoff from intensive pasture systems at various scales: a review. *Journal of Environmental Quality* 33, 1973–1988.

Edwards, P. (1980) A review of recycling organic wastes into fish, with emphasis on the tropics. *Aquaculture* 21, 261–279.

Elbehri, A. and Ford, S.A. (1995) Economic analysis of major dairy forage systems in Pennsylvania: the role of intensive grazing. *Journal of Production Agriculture* 8, 501–506.

Emmick, D.L. and Toomer, L.F. (1991) The economic impact of intensive grazing management on fifteen dairy farms in New York State. In: *Proceeding of Forage and Grassland Conference.* American Forage and Grassland Council, Georgetown, Texas, p. 19

Fales, S.L., McMurry, S.A. and McSweeny, W.T. (1993) The role of pasture in Northeastern dairy farming: historical perspectives, trends and research imperatives for the future. *Agricultural Research* 677, 111–132.

Fraser, M., Cameron, K.C. and Sherlock, R.R. (1994) Lysimeter study of the fate of nitrogen in animal urine returns to irrigated pasture. *European Journal of Soil Science,* 45, 439–447.

Fulhage, C.D. (1997) Manure management consideration for expanding dairy herds. *Journal of Dairy Science* 80, 1872–1879.

Gassman, P.W. and Bouzaher, A. (1995) Livestock pollution: lessons from the European Union. In: Steele, K. (ed.) *Animal Waste and the Land Water Interface.* Lewis Publishers-CRC, New York, pp. 515–522.

Ghebremichael, L.T., Cerosaletti, P.E., Veith, T.L., Hamlett, J.M. and Gburek, W. (2007) Economic and phosphorus-related effects of precision feeding and forage management at a farm-scale. *Journal of Dairy Science* 90(8), 3700–3715.

Gloy, B.A., Tauer, L.W. and Knoblauch, W. (2002) Profitability of grazing versus mechanical forage harvesting on New York dairy farms. *Journal of Dairy Science* 85, 2215–2222.

Goff, J.P. (1998) Phosphorus deficiency. In: Howard, J.L. and Smith, R.A. (eds) *Current Veterinary Therapy 4: Food Animal Practice*. W.B. Saunders, Philadelphia, Pennsylvania, pp. 218–220

Goss, M.J. and Goorahoo, D. (1995) Nitrate contamination of groundwater: measurement and prediction. *Fertilizer Research* 42, 331–338.

Granstedt, A. (1997) Experiences from long term studies in Nordic countries to understanding the conversion process and development of organic farms. In: Proc. 2nd International Workshop of the European Network for Scientific Research Co-ordination in Organic Farming (ENOF). *Steps in the Conversion and Development of Organic Farms*. 3–4 October, 1996, Barcelona.

Hamilton, C.M. and Sims, J.T. (1995) Nitrogen and phosphorus availability in enriched, pelletized poultry litters. *Journal of Sustainable Agriculture* 5, 115–132.

Hanson, G.D., Cunningham, L.C., Morehart, M.J. and Parsons, R.L. (1998) Profitability of moderate intensive grazing of dairy cows in the Northeast. *Journal of Dairy Science* 81, 821–829.

Harris, B.L. and Kolver, E.S. (2001) Review of Holsteinization on intensive pastoral dairy farming in New Zealand. *Journal of Dairy Science* 84(E. Suppl.), E56–E61.

Haygarth, P.M., Chapman, P.J., Jarvis, S.C. and Smith, R.V. (1998) Phosphorus budgets for two contrasting grassland farming systems in the UK. *Soil Use and Management* 14, 160–167.

Herrera-Saldana, R., Gomez-Alarcon, R., Torabi, M. and Huber, J.T. (1990) Influence of synchronizing protein and starch degradation in the rumen on nutrient utilization and microbial synthesis. *Journal of Dairy Science* 73, 142–148.

Hively, W.D., Bryant, R.B. and Fahey, T.J. (2005) Phosphorus concentrations in overland flow from diverse locations on a New York dairy farm. *Journal of Environmental Quality* 34, 1224–1233.

Hunt, P.G., Matheny, T.A. and Szögi, A.A. (2003) Denitrification in constructed wetlands used for treatment of swine wastewater. *Journal of Environmental Quality* 32, 727–735.

James, E., Kleinman, P., Veith, T., Stedman, R. and Sharpley, A. (2007) Phosphorus contributions from pastured dairy cattle to streams of the Cannonsville Watershed, New York. *Journal of Soil and Water Conservation* 62, 40–47.

Jeppsson, K.H. (1999) Volatilization of ammonia in deep-litter systems with different bedding materials for young cattle. *Journal of Agricultural Engineering Research* 73, 49–57.

Kim, Y.J., Geohring, L.D., Jeon, J.H., Collick, A.S., Giri, S.K. and Steenhuis, T.S. (2006) Evaluation of the effectiveness of vegetative filter strips for phosphorus removal with the use of a tracer. *Journal of Soil and Water Conservation* 61, 293–302.

Kleinman, P.J.A. and Sharpley, A.N. (2003) Effect of broadcast manure on runoff phosphorus concentration over successive rainfall events. *Journal of Environmental Quality* 32, 1072–1081.

Knowlton, K.F. and Herbein, J.H. (2002) Phosphorus partitioning during early lactation in dairy cows fed diets varying in phosphorus content. *Journal of Dairy Science* 85, 1227–1236.

Kolver, E.S. (2003) Nutritional limitations to increased production on pasture-based systems. *Proceedings of the Nutrition Society* 62, 291–300.

Kolver, E.S., Muller, L.D., Varga, G.A. and Cassidy, T.W. (1998) Synchronization of ruminal degradation of supplemental carbohydrate with pasture nitrogen in lactating dairy cows. *Journal of Dairy Science* 81, 2017–2028.

Kolver, E.S., Napper, A.J., Copeman, P.J.A. and Muller, L.D. (2000) A comparison of New Zealand and overseas Holstein Friesian heifers. *Proceedings of the New Zealand Society of Animal Production* 60, 265–269.

Kriegl, T. and McNair, R. (2005) Pastures of plenty: financial performance of Wisconsin grazing dairy farms. University of Wisconsin, Madison, Wisconsin.

Krober, T.F., Kulling, D.R., Menzi, H., Sutter, F. and Kreuzer, M. (2000) Quantitative effects of feed protein reduction and methionine on nitrogen use by cows and nitrogen emission from slurry. *Journal of Dairy Science* 83, 2941–2951.

Lazarus, W.E. and Rudstrom, M. (2007) The economics of anaerobic digester operation on a Minnesota dairy farm. *Review of Agricultural Economics* 29, 349–364.

Majer Newman, J. and Clausen, J.C. (1997) Seasonal effectiveness of a constructed wetland for processing milkhouse wastewater. *Wetlands* 17, 375–382.

McDowell, R.W. (2006) Phosphorus and sediment loss in a catchment with winter forage grazing of cropland by dairy cattle. *Journal of Environmental Quality* 35, 575–583.

McDowell, R. and Sharpley, A. (2002) Phosphorus transport in overland flow in response to position of manure application. *Journal of Environmental Quality* 31, 217–227.

McDowell, R.W. and Wilcock, R.J. (2007) Sources of sediment and phosphorus in stream flow of a highly productive dairy farmed catchment. *Journal of Environmental Quality* 36, 540–548.

McDowell, R.W., Nash, D.M. and Robertson, F. (2007) Sources of phosphorus lost from a grazed pasture receiving simulated rainfall. *Journal of Environmental Quality* 36, 1281–1288.

Meals, D. (2000) *Lake Champlain Basin Agricultural Watersheds Section 319 National Monitoring Program Project Executive Summary*. Vermont Department of Environmental Conservation, Waterbury, Vermont.

Meals, D.W. (1993) Assessing nonpoint phosphorus control in the LaPlatte river watershed. *Lake Reservoir Management* 7, 197–207.

Moore, P.A. Jr., Formica, S.J., Van Epps, M. and Delaune, P.B. (2005) Effect of pasture renovation on nutrient runoff from pastures fertilized with manure. In: *Proceedings of the 7th International Symposium on Livestock Environment*. American Society of Agricultural Engineers, 18–20 May 2005, Beijing, China, p. 301.

Moreira, V.R. and Satter, L.D. (2006) Effect of scraping frequency in a freestall barn on volatile nitrogen loss from dairy manure. *Journal of Dairy Science* 89, 2579–2587.

Morken, J. and Sakshaug, S. (1997) New injection technique – Direct ground injection (DGI). In: Voermans, J.A.M. and Monteny, G.A. (eds) *Proceedings of the International Symposium on Ammonia and Odour Control from Animal Production Facilities*, 6–10 October. Research Institute for Pig Husbandry, Rosmalen, the Netherlands, pp. 585–590.

Morken, J. and Sakshaug, S. (1998) Direct ground injection of livestock waste slurry to avoid ammonia emission. *Nutrient Cycling in Ecosystems* 51, 59–63.

Mueller, D.H., Wendt, R.C. and Daniel, T.C. (1984) Phosphorus losses as affected by tillage and manure application. *Soil Science Society of America Journal* 48, 901–905.

Muller, L.D. (1990) Individual concentrate feeding and total mixed rations in meeting nutritional needs of dairy cows. *Proceedings from Dairy Feeding Systems Symposium*. Northeast Regional Agricultural Engineering Service, Ithaca, New York, pp. 113–117.

Muller, L.D. (1992) Feeding management strategies. In: Van Horn, H.H. and Wilcox, C.J. (eds) *Large Dairy Herd Management*. Management Services, American Dairy Science Association, Champaign, Illinois, pp. 326–335.

Neeteson, J.J., Schroder, J.J. and Hassingk, J. (2001) The Netherlands. In: DeClerq, P. *et al.* (eds) *Nutrient Management Legislation in European Countries*. Wageningen Pers., Wageningen, the Netherlands, pp. 283–293.

National Research Council. (2001) *Nutrient Requirements of Dairy Cattle*, 7th rev. edn. National Academy of Sciences, Washington, DC.

Oenema, O., Bannink, A., Sommer, S.G. and Velthof, G.L. (2001) Gaseous nitrogen emissions from livestock farming systems. In: Follett, R.F. and Hatfield, J.L. (eds) *Nitrogen in the Environment: Sources, Problems and Management*. Elsevier Science, Amsterdam, pp. 255–289.

Osei, E., Gassman, P.W., Jones, R.D., Pratt, S.J., Hauck, L.M., Beran, L.J., Rosenthal, W.D. and Williams, J.R. (2000) Economic and environmental impacts of alternative practices on dairy farms in an agricultural watershed. *Journal of Soil and Water Conservation* 55, 466–472.

Parker, W.J., Muller, L.D., Fales, S.L. and McSweeney, W.T. (1991) A physical and financial comparison of dairy farming in Pennsylvania and New Zealand. Staff paper 195, Department of Agricultural Economy and Rural Sociology, Pennsylvania State University, University Park, Pennsylvania.

Parker, W.J., Muller, L.D. and Buckmaster, D.R. (1992) Management and economic implications of intensive grazing on dairy farms in the Northeastern States. *Journal of Dairy Science* 75, 2587–2597.

Parker, W.J., Muller, A., Fales, S.L. and McSweeney, W.T. (1993) A survey of dairy farms in Pennsylvania using minimal or intensive grazing systems. *Professional Animal Scientist* 9, 77–85.

Peterson, P.R. and Gerrish, J.R. (1996) Grazing systems and spatial distribution of nutrients in pastures: livestock management considerations. In: Joost, R.E. and Roberts, C.A. (eds) *Nutrient Cycling in Forage Systems.* Proceedings of Potash and Phosphate Institute and the Foundation for Agronomy Research, Columbia, Missouri. 7–9 March. PPI, Norcross, Georgia, pp. 203–212.

Petersen, S.O., Lind, A.M. and Sommer, S.G. (1998) Nitrogen and organic matter losses during storage of cattle and pig manure. *Journal of Agricultural Science* 130, 69–79.

Powell, J.M., Satter, L.D. and Misselbrook, T.A. (2004) Dairy manure and air quality: the issues. Available at: http://ipcm.wisc.edu/Publications/tabid/54/Default.aspx

Preedy, N., McTiernan, K., Mathews, R., Heathwaite, L. and Haygarth, P. (2001) Rapid incidental phosphorus transfers from grassland. *Journal of Environmental Quality* 30, 2105–2112.

Rauniyar, G.P. and Parker, W.J. (1999) Longitudinal analysis of dairy farm income and expenditure in New Zealand: a review of 25 years of adjustment. Australasian Agribusiness Review, vol. 7. Available at: http://www.agrifood.info/review/1999/Rauniyar.html

Rook, A.J., Tallowin, J.R.B., Rutter, S.M., Orr, R.J., Hopkins, A. and Gibb, M.J. (2005) Behavioural and community ecology: mechanisms driving biodiversity in pasture-based systems. In: Rook, A.J. and Penning, P.D. (eds) *In Grazing Management, the Principles and Practice of Grazing, for Profit and Environmental Gain, Within Temperate Grassland Systems.* British Grassland Society, Reading, UK, pp. 54–65.

Rotz, C.A. (2004) Management to reduce nitrogen losses in animal production. *Journal of Animal Science* 82(E. Suppl.), E119–E137.

Rotz, C.A., Sharpley, A.N., Satter, L.D., Gburek, W.J. and Sanderson, M.A. (2002) Production and feeding strategies for phosphorus management on dairy farms. *Journal of Dairy Science* 85, 3153.

Saporito, L.S. and Lanyon, L.E. (1998) Evaluating management and soil and weather contributions to potential nitrate leaching from a Pennsylvania dairy farm using NLEAP. *Journal of Environment Quality* 27, 1367–1375.

Schellinger, G.R. and Clausen, J.C. (1992) Vegetative filter treatment of dairy barnyard runoff in cold regions. *Journal of Environmental Quality* 21, 40–45.

Schmit, T.M., Boisvert, R.N. and Tauer, L.W. (2001) Measuring the financial risks of New York dairy producers. *Journal of Dairy Science* 84, 411–420.

Schwer, C.B. and Clausen, J.C. (1989) Vegetative filter treatment of dairy milkhouse waste water. *Journal of Environmental Quality* 18, 446–451.

Serodes, J.B. and Normand, D. (1999) Phosphorus removal in agricultural wastewater by a recently constructed wetland. *Canadian Journal of Civil Engineering* 26, 305–311.

Shamir, E., Thompson, T.L., Karpiscak, M.M., Freitas, R.J. and Zauderer, J. (2001) Nitrogen accumulation in a constructed wetland for dairy wastewater treatment. *Journal of the American Water Resources Association* 37, 315–325.

Sharpley, A.N. (1985) Depth of surface soil-runoff interaction as affected by rainfall, soil slope and management. *Soil Science Society of America Journal* 49, 1010–1015.

Sharpley, A.N. and Halvorson, A.D. (1994) The management of soil phosphorus availability and its impact on surface water quality. In: Lal, R. and Stewart, B.A. (eds) *Soil Processes and Water Quality.* Lewis Publishers, Boca Raton, Florida, pp. 7–90.

Sharpley, A.N. and Syers, J.K. (1976) Phosphorus transport in surface runoff as influenced by fertilizer and grazing cattle. *New Zealand Journal of Science* 19, 277–282.

Sharpley, A.N. and Syers, J.K. (1979) Loss of nitrogen and phosphorus in tile drainage as influenced by nitrogen fertilizer application and grazing animals. *New Zealand Journal of Agricultural Research* 22, 123–217.

Silva, R.G., Cameron, K.C., Di, H.J. and Hendry, T. (1999) A lysimeter study of the impact of cow urine, dairy shed effluent and nitrogen fertiliser on nitrate leaching. *Australian Journal of Soil Research* 37, 357–369.

Smith, S.F. (1994) Moderate size farms can be successful. In: *Agricultural Update: Farm Business and Financial Management. Vol. 4, No. 4*. Cornell Cooperative Extension, Ithaca, New York, p. 2

Smith, K.A., Chalmers, A.G., Chambers, B.J. and Christie, P. (1998) Organic manure phosphorus accumulation, mobility and management. *Soil Use and Management* 14, 154–159.

Smith, E., Gordon, R., Madani, A. and Stratton, G. (2006) Year-round treatment of dairy wastewater by constructed wetlands in Atlantic Canada. *Wetlands* 26, 349–357.

Soder, K.J. and Rotz, C.A. (2001) Economic and environmental impact of four levels of concentrate supplementation in grazing dairy herds. *Journal of Dairy Science* 84, 2560–2572.

Soder, K.J. and Rotz, C.A. (2003) Economic and environmental impact of utilizing a total mixed ration in Pennsylvania grazing dairy herds. *Professional Animal Scientist* 19, 304–311.

Soder, K.J. and Muller, L.D. (2007) Case study: a survey of total mixed ration use on grazing dairy farms in Pennsylvania and New York. *Professional Animal Scientist* 23, 300–307.

Soriano, F.D., Polan, C.E. and Miller, C.N. (2001) Supplementing pasture to lactating Holsteins fed a total mixed ration diet. *Journal of Dairy Science* 84, 2460–2468.

Stout, W.L. (2003) Effect of urine volume on nitrate leaching in the northeast USA. *Nutrient Cycling in Agroecosystems* 67, 197–203.

Stout, W.L., Fales, S.L., Muller, L.D., Schnabel, R.R., Elwinger, G.F. and Weaver, S.R. (2000) Assessing the effect of management intensive grazing on water quality in the northeast USA. *Journal of Soil and Water Conservation* 55, 238–243.

Tallam, S.K., Ealy, A.D. Bryan, K.A. and Wu, Z. (2005) Ovarian activity and reproductive performance of dairy cows fed different amounts of phosphorus. *Journal of Dairy Science* 88, 3609–3618.

Tallowin, J.R.B., Rook, A.J. and Rutter, S.M. (2005) Impact of grazing management on biodiversity of grasslands. *Animal Science* 81, 193–198.

Texas Water Resources Institute. (2006) Dairy compost utilization by the public and public entities. Available at: http://compost.tamu.edu/index.php

Tozer, P.R., Bargo, F. and Muller, L.D. (2003) Economic analyses of feeding systems combining pasture and total mixed ration. *Journal of Dairy Science* 86, 808–818.

Tucker, W.B., Rude, B.J. and Wittayakun, S. (2001) Case Study: performance and economics of dairy cows fed a corn silage-based total mixed ration or grazing annual ryegrass during mid to late lactation. *Professional Animal Scientist* 17, 195–201.

Uusi-Kamppa, J., Braskerud, B., Jansson, H., Syversen, N. and Uusitalo, R. (2000) Buffer zones and constructed wetlands as filters for agricultural phosphorus. *Journal of Environmental Quality* 29, 151–158.

Vadas, P.A., Harmel, R.D. and Kleinman, P.J.A. (2007) Transformations of soil and manure phosphorus after surface application of manure to field plots. *Nutrient Cycling in Agroecosystems* 77, 83–99.

Van Vuuren, A.M., Tamminga, S. and Ketelaar, R.S. (1991) *In sacco* degradation of organic matter and crude protein of fresh grass (*Lolium perenne*) in the rumen of grazing dairy cows. *Journal of Agricultural Science Cambridge* 116, 429–436.

Vietor, D.M., Griffith, E.N., White, R.H., Provin, T.L., Muir, J.P. and Read, J.C. (2002) Export of manure phosphorus and nitrogen in turfgrass sod. *Journal of Environmental Quality* 31, 1731–1738.

Watson, C.A., Bengtsson, H., Ebbesvik, M., Loes, A.K., Myrbeck, A., Salomon, E., Schroder, J. and Stockdale, E.A. (2002) A review of farm-level nutrient balances for organic farms as a tool for management of soil fertility. *Soil Use and Management* 18, 264–273.

Webb, J. (2001) Estimating the potential for ammonia emissions from livestock extreta and manures. *Environmental Policy* 111, 395–406.

White, S.L. (2000) *Investigation of Pasture and Confinement Dairy Feeding Systems Using Jersey and Holstein Cattle*. North Carolina State University, Raleigh, North Carolina.

White, S.L., Sheffield, R.E., Washburn, S.P., King, L.D. and Green, J.T., Jr. (2001) Spatial and time distribution of dairy cattle excreta in an intensive pasture system. *Journal of Environmental Quality* 30, 2180–2187.

Whitehead, D.C. and Bristow, A.W. (1990) Transformations of nitrogen following the application of ^{15}N-labelled cattle urine to an established grass sward. *Journal of Applied Ecology* 27, 667–678.

Withers, P.J.A., Peel, S., Mansbridge, R.J., Chalmers, A.G. and Lane, S.P. (1999) Transfers of phosphorus within three dairy farming systems receiving varying inputs in feeds and fertilizers. *Nutrient Cycling in Agroecosystems* 55, 63–75.

Wright, P. (1996) Prevention, collection and treatment of concentrated pollutions sources on farms. In: *Animal Agriculture and the Environment: Nutrients, Pathogens and Community Relations*. Northeast Regional Agricultural Engineering Service, Ithaca, New York, pp. 142–158.

Wu, Z. and Satter, L.D. (2000) Milk production and reproductive performance of dairy cows fed two concentrations of phosphorus for two years. *Journal of Dairy Science* 83, 1063.

Wu, Z., Satter, L.D. and Sojo, R. (2000) Milk production, reproductive performance, and fecal excretion of phosphorus by dairy cows fed three amounts of phosphorus. *Journal of Dairy Science* 83, 1028–1041.

Wulf, S., Maeting, M. and Clemens, J. (2002) Application technique and slurry co-fermentation effects on ammonia, nitrous oxide and methane emissions after spreading: I. Ammonia volatilization. *Journal of Environmental Quality* 31, 1789–1794.

Xue, Y., Kovacic, D.A., David, M.B., Gentry, L.E., Mulvaney, R.L. and Lindau, C.W. (1999) *In situ* measurement of denitrification in constructed wetlands. *Journal of Environmental Quality* 28, 263–269.

Index